Sensors and Transducers
Second Edition

D. PATRANABIS

Formerly, Professor and Head
Department of Instrumentation and Electronics Engineering
Jadavpur University, Kolkata

Professor and Head
Department of Applied Electronics and Instrumentation
Engineering Heritage Institute of Technology, Kolkata

PHI Learning Private Limited

Delhi-110092

2026

*In fond memory of **Shri Asoke K. Ghosh** (October 1942 – February 2024), Founder Chairman and Managing Director of PHI Learning, whose vision endlessly inspires.*

The Legacy Continues....

Published by Pushpita Ghosh, PHI Learning Private Limited, Rimjhim House, 111, Patparganj Industrial Estate, Delhi-110092 and Printed by Mudrak, C-55, Sector 65, Noida, U.P.-201301.

₹725.00

SENSORS AND TRANSDUCERS, Second Edition
D. Patranabis

ISBN-978-81-203-2198-4 (Print Book)
ISBN-978-93-90544-58-5 (e-Book)

Dedicated to **My Parents**

Contents

Preface

Sensors and transducers are used in automation in construction, domestic appliances, industries, transport, space exploration, defence equipment, health services, and other applications. Advances in processing and computation have opened up opportunities for very accurate control of plants, processes, and systems. Sensors/Transducers have helped achieve substantial accuracy and control as automation of any kind begins with the measurement of certain system parameters of which sensors and transducers form an essential and indispensable part. Industrial process monitoring has become possible by appropriate and accurate sensing of the relevant variables. In fact, industrial growth moves hand-in-hand with the growth of the measurement science and technology. Growing technologies require increased accommodation and as a consequence, conventional disciplines in the engineering and science faculties have accommodated new developments.

In India, there are more than two hundred engineering institutes teaching instrumentation as a separate discipline and at the heart of instrumentation, lie sensors/transducers. With the developments in sensing systems and the plethora of courses offered, it becomes necessary to have a comprehensive compilation of the principles, analyses, and applications of sensors in a single volume. The students of this and associated disciplines, the teachers, the researchers, and the professionals also require some literature to begin with. Therefore, the experience in teaching and research in the discipline of over four decades has encouraged me to accept the challenge of bringing out such a treatise.

This book opens with classification and characterization of sensors and transducers in Chapter 1 while Chapter 2 makes a thorough study of mechanical and electromechanical sensors. Thermal sensors are comprehensively covered in Chapter 3 while Chapter 4 describes all types of magnetic sensors. Radiation sensors including optical types have been given in-depth consideration in Chapter 5. Sensors for analysis, the electroanalytical sensors, are included in Chapter 6 where all types of electrochemical as well as electroceramic sensors have been dealt with in detail. In today's world where technologies grow at a fast pace, automation demands intelligent/smart sensors. Chapter 7 covers this aspect by explaining their advantages and operating principles. Sensor technology is in the process of undergoing sea-change specially when micro- and nanotechnologies are supporting its growth and development. Sensors based on such technologies have therefore been discussed in Chapter 8. In the end, Chapter 9 discusses certain areas of application with the types of sensors as used in contemporary times. The theoretical explanation, discussion, and analysis in the text are well-supported by review questions and elaborate diagrams in each chapter.

I wish to extend my sincere thanks to my students for persistent demand and support to write such a text, and to my family for the trust and understanding they have shown.

D. PATRANABIS

Chapter 1

Getting Started!

1.1 WHAT ARE SENSORS/TRANSDUCERS?

Instrument Society of America defines a sensor or transducer as *a device which provides a usable* **output** *in response to a specified* **measurand**. Here, the output is defined as an 'electrical quantity' and measurand as a 'physical quantity, property, or condition which is measured.'

This definition can now be generalized by extending 'electrical quantity' to any type of signal such as mechanical and optical and extending 'physical quantity, property, or condition being measured' to those of nature—chemical and biochemical and so on.

1.2 PRINCIPLES

Different views exist over a common definition of both sensors and transducers. As a result, different definitions have been adopted for an easy distinction. One set of definitions holds as— an element that senses a variation in input energy to produce a variation in another or same form of energy is called a *sensor* whereas a *transducer* uses transduction principle to convert a specified measurand into usable output. Thus, a properly cut piezoelectric crystal can be called a sensor whereas it becomes a transducer with appropriate electrodes and input/output mechanisms attached to it.

In general, however, the sensing principles are physical or chemical in nature and the associated gadgets are only secondary and hence, the distinction is gradually being ignored. The principles can be grouped according to the form of energy in which the signals are received and generated. A matrix-like arrangement can thus be obtained for elaborating the principles. Signals can be divided into six categories on the basis of energies generated or received, namely (i) mechanical, (ii) thermal, (iii) electrical, (iv) magnetic, (v) radiant, and (vi) chemical.

Table 1.1 enlists physical and chemical transduction principles alongwith some elaborations.

Table 1.1 Physical and chemical transduction principles

Input \ Output	Mechanical	Thermal	Electrical	Magnetic	Radiant	Chemical
Mechanical	Mechanical including acoustic effects. eg: diaphragm.	Friction effects, cooling effects. eg: thermal flowmeter.	Piezoelectricity, piezoresistivity, resistive, inductive, and capacitive changes.	Piezomagnetic effects.	Photoelasticity, interferometry, Doppler effect.	—
Thermal	Thermal expansion. eg: xpansion thermometry.	—	Seebeck effect, pyroelectricity, thermoresistance. eg: Johnson noise.	—	Thermo-optical effects. eg: liquid crystals, thermo-radiant emission.	Thermal dissociation, thermally induced reaction.
Electrical	Electrokinetic effects. eg: inverse piezoelectricity.	Peltier effect, Joule heating.	Charge controlled devices, Langmuir probe.	Biot-Savart's electromagnetic law.	Electroluminescence, Kerr effect.	Electrolysis, electrically induced reaction. eg: electromigration.
Magnetic	Magnetostriction, magnetometers.	Magnetothermal effects (Righi-Leduc effect).	Ettinghaussen-Nernst effect, Galvanomagnetic effect. eg: Hall effect, magnetoresistance.	—	Magneto-optical effects. eg: Faraday effect, Cotton-Mouton effect.	—
Radiant	Radiation pressure.	Bolometer, thermopile.	Photoelectric effects. eg: photovoltaic cell, LDR's.	—	Photorefractivity, photon induced light emission.	Photodissociation, photosynthesis.
Chemical	Photoacoustic effect, hygrometry.	Thermal conductivity cell, calorimetry.	Conductimetry, potentiometry, voltametry, flame-ionization, chem FET.	Nuclear magnetic resonance.	Spectroscopy. eg: emission and absorption types, Chemiluminescence.	—

Table 1.1 produces a matrix of 6 × 6 entries, each of which can be another submatrix whose rows and columns are designated by the signal types within each major domain as shown in Table 1.2. There are some entries that are involved with more than one input or output as in the

Table 1.2 Energy types and corresponding measurands

Energy	Measurands
Mechanical	Length, area, volume, force, pressure, acceleration, torque, mass flow, acoustic intensity, and so on.
Thermal	Temperature, heat flow, entropy, state of matter.
Electrical	Charge, current, voltage, resistance, inductance, capacitance, dielectric constant, polarization, frequency, electric field, dipole moment, and so on.
Magnetic	Field intensity, flux density, permeability, magnetic moment, and so forth.
Radiant	Intensity, phase, refractive index, reflectance, transmittance, absorbance, wavelength, polarization, and so on.
Chemical	Concentration, composition, oxidation/reduction potential, reaction rate, pH, and the like.

case of the thermomagnetic or galvanomagnetic effects, both of which have two inputs. Thus, Hall effect has electrical and magnetic inputs. They can, therefore, go to a different location in the matrix. In such a situation, application aspect is given more importance which fixes the entry position. However, the above classification is only to provide an illustration to represent the physical and chemical effects that form a variety of sensors.

1.3 CLASSIFICATION

It is very difficult to classify sensors under one criterion and hence, different criteria may be adopted for the purpose. Some of these include:

1. transduction principles using physical or chemical effects,
2. primary input quantity, that is, the measurand,
3. material and technology, that have acquired more importance lately,
4. application, and
5. property.

Others of consequence are cost and accuracy.

People would choose their own criterion to suit their areas of activity. For example, development engineering groups prefer material and technology as the basis. However, the *transduction principle* is the basic criterion which should be followed for a systematic approach. Table 1.1 presents such a grouping.

Another very preliminary classification or subclassification as it may be, is based on the energy or power supply requirements of the sensors. This means that some sensors require power supply and there are some that do not. As a result, the sensors are called *active* and *passive* respectively. Sometimes, terms such as *self-generating* and *modulating* are used to qualify these.

Conventional sensors are now aptly supported by technologies which have yielded Micro Electro Mechanical Sensors (MEMS), CMOS image sensors, displacement and motion detectors and biosensors. Similarly, Coriolis, magnetic and ultrasonic flowmeters, photoelectric, proximity, Hall effect, infrared, integrated circuit (IC), temperature, radar-based level sensors are also relatively modern.

Application based classification is another very convenient way to show the segmentation in a very broad manner. But new technologies coupled with the existing ones can display a good segmentation when classified based on some property. Application based classification is represented as

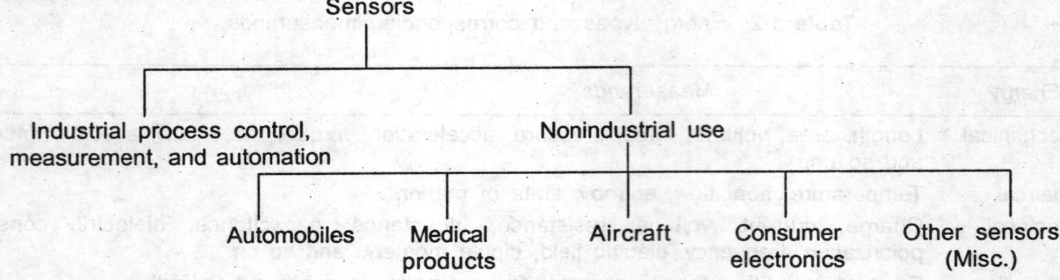

Property based classification is much more elaborate and to a certain extent exhaustive. Here, the subdivision is in technology scale. A brief presentation of this segmentation is given in Table 1.3.

Table 1.3 Property based classification

	Property				
	Flow	*Level*	*Temperature*	*Pressure*	*Proximity and displacement*
Technology	Differential pressure, positional displacement, vortex, thermal mass, electromagnetic, Coriolis, ultrasonic, anemometer, open channel.	Mechanical, magnetic, differential pressure, thermal displacement, vibrating rod, magnetostrictive, ultrasonic, radio frequency, capacitance type, microwave/radar, nuclear.	Filled-in systems, RTDs, thermistors, IC, thermocouples, inductively coupled, radiation (IR).	Elastic, liquid-based manometers, inductive/LVDT, piezoelectric, electronic, fibre optic, MEMS, vacuum.	Potentiometric, inductive/LVDT, capacitive, magnetic, photoelectric, magnetostrictive, ultrasonic.

Table 1.3 *(cont.)*

	Property				
	Acceleration	*Image*	*Gas and chemical*	*Biosensors*	*Others*
Technology	Accelerometers, gyroscopes.	CMOS, CCDs (charge coupled devices).	Chemical bead, electrochemical, thermal conductance, paramagnetic, ionization, infrared, semiconductor.	Electrochemical, light-addressable potentiometric (LAP), surface plasmon resonance (SPR), resonant mirror	Mass, force, load, humidity, moisture, viscosity.

CMOS image sensors have low resolution compared to earlier developed charge coupled devices but their small size, less cost, and low power consumption are considered better substitutes for CCDs as camera-on-a-chip sensors.

In biosensors group, SPR and LAP are relatively new optical technology-based sensors. Accelerometers are separately grouped mainly because of their role in the development of future automobiles, aircrafts, in industrial sector and in lesser developed areas of toys, videograms, physical therapy and so on, which is increasing sharply specially when the micromachining processes are decreasing the size considerably retaining the usual level of performance. Table 1.4 shows the emerging sensor technologies with current and future application schedules as a chart.

Table 1.4 Emerging sensor technologies

	Image sensors	Motion detectors	Biosensors	Accelerometers
Sensors	Technology: CMOS-based	Technology: IR, ultrasonic, microwave/radar	Technology: electrochemical	Technology: MEMS-based
Applications	Traffic and security surveillance, blind-spot detection as autosensors (robots etc.), video conferencing, consumer electronics, biometrics, PC imaging	Obstruction detection (robots, auto), security detection (intrusion), toilet activation, kiosks videograms and simulations, light activation	Water testing, food testing (contamination detection), medical care device, biological warfare agent detection	Vehicle dynamic system (auto), patient monitoring (including pace makers etc.)

1.4 PARAMETERS

The normal environmental conditions from where the data are made available through sensors are noisy and keep changing. The high fidelity mapping of such a varying reality requires extensive studies of 'fidelity' of the sensors themselves. Or, in other words, sensors are required to be appropriately characterized. These are done in terms of certain parameters and characteristics of the sensors.

1.4.1 Characteristics

Sensors like measurement systems have two general characteristics, namely 1. static, and 2. dynamic.

Static characteristics

(a) *Accuracy specified by inaccuracy* or *usually error:* which is given by

$$\varepsilon_a \% = \frac{x_m - x_t}{x_t} \times 100 \tag{1.1}$$

where

    t stands for true value,

    m for measured value, and

    x stands for measurand.

This is often expressed for the full scale output (fso) and is given by

$$\varepsilon_{\text{fso}}\% = \frac{x_m - x_t}{x_{\text{fso}}} \times 100 \tag{1.2}$$

Obviously,

$$|\varepsilon_{\text{fso}}| \leq |\varepsilon_a|$$

For multi-error systems, the overall performance in terms of error can be assessed either through (i) the worst case approach which assumes that all errors add up in the same direction so that the overall error is very high, being the linear sum of all the performance errors, or through (ii) the root mean square approach which is optimistic as well as practical, when the total performance error is assessed as

$$\varepsilon_0 = \left[\sum_i (\varepsilon_i)^2 \right]^{1/2} \tag{1.3}$$

(b) *Precision:*   describes how far a measured quantity is reproducible as also how close it is to the true value.

The term 'repeatability' is close to precision which is the difference in output y at a given value of the input x when obtained in two consecutive measurements. It may be expressed as % FSO. Figure 1.1 shows the plot of repeatability.

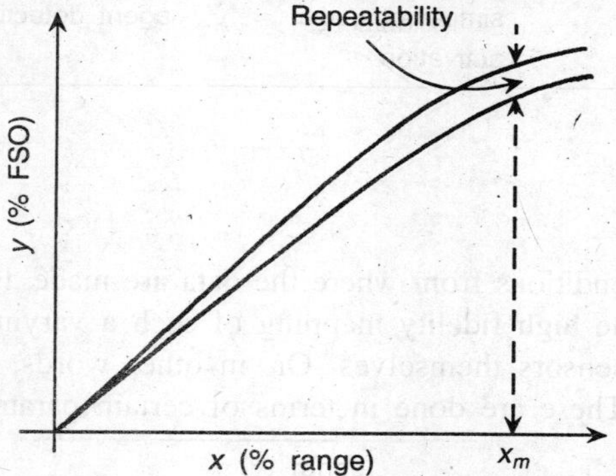

Fig. 1.1 Repeatability in y–x coordinates.

(c) *Resolution:* is defined as the smallest incremental change in the input that would produce a detectable change in the output. This is often expressed as percentage of the measured range, *MR*. The measured range is defined as the difference of the maximum input and the minimum input, that is, $MR = x_{\max} - x_{\min}$. For a detectable output Δy, if the minimum change in x is $(\Delta x)_{\min}$, then the maximum resolution is

$$R_{\max}(\%) = \frac{100\,(\Delta x)_{\min}}{MR} \tag{1.4}$$

Over the range of operation, an average resolution has also been defined as

$$R_{av}(\%) = 100 \, \frac{\sum\limits_{1}^{n} \Delta x_i}{n \cdot MR} \tag{1.5}$$

(d) *Minimum Detectable Signal (MDS):* Noise in a sensor occurs because of many reasons—internal sources or fluctuations due to externally generated mechanical and electromagnetic influences. Noise is considered in detail, on individual merits and often an equivalent noise source is considered for test purposes.

If the input does not contain any noise, the minimum signal level that produces a detectable output from the sensor is determined by its noise performance or noise characteristics. For this, the equivalent noise source is connected to the input side of the ideal noiseless sensor to yield an output which is the actual output level of the sensor. The MDS is then taken as the RMS equivalent input noise. When signal exceeds this value, it is called a detectable signal.

(e) *Threshold:* At the zero value condition of the measurand, the smallest input change that produces a detectable output is called the threshold.

(f) *Sensitivity:* It is the ratio of the incremental output to incremental input, that is

$$S = \frac{\Delta y}{\Delta x} \tag{1.6}$$

In normalized form, this can be written as

$$S_n = \frac{\Delta y/\Delta x}{y/x} \tag{1.7}$$

If sensitivity or the output level changes with time, temperature and/or any other parameters without any change in input level, drift is said to occur in the system which often leads to instability.

(g) *Selectivity and specificity:* The output of a sensor may change when afflicted by environmental parameters or other variables and this may appear as an unwanted signal. The sensor is then said to be *non-selective*. It is customary to define selectivity or specificity by considering a system of n sensors each with output $y_k (k = 1, 2, ..., n)$. The partial sensitivity S_{jk} is defined as the measure of sensitivity of the kth sensor to these other interfering quantities or variables x_j as

$$S_{jk} = \frac{\Delta y_k}{\Delta x_j} \tag{1.8}$$

A selectivity matrix would thus be obtained with S_{jk} as the jkth entry. Obviously, an ideally selective system will have only diagonal entries S_{jj} in the selectivity matrix. An ideally specific system is characterized by having a matrix with a single entry in the diagonal. Following relationship describes selectivity, λ;

$$\lambda = \min\left[\frac{S_{jj}}{\sum\limits_{k=1}^{n} |S_{jk}| - |S_{jj}|}\right] \qquad j = 1, 2, ..., n \tag{1.9}$$

Thus, for a selective group, denominator tends to zero and $\lambda \to \infty$. Also, specificity is a special case of selectivity.

(h) *Nonlinearity:*   Deviation from linearity, which itself is defined in terms of superposition principles, is expressed as a percentage of the full scale output at a given value of the input. Nonlinearity can, however, be specified in two different ways, namely (i) deviation from best fit straight line obtained by regression analysis, and (ii) deviation from a straight line joining the end points of the scale. These are shown in Figs. 1.2(a) and (b). The maximum nonlinearity in the first method is always less than the maximum nonlinearity in the second one. The figure is actually half.

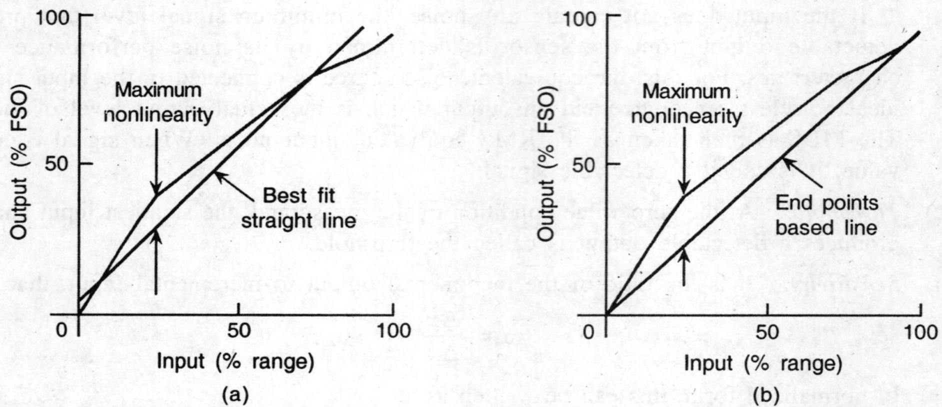

Fig. 1.2   Nonlinearity with (a) best-fit characteristics and (b) terminal-based characteristics.

A consequence of nonlinearity is *distortion* which is defined as the deviation from an expected output of the sensor or transducer. It also occurs due to presence of additional input components. If deviation at each point of the experimental curve is negligibly small from the corresponding point in the theoretical curve or from a curve made by using least square or other standard fits, the sensor is said to have *conformance* which is quantitatively expressed in % FSO at any given value of the input.

(i) *Hysteresis:*   It is the difference in the output y of the sensor for a given input x when x reaches this value in upscale and downscale directions as shown in Fig. 1.3. The causes

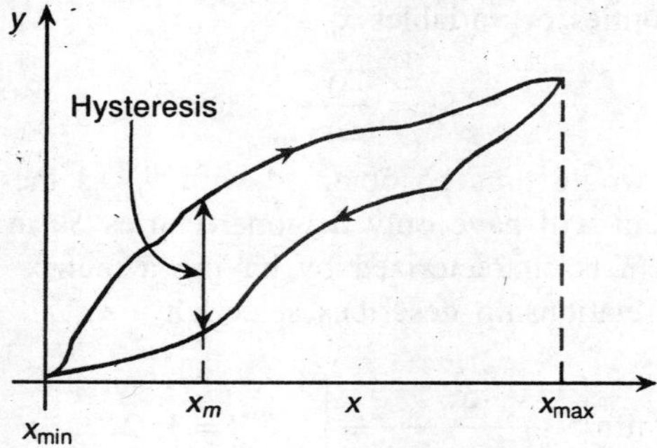

Fig. 1.3   The hysteresis curve.

are different for different types of sensors. In magnetic types, for example, it is the lag in alignment of the dipoles, in semiconductor types it is the injection type slow traps producing the effect, and so on.

(j) *Output impedance:*   It is a characteristic to be considered on individual merit. It causes great restriction in interfacing, specifically in the choice of the succeeding stage.

(k) *Isolation and grounding:* Isolation is necessary to eliminate or at least reduce undesirable electrical, magnetic, electromagnetic, and mechanical coupling among various parts of the system and between the system and the environment. Similarly, grounding is necessary to establish a common node among different parts of the system with respect to which potential of any point in the system remains constant.

Dynamic caracteristics

These involve determination of transfer function, frequency response, impulse response as also step response and then evaluation of the time-dependent outputs. The two important parameters in this connection are (a) fidelity determined by dynamic error and (b) speed of response determined by lag.

For determining the dynamic characteristics, different specified inputs are given to the sensor and the response characteristics are studied. With step input, the specifications in terms of the time constant of the sensor are made. Generally, the sensor is a single time constant device and if this time constant is τ, then one has the specifications as given in Table 1.5.

Table 1.5   % Response time of the sensors

% Response time	*Value in terms of τ*
$t_{0.1}$ or 10	0.104τ
$t_{0.5}$ or 50	0.693τ
$t_{0.9}$ or 90	2.303τ

This gives $t_{0.9}/t_{0.5} = 3.32$ which is taken as a quick check relation.

Impulse response as well as its Fourier transform are also considered for time domain as well as frequency domain studies.

1.5   ENVIRONMENTAL PARAMETERS (EP)

These are the external variables such as temperature, pressure, humidity, vibration, and the like which affect the performance of the sensor. These parameters are not the ones that are to be sensed.

For non-temperature transducers, temperature is the most important environmental parameter (EP). For any EP, the performance of the transducer can be studied in terms of its effect on the static and dynamic characteristics of the sensor as has already been discussed. For this study, one EP at a time is considered variable while others are held constant.

1.6 CHARACTERIZATION

Characterization of the sensors can be done in many ways depending on the types of sensors, specifically microsensors. These are electrical, mechanical, optical, thermal, chemical, biological, and so on.

Electrical characterization

It consists of evaluation of electrical parameters like (a) impedances, voltage and currents, (b) breakdown voltages and fields, (c) leakage currents, (d) noise, (e) cross talk, and so on.

The knowledge of the sensor 'output impedance' is very important for coupling the measuring equipment to it. For voltage sensitive sensors, the ratio of the input impedance of the measuring equipment to the output impedance of the transducer/sensor should be very high while for current sensitive sensors, reverse is true.

'Breakdown' of the insulating parts of the sensor is very critical as the health of the system depends on it. For metal-insulator-metal (MIM) or for metal-insulator-semiconductor (MIS) structures, the breakdown of the insulating film is studied by the system of Fig. 1.4. Three different types of breakdown are of interest for such a film: (i) dielectric strength, (ii) wear out, and (iii) current induced breakdown. Tests are also different and are performed with a particular independent parameter for a particular case. The three case studies are illustrated in Figs. 1.5(a), (b), and (c).

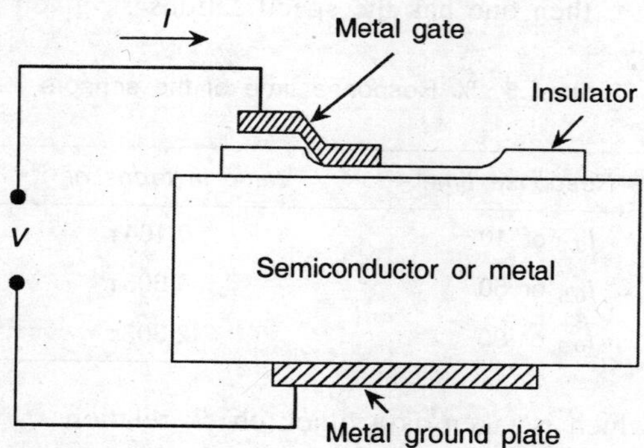

Fig. 1.4 Structure of a metal oxide semiconductor.

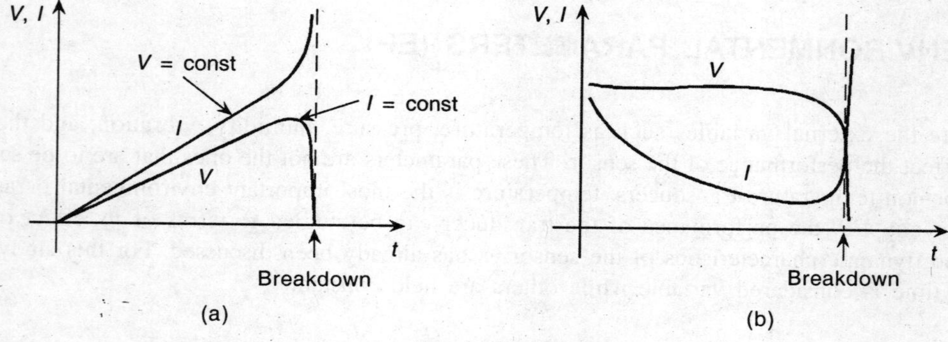

Fig. 1.5 Cont.

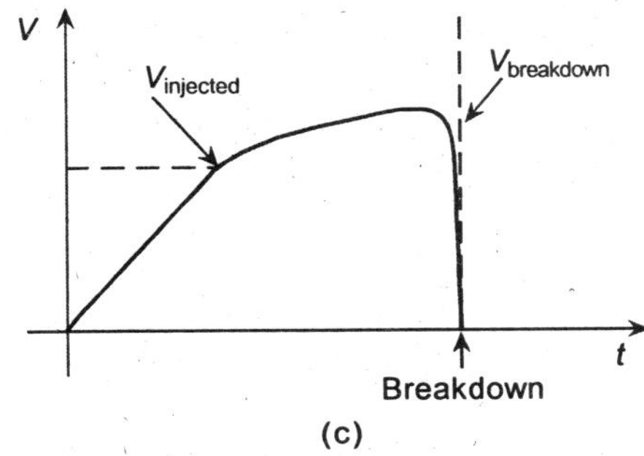

Fig. 1.5 Breakdown characteristics: (a) dielectric strength, (b) wear out, (c) current induced type.

Breakdown generally implies a sudden or 'avalanche' change in the voltage or current—voltage dropping to a negligible value and current rising to a very high value. Breakdown may be extrinsic or intrinsic though the mechanism in either case is basically the same. There occurs a high local field in the material which may be defect-induced which then is called *extrinsic*. However, if this is high field-induced, it is called *intrinsic* type. In the latter case, the high field induces microvoids to generate defects leading it to behave as the extrinsic type.

'Leakage current' measurement specifies the sensor quality, specifically its insulating quality as also the quality of p–n junctions wherever it exists.

'Noise' comes from electromagnetic interference, ac magnetic fluctuation, 50 Hz supply pick up, mechanical or acoustical vibration, or photon-induced output. Sensors are to be characterized for noise testing for immunity to such noise. For testing purposes, different noise sources are developed.

In multichannel or array sensors, 'crosstalk' may occur due to overlapping of signals between the two adjacent transducer elements. It may, however, occur in a single transducer system because of inductive or capacitive coupling or coupling through the common voltage source during transduction inside the element. It is measured using correlation techniques.

Mechanical and thermal characterization

It involves mechanical and thermal properties related to the overall reliability and integrity of the transducer, as well as relevant transduction process. Reliability is an important aspect of characterization. By means of testing, the functional and reliable portion of a batch of sensors or transducers is identified. Basically, failure analysis is performed and the mechanism of failure is attempted to be eliminated and thereby reduce the subsequent failures. In fact, the above two approaches are supplementary to each other.

Failure of transducers can be divided into three different categories:

   (i)   Catastrophic early life failures, often called infant mortality,
  (ii)   Short term drifts in the sensor parameters, and
 (iii)   Long term drifts and failures.

Catastrophic failure of the sensors is the complete failure in the normal operation. It is called *wear out* if it occurs in later life.

Short term and long term drifts are, in effect, changes in sensor parameters and are, therefore, to be studied more intensely for the sensor characterization.

The reliability of an item is given by what is known as reliability function, $R(x)$ which is the probability that the item would survive for a stated interval, say, between 0 and x. If $F(x)$ is the probability of failure, then

$$R(x) = 1 - F(x) = \int_{x}^{+\infty} f(t)\,dt \tag{1.10}$$

$$= \frac{\text{No. of 'sound' components at instant } x}{\text{Total no. of components at } x = 0} \tag{1.11}$$

The probability of failure $F(x)$ is actually a cumulative distribution function and in reliability statistics, the distribution functions that are used may have the following characteristics (i) normal, (ii) exponential, (iii) log-normal, (iv) gamma, and (v) Weibull, depending on the usage.

In general, the failure rate has a time dependency as shown in Fig. 1.6. The curve between failure rate and time appears like a bath tub which can be divided into three distinct zones— Zone 1 is the infant mortality zone, Zone 2 is the working zone with a constant failure rate, and Zone 3 is the wear out zone. As the normal working life is usually very long, estimating failure is tested by 'accelerated ageing test'. Before this ageing test, screening steps are taken to isolate the defective transducers. These steps vary depending on the types of the transducers. For standard silicon integrated type sensors (SITS), the typical tests that are performed are briefly discussed here.

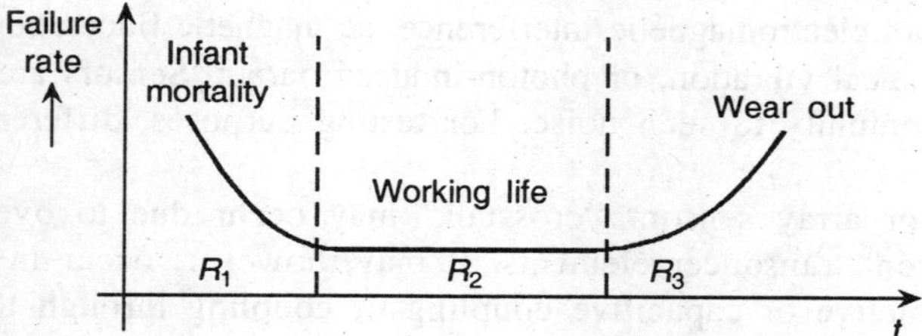

Fig. 1.6 The bath-tub curve.

High temperature burn in: The sensors are subjected to a high temperature over a stipulated period, usually at 125°C for 48 hours for SITS, when the defective units are burnt out and the remaining ones are expected to run for the expected life.

High temperature storage bake: The units are baked at a high temperature, usually at 250°C for SITS, for several hours when the instability mechanisms such as contamination, bulk defects, and metallization problems are enhanced in some units which were initially defective. These units are then screened out.

Electrical overstress test: Where progressively larger voltages upto 50% in excess of specification are applied over different intervals of time so that failures due to insulation, interconnection or oxide formation can occur in some units which were originally defective and are screened out.

Thermal shock test: Mainly done for packaging defects where the units are subjected to a temperature between −65° and 125°C for about 10 seconds for every temperature. The time is gradually increased to 10 minutes and the cycle is repeated 10 times. The failed units are rejected.

Mechanical shock test: Also for packaging, this test is performed by dropping the units from a specified height that varies from 3 to 10 m. Alternately, the unit is shaken by attaching it to a shaking table for a specified period of time.

As has already been mentioned, real-time operational test for reliability is difficult to perform so that accelerated ageing test has been proposed. The test should simulate the real ageing process in a much shorter time. High stress is imposed on the sensor and results from such a test are used to predict the performance in the normally stressed condition. The results should be interpreted for (i) true accelerated ageing, (ii) valid extrapolation to obtain expected performance under normal conditions, and (iii) determining the acceleration factor for the scaling, that is, how many hours of normal operation correspond to 1 hour of accelerated operation.

Appropriate models have been developed for the purpose and failures with respect to specified parameters such as leakage current, temperature, and so forth are predicted.

Optical characterization

It is usually done by ascertaining absorption coefficient, refractive index, reflectivity and the like. Here, again the consideration of the individual merit comes in.

Chemical/biological characterization

This is basically a test of the sensor with respect to its resistance to chemicals or corrosion in industrial as well as biological environment. Safety is an important aspect here particularly in case of biomedical sensors which should be tested against toxic or harmful effects in the prescribed environment.

REVIEW QUESTIONS

1. What are primary and secondary signals in sensor or transducer classification? Give examples of some magnetic-electric sensors and chemical-electrical sensors.

2. (a) What do you mean by minimum detectable signal? If the input noise of a sensor is sinusoidal in nature with a peak-to-peak value of 0.1 mV, what would be the MDS? [*Hint:* The rms value of the noise is the MDS which is $0.05/(2)^{1/2} = 0.035$ mV]

   (b) Define selectivity and specificity. How are they related?

3. Discuss the sensor characterization methods. How is a sensor electrically characterized? Support your answer with diagrams.

4. What are the different types of failures possible in a sensor? How do you define reliability function? If m units of produced items have been checked n times and the average failure at an instant of time, t, is found to be 1%, what is the value of the reliability function?
   [*Hint:* As per definition $R(t) = 1 - 0.01 = 0.99$]

5. How is a 'bath tub' curve associated with failures of transducers? What are the screening steps taken in standard silicon integrated sensors?

2

Mechanical and Electromechanical Sensors

2.1 INTRODUCTION

The controversy associated with formal definitions of sensors and transducers has apparently been resolved but not as yet the one with the classification perhaps, although, a generalized concept has been introduced in Chapter 1 in terms of input-output or primary and secondary signals.

Mechanical sensors, are those which have a mechanical quantity as the input and the output may be a quantity such as an electrical, magnetic, optical, thermal, and so on. In such a case, motion, displacement, speed, velocity, force, acceleration, torque and other such quantities should be measured by mechanical sensors. Process variables like pressure, flow, and level should also be considered as mechanical inputs and sensors for measurement of such variables should also be considered as mechanical sensors.

In many such sensors 'electromechanical coupling' is involved. As such, the primary objective is to convert the input form into an electrical output form for convenience of processing and display. In this respect, mechanical sensors are also termed as electromechanical or mechano-electrical sensors. It would, however, be seen that many sensors may be categorized under more than one category without being inappropriate. In the proceedings, appropriate references may be made for such multiplacement.

Further, same sensor is often used for measuring different variables by appropriate adaptation. In this way, it is not always possible to uniquely identify a sensor for a specific function. It is, therefore, perhaps more appropriate to discuss the sensors in the way they are formed and developed and not in the way a specific variable is measured.

Many mechanical variables are secondary in nature such as the 'motion' of the tip of a bimetallic element (thermal sensor) which is the result of the temperature variation of the element. Temperature, here, is the primary variable or input. The tip motion is angular or rotational. Similarly, a diaphragm actuated by pressure has a linear motion in its central part, or a bellows element actuated by pressure has a linear displacement of its free end. In contrast, a shaft under a torque with one end rigidly held has its free end in rotation.

Such rotational or translational displacements are measurable by various means. But direct measurement by using a pointer attached to the 'moving' end often leads to poor accuracy because of small movement and/or low resolution. Instead, resistive potentiometers, LVDT's, capacitive sensors, and so forth are used, not only for displacements alone but also for various other related variables, as has already been mentioned.

2.2 RESISTIVE POTENTIOMETERS

Resistive potentiometer is a kind of variable resistance transducer. Others in this category are strain gauges, RTD, thermistor, wire anemometer, piezoresistor and many more. A typical scheme of the potentiometer is shown in Fig. 2.1. This is a precision wire-wound potentiometer which is used as a sensor. A major advantage with this type is its large output. Resolution and noise are important aspects to be discussed in connection with it. For the former, the cross-section of the n-turn winding is shown in Fig. 2.2(a) with the wiper in two different possible positions: (i) touching only one wire and (ii) touching two turns is important as is obvious. In case (i) for a voltage supply V to the potentiometer, the voltage resolution would be

$$\Delta V = \frac{V}{n} \tag{2.1}$$

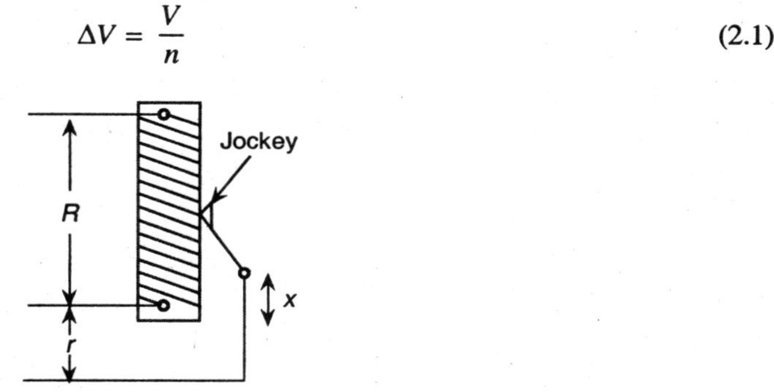

Fig. 2.1 Wire-wound potentiometer.

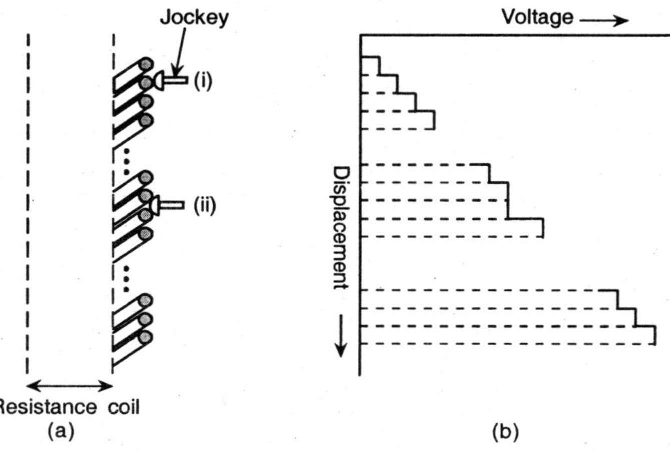

Fig. 2.2 (a) Jockey contact schemes: (i) single wire contact, (ii) two-wire shorting, (b) corresponding voltage levels.

In Fig. 2.2(b), the solid line stairs show the output voltage steps each of which is equivalent to a value V/n. But during transit, two adjacent wires are likely to be shorted as shown in (ii) of Fig. 2.2(a) and a minor resolution pulse of magnitude

$$\Delta V_m = V_p\left[\frac{1}{n-1} - \frac{1}{n}\right] \tag{2.2}$$

is obtained, where pth and $(p + 1)$th wires are shorted. This shows that with increasing value of p, minor pulse magnitude also increases and the loss in resolution due to this shorting leads to an actual resolution value

$$\Delta V - \Delta V_m = \frac{V}{n} - V_p\left[\frac{1}{n-1} - \frac{1}{n}\right] \tag{2.3}$$

The jockey shape/profile or the ratio of jockey radius to wire radius and geometry of wire winding should be considered for reducing ΔV_m. If jockey radius is small, with the jockey in use for some time with pressure, the wire gets its round surface worn out to develop a flat surface and finally gets torn. With a large radius of wire and close winding, this effect is small but may short more than two wires during the movement of the jockey and hence, precision of measurement is affected. For circular wire and circular jockey, it is recommended that the ratio of their radii be around 10, that is, $r_{\text{jockey}}/r_{\text{wire}} \approx 10$.

Also, materials of resistance wire and jockey are equally important, particularly from the wearing point of view and 'noise'. For noise, among other things, the jockey construction is to be considered seriously. A few types of the jockeys are shown in Figs. 2.3(a), (b), (c), and (d). The pressure at contact with the wire is provided by giving an adequate flexibility to the arm in relation to its mass. However, the required pressure is dependent on the materials, jockey to wire radius ratio, and the proposed lifespan of the potentiometer. A value of 10–50 mN is quite common.

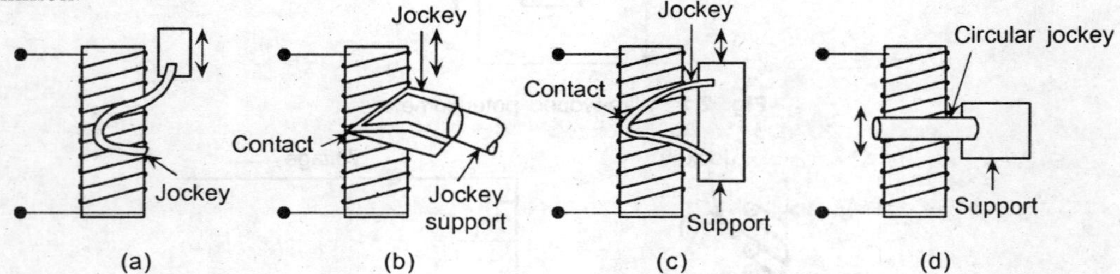

Fig. 2.3 Different designs of jockeys.

Noise is contributed by

1. irregularities in resolution—a random type noise,
2. thermal motions of molecules that come in equilibrium with random motions of electrons giving rise to white/Johnson noise with equivalent voltage output as

   $$\{\langle V^2\rangle\}^{1/2} = \sqrt{4kTR\Delta f}\,,$$

3. contact non-uniformity mainly produced due to changing contact area and hence, contact resistance—aggravated by the presence of foreign particles in the area of

contact (contact area changes with use, also contamination and oxidation change the resistance and hence, noise),

4. rubbing action between the jockey and the wire—an equivalent of 100–300 μV is easily obtained with this rubbing action, and
5. thermoelectric action specifically at high temperatures and dc operations.

Sensitivity, under ideal unloaded condition of the potentiometer is the output voltage per unit travel of the jockey. Irregularities occur (i) at the potentiometer ends and (ii) due to power dissipation and corresponding rise in resistance of the potentiometer. Adequate corrections are to be made for these. A proper choice of the wire material with safety limit extended in current carrying capacity can minimize these errors to a certain extent.

As discussed earlier, with $n\%$ resolution of full scale (FS), the linearity of measurement in the scale is limited and the error on this count is smaller than $\pm(n/2\%)$ FS. Other factors that contribute to nonlinearity are (i) irregularities in winding pitch, (ii) mechanical uncertainties in jockey's movements, and (iii) tolerance/variation in wire and former dimensions and diameters. Linearity, better than the apparently calculated value can be obtained by using more number of turns than the theoretically calculated value.

The performance of the potentiometer changes in the loaded condition. Specifically, linearity is badly affected. Considering the circuit of Fig. 2.4, if R_L is the load resistance, V_i and V_o are input and output voltages respectively, R_i is the instantaneous tapped resistance across which V_o is obtained, and if the jockey begins movement from the bottom end, so that minimum $R_i = 0$ and maximum $R_i = R$, then

$$\frac{V_o}{V_i} = \frac{R_i/R}{1 + \dfrac{R_i}{R_L}\left(1 - \dfrac{R_i}{R}\right)} \tag{2.4}$$

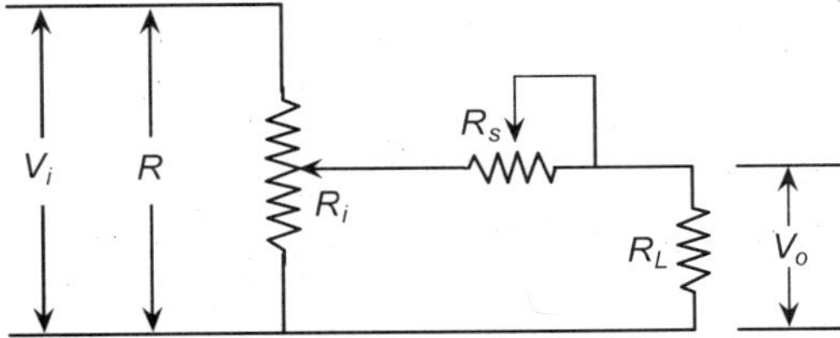

Fig. 2.4 Circuital method of drawing output from the potentiometer (for better linearity).

The figure also shows a variable series resistance R_s which is, in fact, optional and Eq. (2.4) has been obtained with $R_s = 0$. Generally, for ideal condition, $V_{oi}/V_i = R_i/R$. Representing R_i/R by ρ and R/R_L by λ, the percentage error in output–input voltage ratio is given as

$$\varepsilon = \frac{(V_{oi}/V_i - V_o/V_i)}{(V_{oi}/V_i)} \times 100$$

$$= \left(1 - \frac{1}{(1 + \lambda\rho\,(1 - \rho))}\right) \times 100 \tag{2.5}$$

Plots of ε versus ρ can be drawn with λ as a parameter, using Eq. (2.5), to show that the percentage deviation from linearity may be as high as 20% at $R_i = R/2$ for $R_L = R$. However, this is kept to within 1% by making $R_L \geq 20R$.

Alternate methods make use of

(i) a potentiometer which itself has nonlinear characteristics or

(ii) a nonlinear variable resistance R_s in series with the load.

The first method, in effect, proposes a design of the former, on which the winding is made, to have a nonlinear profile on the side the jockey moves. This nonlinear profile is such that the resistance ratio R_i/R curve drawn against the jockey movement (travel) is complementary to that of V_o/V_i.

In the second method, since R_s is also variable, a double jockey system—one for R and the other for R_s with equal lengths to move should be used. It can be shown that a resistance $R_s = R/4$ with parabolic resistance characteristics about an axis of symmetry at $x = 0.5$ are necessary for the purpose, where x is the normalized movement from 0 to 1.

As has already been mentioned, materials, both of the wire and the jockey are equally important. Table 2.1 shows a list of materials for the wire and the jockey which can be used in correspondence.

Table 2.1   Materials for wire and jockey

Wire	*Jockey*
1. Copper–nickel alloys like constantan (Cu 55–Ni 45), advance, ferry alloy, eureka and so on.	(a) Gold, gold–silver, (b) Ni 40–Ag 60, 10% graphite in Cu or 2–5% graphite in Ag.
2. Nickel–chromium alloys such as nichrome (Ni 80, Cr 20), Karma and so forth.	Group (b) above, and/or Rh or Rh-plated metals, gold–silver, osmium–iridium, Cu 40–Pd, ruthenium 10–Pt, Gold.
3. Silver–palladium alloys	Pt–iridium, Au 10–Cu 13–Ag 30–Pd 47.
4. Platinum–iridium	Pt–iridium

The wire is precision-drawn and annealed in a reducing atmosphere. The resistance per unit length varies from 0.25–1.5 $\mu\Omega$. The temperature coefficient of resistance is material-dependent and lies between $2 \times 10^{-5}/^\circ\text{C}$ and $10^{-4}/^\circ\text{C}$. Wire diameter tolerance is prescribed to be less than 5% at 0.025 mm.

2.3   STRAIN GAUGE

Although the basic principle of change in resistance of a metallic wire in response to strain produced in it was known as early as the mid-nineteenth century, its application in areas of commercial importance for measurement started becoming popular only about ninety years after that. Presently, the literature in strain gauges and their applications is so vast that it is difficult to prepare even a gist of all these in the folds of a section as proposed here. Strain gauges are of two types, namely the resistance type and the semiconductor type—the latter being of more recent origin.

2.3.1 Resistance Strain Gauge

Resistance strain gauges can be divided into two categories—(a) unbonded and (b) bonded—the former, being of limited use has received less attention than the latter. Unbonded strain gauge consists of a piece of wire stretched in multiple folds between a pair or more of insulated pins fixed to movable members of a 'body' or even a single flexible member whose strain is to be measured. There occurs a relative motion between the two members on strain and the wire gets strained as well with a corresponding change in its resistance value. The scheme of such a system is shown in Fig. 2.5.

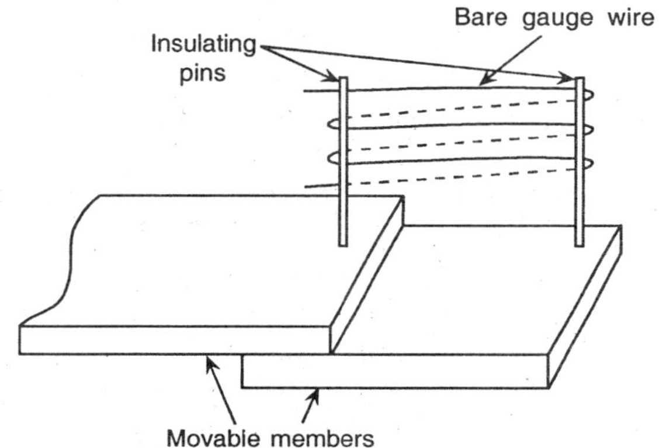

Fig. 2.5 Mounting an unbonded strain gauge.

The bonded type is more common and in its simplest form consists of wire/strip of resistance material arranged usually in the form of a grid for larger length and resistance value. The grid is bonded to the test specimen with an insulation layer between the gauge material and the specimen as shown in Fig. 2.6. If the insulation and the bonding material thickness is h which also is the height of the wire above the specimen surface and H is the distance of the neutral axis of the specimen from its surface, then the actual strain ε, in terms of measured strain ε_m, is given by

$$\varepsilon = \varepsilon_m \frac{H}{h + H} \tag{2.6}$$

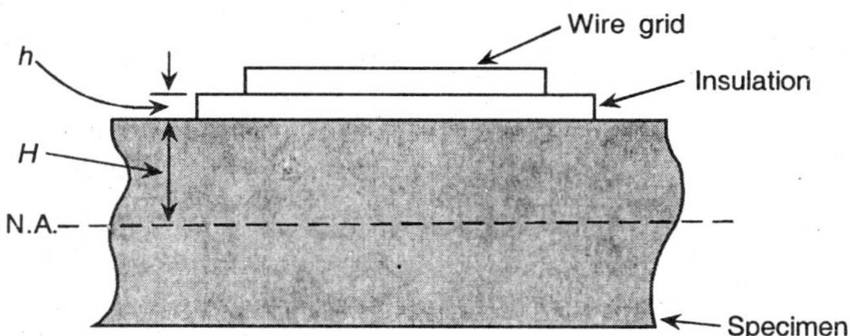

Fig. 2.6 Mounting a bonded strain gauge.

Depending upon the implementation, the resistance gauges can be classified as:

(a) Unbonded metal wire,
(b) Bonded metal wire,
(c) Bonded metal foil,
(d) Thin metal film by vacuum deposition, and
(e) Thin metal film by sputter deposition.

Considering a circular cross-section metal resistance wire of length l and cross-sectional area A with resistivity ρ of the material, the unstrained resistance of the wire is given by

$$R = \frac{\rho l}{A} \tag{2.7}$$

If the wire is uniformly stressed along its length (Fig. 2.7) and if the stress is given by σ, then

$$\frac{dR}{d\sigma} = \frac{d}{d\sigma}\left(\rho \frac{l}{A}\right) \tag{2.8}$$

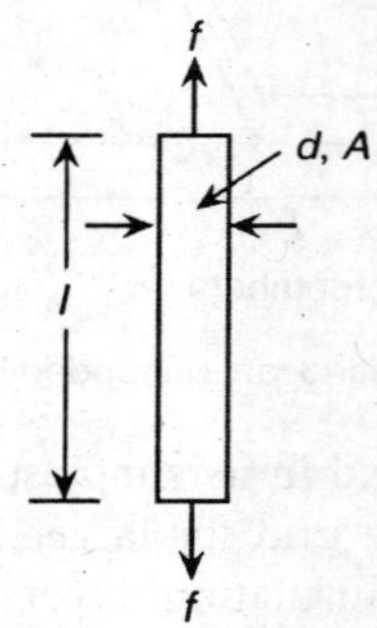

Fig. 2.7   Straining of an elastic member.

which gives

$$\left(\frac{1}{R}\right)\frac{dR}{d\sigma} = \left(\frac{1}{l}\right)\left(\frac{\partial l}{\partial \sigma}\right) - \left(\frac{1}{A}\right)\left(\frac{\partial A}{\partial \sigma}\right) + \left(\frac{1}{\rho}\right)\left(\frac{\partial \rho}{\partial \sigma}\right)$$

Eliminating all σ terms, we get

$$\frac{dR}{R} = \frac{\partial l}{l} - \frac{\partial A}{A} + \frac{\partial \rho}{\rho} \tag{2.9}$$

If the wire has a diameter d then the lateral contraction of the wire, $\Delta d/d = (1/2)\,(dA/A)$, is related to the fractional extension of the length, $\varepsilon = \Delta l/l$ by the Poisson's ratio μ as

$$\frac{\Delta d}{d} = -\frac{\mu \Delta l}{l} \tag{2.10}$$

so that Eq. (2.9) changes to

$$\frac{\Delta R}{R} = \left(1 + 2\mu\right)\frac{\Delta l}{l} + \frac{\Delta \rho}{\rho} \tag{2.11}$$

The strain sensitivity or the gauge factor λ is now defined as the ratio $(\Delta R/R)/(\Delta l/l)$ and is given by

$$\lambda = \frac{\Delta R/R}{\Delta l/l} = 1 + 2\mu + \frac{\Delta \rho/\rho}{\Delta l/l} \qquad (2.12)$$

It is generally assumed that resistivity of a metallic material is usually constant implying that the gauge factor λ is constant at 1.6 as most metal has a Poisson ratio of 0.3. It can have a maximum value of 0.5. But it is known that λ varies from metal to metal and under elastic strain its value is, in general, different from 1.6 meaning thereby that the resistivity also changes with strain. The last term on the right hand side of Eq. (2.12) is due to piezoresistance effect or Bridgeman effect and is often expressed as

$$\frac{\Delta \rho/\rho}{\Delta l/l} = \psi E \qquad (2.13)$$

where ψ is the Bridgeman or longitudinal piezoresistance coefficient and E is the modulus of elasticity.

As mentioned, μ has a maximum value of 0.5 which occurs in the plastic constant volume case so that when the change from essentially elastic to essentially plastic strain occurs, strain sensitivity also changes as shown in Figs. 2.8(a), (b), and (c). In Fig. 2.8(d), there is no change in strain sensitivity and the constant value is around 2 indicating that $(\partial \rho/\rho)/(\partial l/l)$ compensates for the gauge factor in the elastic strain region. Hard drawn nickel shows that its gauge factor is initially negative changing gradually to positive value. Minalpha has a slow and smooth transition and the change is not sharp. Obviously, the curve of Fig. 2.8(d) (i) is the most suitable one.

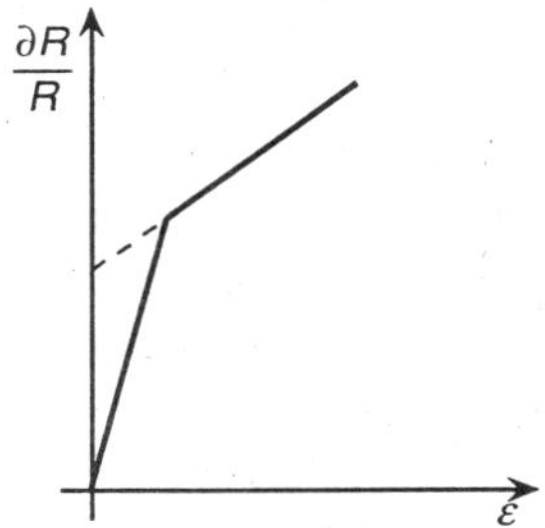

(a)   Fe, hard Cu, Ag, Pt, 10% Ir + Pt, 10% Rh + Pt

(b)   Ag (40%), Pd

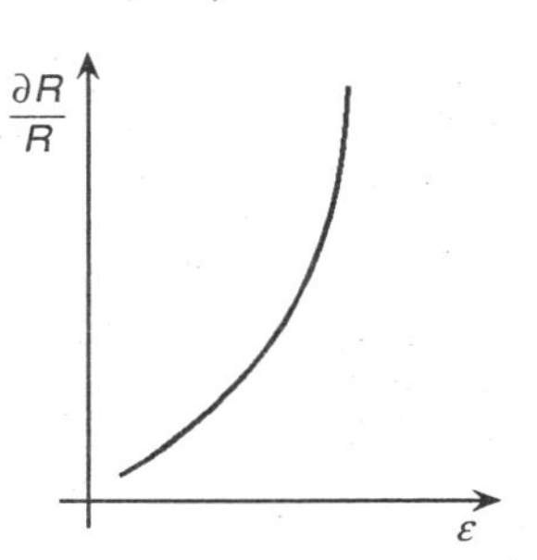

(c)   Minalpha

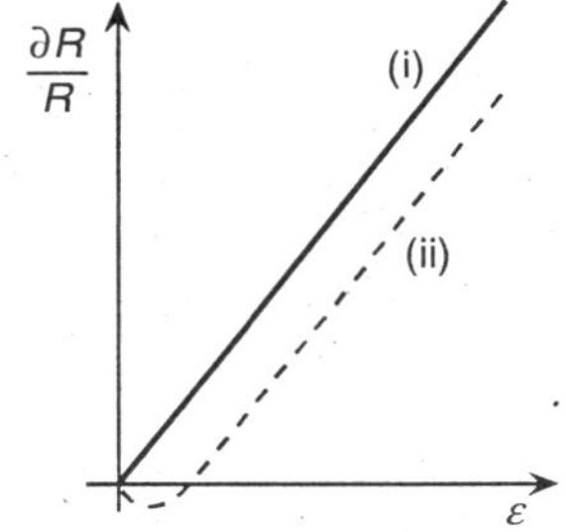

(d)   (i) Annealed Cu/Ni,   (ii) Hard drawn Ni

Fig. 2.8   Strain sensitivity for different materials.

Unbonded strain gauges are used in preloaded conditions not to allow the 'strings' to go slack. The wires are nickel alloys such as Cu–Ni, Cr–Ni, or Ni–Fe with gauge factor between 2 and 4 and diameters varying from 0.02–0.03 mm.

The bonded strain gauges are of a few types. When wire is used, the possibilities are (i) flat grid type, (ii) wrap around type, and (iii) woven type, although the flat grid type is more popular of all the three. Etched foil type resistance strain gauge is one variety that, in recent years, has most extensively been used.

A gauge consists of the resistance element of proper design/shape, the gauge backing, cement, connection leads, and often protective coating or other protective means.

The construction of the flat grid bonded strain gauge is shown in Fig. 2.9. Such a construction has the advantage of better strain transmission from the member to the wire grid, small hysteresis and creep, and is more accurate when the strain member is thin.

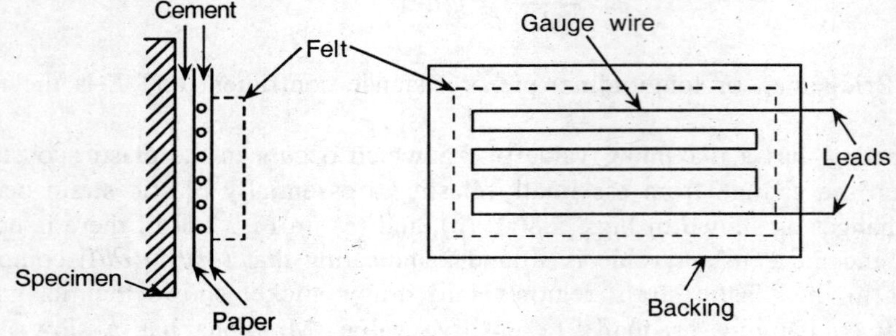

Fig. 2.9 Grid type gauge.

The foil gauges are etched out from deposited films or sheets and have higher surface area to cross-section ratio than wire gauges, and hence, have better heat transfer property so that they can handle higher current.

For wire gauges, the wires are usually drawn and often annealed, while bonded foil gauges consist of sensing elements which are formed from sheets of thickness less than 5×10^{-4} cm by photoetching processes so that any arbitrary shape can be given to these elements.

Because the wire grid in the grid type structure has a finite width, the gauge has a sensitivity to transverse strain which may be as large as 2% of the longitudinal sensitivity. In foil grid structure, the end turns can be made wider or fat enough so that the transverse strain sensitivity is lesser. A typical grid structure foil gauge is shown in Fig. 2.10.

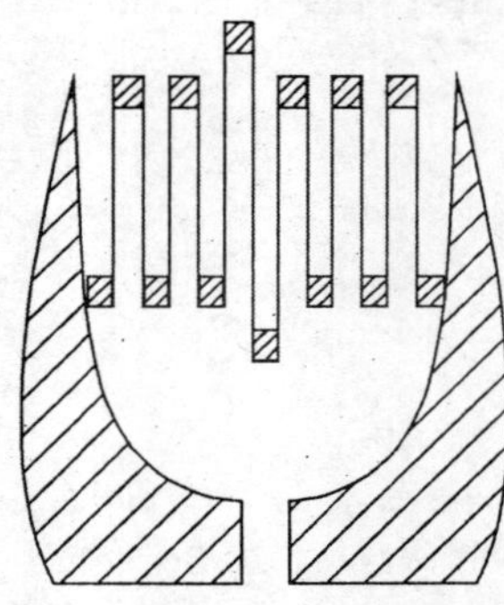

Fig. 2.10 Grid structure gauge with reduced transverse strain sensitivity.

Vacuum deposition and sputter deposition thin film gauges are produced where cement between the elastic element and the gauge is not necessary for bonding. In the former case, a suitable elastic metal element which can be adapted for strain generation such as a diaphragm for pressure measurement, is placed in a vacuum chamber with a suitable dielectric material of much lower vapour point than the metal. With application of requisite amount of heat, this dielectric material vapourizes and then condenses and finally forms a thin layer on the metallic member. A template of a suitable shape is now placed over it and the evaporation-deposition process is repeated with the gauge material. Thus, the gauge is formed over the insulator substrate.

In the sputtering-deposition process, the first step is nearly the same to form an insulating layer on the strain member. In the second step, without using a template, the metallic gauge material is sputtered over the entire substrate and the gauge pattern is defined by using micro-imaging techniques and photosensitive masking materials from outside the chamber and finally sputter-etching is used to remove all unmasked layers inside the vacuum chamber.

If strain members are not available or bonded metal foil gauges are required to be produced for 'general purpose' uses, the gauges are produced on flexible insulating carrier films such as polyimide and glass-reinforced phenolic having a thickness of about 0.002 cm.

One important aspect of resistance gauges is its temperature coefficient of resistance. The temperature at which the strain is measured may be different from the temperature when the strain member is bonded. This gives rise to a differential expansion between the strain member and the gauge resulting in error in strain measurement. Table 2.2 shows the list of resistance type strain gauge materials with corresponding gauge factors and temperature coefficient of resistance along-with the resistivity values.

Table 2.2 Strain gauge materials and their properties

Material	Approx. nominal composition (%)	Gauge factor	Thermal coefficient of resistance (%/°C)	Nominal resistivity ($\mu\Omega$ cm)
Constantan, Advance, Ferry	Ni 45, Cu 55	2.1–2.2	2×10^{-3}	0.45–0.48
Karma	Ni 74, Cr 20, Fe 3 Cu 3	2.1	2×10^{-3}	1.25
Nichrome V	Ni 80, Cr 20	2.2–2.6	10^{-2}	1.00
Isoelastic	Ni 36, Cr 8, Fe 52, Mn–Si–Mo 4	3.5–3.6	1.75×10^{-2}	1.05
Pt–W alloy	Pt 92, W 8	3.6–4.5	2.4×10^{-2}	0.62
Nickel	Ni 100	12	0.68	0.65
Manganin	Cu 84, Mn 12, Ni 4	0.3–0.48	2×10^{-3}	—
Platinum	Pt 100	4.8	0.4	0.1

The adhesives used to bond the gauge (backings) to the elastic member to be strained should be carefully selected. They must

    (a)   transmit the strain fully from the member surface to the gauge,
    (b)   have high insulation property,
    (c)   have high mechanical strength,
    (d)   have low thermal insulation,
    (e)   be as thin as possible yet provide strong bonding, and
    (f)   be suited to the environment, specifically the metal–paper and metal–dielectric interfacing.

Table 2.3 gives the properties of a few adhesives specially made for bonding strain gauges.

Table 2.3 Properties of adhesives

Material-base	Temperature range (°C)	Cure-time (hrs)	Cure pressure kg/cm^2	Max. strain at room temp. (%)	Recommended lifetime (yrs)
Acrylic	−75–65	1/12	Normal	10–15	1/2
Nitrocellulose	−75–65	24–48	1/2–1	10–15	2
Epoxy	0–200	12–24	1–3	6	1
Epoxy-phenolic	0–220	2	2–3	3–4	1
Polyimide	0–400	2–3	2.5–3	2–3	1/3
Ceramic	0–700	1	—	1/2	1

Acrylic has long term instability, nitrocellulose is a general purpose adhesive. Epoxy is resistant to moisture and has long term stability while epoxy-phenolic can be used in a thinner layer than the others. Polyimide and ceramic-base cements can be used at high temperatures though the latter is not very commonly used.

The recommended value of electrical insulation is of the order of 10^9 ohm at 50 V dc. If this value is not complied with, the gauge is likely to be 'shorted' and reading is susceptible to error. Most of the adhesives are vulnerable to high temperature and moisture/humidity which deteriorate their insulating as well as mechanical properties. Epoxy-base adhesives have been produced in various combinations with resins and hardeners for improving their properties.

Other than the adhesives given in Table 2.3, flame-spray and welding techniques have also been developed and are specifically used in some cases of free filament wire gauges. In the flame-spray, a solid rod is atomized to produce a ceramic spray which solidifies on the wires of the strain gauge making a bond without damaging the gauge or the strain member. This can be used upto about 800°C from near absolute zero while in the welding technique, the gauge is first epoxied to a thin metal shim. With low energy spot welder, the shim is then attached to the specimen. The foil gauges are specifically suitable with shim of thickness varying between 0.1–0.12 mm.

Gauges are made available in combinations often called 'rosettes' and these are designed in various configurations for specific stress–strain analysis and/or for transducer applications. A number of gauges are given relative orientations following certain pattern for the purpose. Thus, a three-gauge rosette used in stress analysis solves problems of a surface stress in magnitude and direction. Since the stress/strain is necessary to be measured at a point, it is best to stack these three gauges to form a rosette on that point. In fact, this sandwich pattern rosette is available from the manufacturers under the name 'stacked rosette'. Figure 2.11(a) shows such a three element rosette stacked at 45° to each other. In this, the topmost gauge is farthest from the specimen and all the gauges are insulated from each other, the topmost gauge gets heated up more compared to the bottommost which use the specimen as the heat sink. Two element stack type design is also commercially available. Such a design has an advantage that the strain/stress at the same point is sensed by all the gauges.

The alternative to the stack type design is the planar design which covers a small area rather than a point. Rosettes with such a design are available in two element 90° planar—usual and shear, three element 45°, 60° planar. They can be generated on the specimen as well. Figures 2.11(b), (c), and (d) show some of the types.

In fact, the technology of generating gauges on the specimen itself or on substrate as mentioned earlier by vacuum process has lead to wide scope of gauge pattern variation. It can be of any type depending on the specific requirements. The number of gauges at a location can also be changed as per this practice. Figure 2.11(e) shows a gauge pattern variation for measurement of strain in a diaphragm. Gauges 1 and 3 are subjected to tensile tangential stress while gauges 2 and 4 are subjected to compressive radial stress. 1′, 2′, 3′, and 4′ are contact terminals.

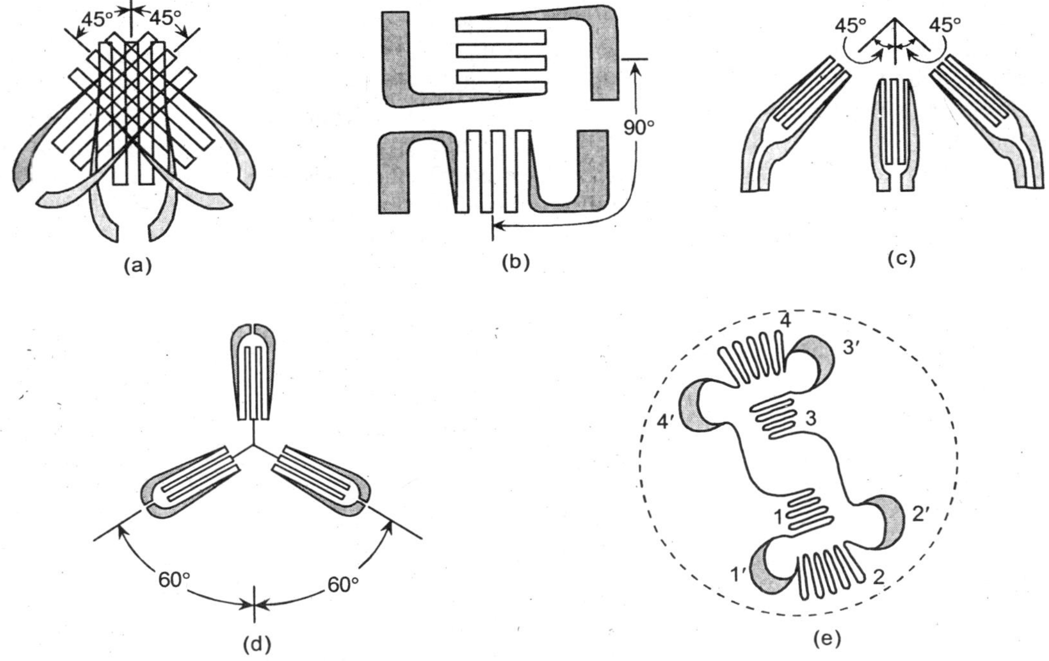

Fig. 2.11   Rosetted strain gauges: (a) three-element stacked type, (b) two-element right-angled, (c) three elements at 45° to each other, (d) three elements at 120° to each other, (e) gauge pattern on a diaphragm.

2.3.2 Semiconductor Strain Gauges

First lot of semiconductor strain gauges were produced early in mid-thirties from single crystal silicon or germanium by cutting thin strips. Lot of work has since been done and is still being done on the improvement of their performance and manufacturing ease because it has been known that although the semiconductor gauges have higher gauge factors, they are much inferior to the resistance types in so far as linearity and temperature stability are concerned (specifically the latter). But the discovery of semiconductor strain gauges has cleared the path of smart sensors, including production of strain sensitive cantilevers and diaphragms by doping selected small areas of monolithic silicon slice. Semiconductor strain gauges can be divided into two classes—(i) bonded semiconductor and (ii) diffused semiconductor—depending on their implementation.

Strain sensitivity of semiconductor material depends, among other things, on the crystal material such as Si or Ge, doping levels (if any), type of doping materials, crystal cut-axis orientation, and so on. Because the bandgaps both in intrinsic and extrinsic semiconductors are affected by temperature variation, semiconductor gauges are more prone to temperature variations. For intrinsic semiconductors, gauge factors are larger decreasing with increasing degrees of doping, the thermal coefficients of resistivity also decrease correspondingly.

As has been shown, the gauge factor of strain gauge is given by the relation

$$\lambda = 1 + 2\mu + \psi E \tag{2.14}$$

The strain sensitivity of a semiconductor gauge is high and the large value is due to the large value of ψE, that is, $(\Delta\rho/\rho)/(\Delta l/l)$, specifically ψ. The value of Poisson's ratio for semiconductors is less than that of metals although it is more in Si than Ge. Table 2.4 shows different values of Young's moduli (E), μ's, ρ's, and λ's for different Si and Ge crystals.

There are a number of piezoresistive coefficients in a semiconductor material, they are called 'fundamental'. The longitudinal piezoresistive coefficients, in which the stress and current are in the same direction and the transverse piezoresistive coefficients, in which the stress and current are perpendicular to each other, are computed from these fundamental coefficients and the direction cosines of the current with respect to the crystallographic axes.

Table 2.4 Properties of semiconductor gauges

Material with crystal orientation	μ	E (10^{10}N/m^2)	ρ $(10^{-3}\ \Omega\text{m})$	λ (longitudinal)	Thermal coefficient of resistance β $(10^{-5}/°C)$
p-Si (111)	0.180	18.7	78	175	$70 \le \beta \le 700$
n-Si (100)	0.275	13.0	118	−135	$70 \le \beta \le 700$
p-Ge (111)	0.155	15.5	150	105	$70 \le \beta \le 700$
n-Ge (111)	0.156	15.5	160	−155	$70 \le \beta \le 700$

Practical aspect of using a semiconductor strain gauge is governed by λ, R, gauge length, encapsulation/backing, bonding, leads geometry, and means of temperature compensation. Size and shape of the gauge are equally important. Some possible and useful shapes are given in Figs. 2.12(a), (b), (c), and (d). Sizes are determined by the specimen size as also resistance value R.

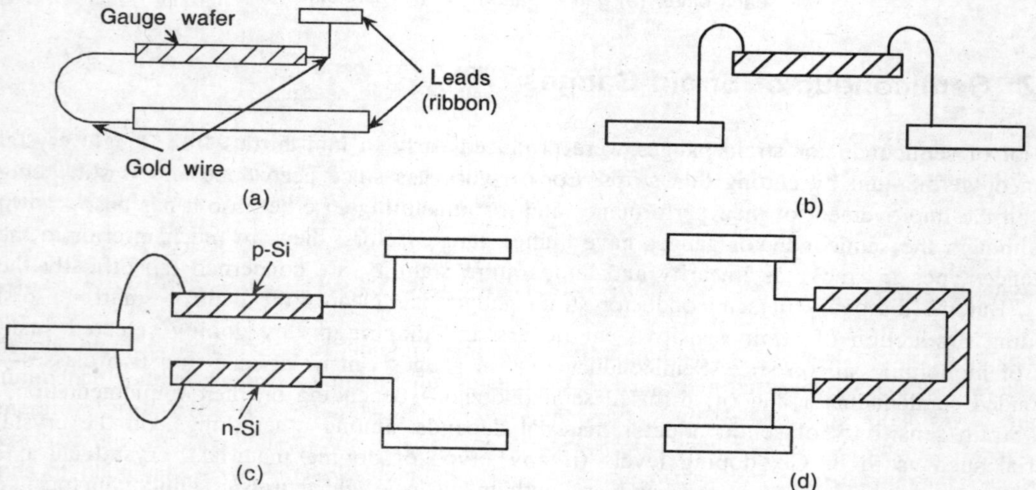

Fig. 2.12 Semiconductor gauges of different shapes and mountings.

Semiconductor gauges with/without backing are bonded to the specimen with epoxy-based adhesives, or for better, diffused semiconductor gauges are attached to the specimen by semiconductor diffusion process. The gauge is diffused directly on to the surface of the specimen such as a diaphragm, using photolithographic masking technique and an impurity such as boron is diffused into it. No separate bonding is necessary here. In recent times, the specimen, that is the strained member such as a cantilever or a diaphragm itself is also made from Si and the whole unit is developed into a smart sensor. A diaphragm of 2.5–25 mm diameter or cantilever of appropriate size is obtained in the main substrate of Si which is 50–750 mm in diameter. The four arm bridge is developed on this diaphragm as also the circuit of measurement by diffusion process.

The semiconductor strain gauge is basically nonlinear and an empirical relation between $\Delta R/R$ and ε

$$\frac{\Delta R}{R} = \sum_{j=1}^{n} k_j \varepsilon^j \tag{2.15}$$

is suggested, where k_j's are constants that depend on the materials and doping levels. Also, at high stress conditions temperature dependence of these coefficients are observed. Nonlinearity has been found to be improved by heavily doping the basic material of lower resistivity but then strain sensitivity is less. Often approximation by truncating the series upto $j = 2$ is good enough for practical use. Thus, an n-Si gauge of $\rho = 3.1 \times 10^{-4}$ ohm m would have

$$\lambda = -110 + 10^5 \varepsilon$$

and a p-Si with $\rho = 0.2 \times 10^{-3}$ ohm m would have

$$\lambda = 120 + 4 \times 10^4 \varepsilon$$

However, with higher resistivity such as $\rho = 78 \times 10^{-3}$ ohm m, a p-Si has a gauge factor (see Table 2.4)

$$\lambda = 175 + 7.26 \times 10^4 \varepsilon$$

As has been mentioned already, increasing doping decreases sensitivity towards temperature as well. Figure 2.13 shows the temperature–gauge factor curves for varying degrees of doping of a semiconductor gauge.

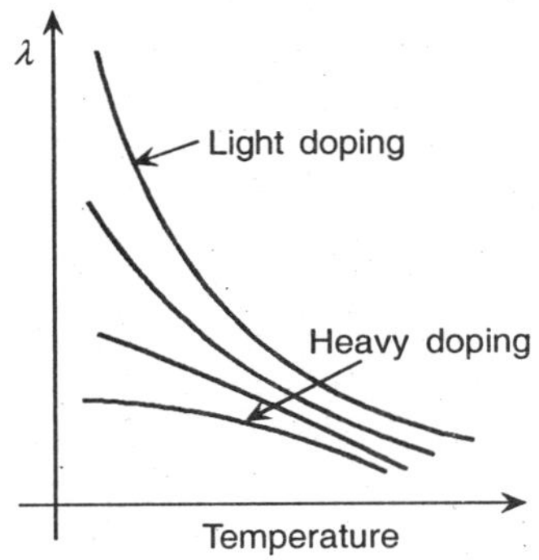

Fig. 2.13 Gauge factor versus temperature plots for different doping levels.

In fact, doping changes the gauge resistance as well, decreasing it with high doping level. Figure 2.14 shows the ρ–T characteristics with doping as a parameter. Figure shows that higher doping gives high value of ρ and β, the temperature coefficient of resistance—positive as well as high. But this occurs only upto a certain temperature above which the material behaves as in intrinsic conduction mode with negative temperature coefficient also of a very high value. However, with heavy doping, ρ is moderate and β quite small, and this condition persists over a wider temperature range.

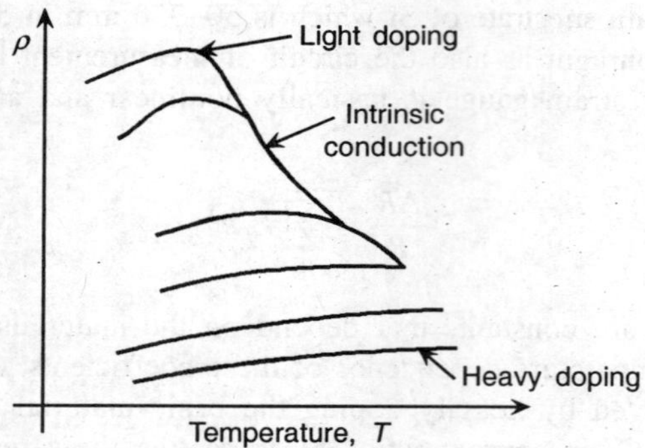

Fig. 2.14   Resistivity versus temperature for doping level variations.

As has been discussed already, semiconductors under strain show piezoresistive effect which is so predominant over other effects related to Poisson's ratio and so on that based only on this dominant effect, pressure transducers have been produced and within the elastic limits of silicon, electrical output is found proportional to mechanical strain or stress. The scheme consists of a cantilever beam of silicon about 0.1 mm thick on to both sides of which planar resistors are produced by diffusion. Figure 2.15 shows the scheme with the header with connecting terminals. With the beam under stress, the resistors on the two sides of the beam undergo different changes because of compression on one side and extension on the other. The difference is measured by a bridge. The length l can be inserted in a pressure cell where a diaphragm actuated by the inlet pressure is so mounted and attached to the cantilever that the deformation of the diaphragm is transmitted to the cantilever and hence, to the diffused resistance gauges.

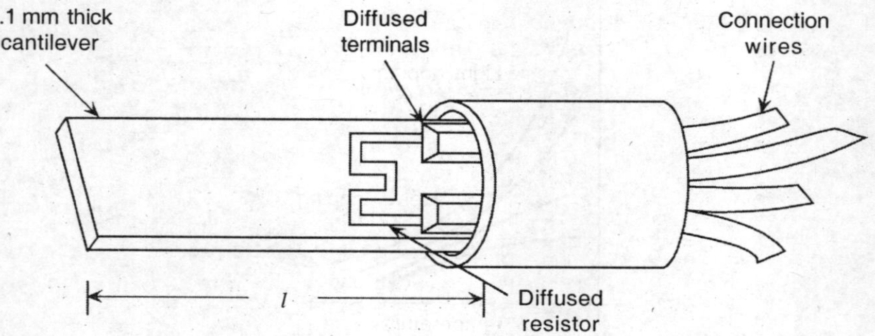

Fig. 2.15   Sensor using semiconductor piezoresistive effect.

For pressure measurement, thin silicon diaphragms with diffused resistors have been developed. A typical scheme is shown in Fig. 2.16. The piezoresistors are usually embedded in the diaphragm so that they get the strain of the diaphragm unabated.

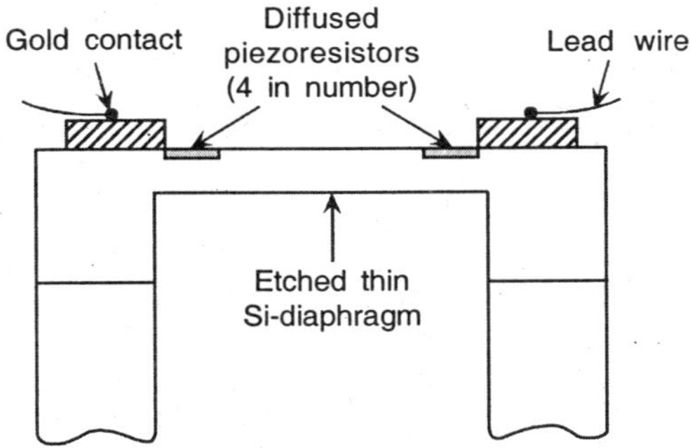

Fig. 2.16   Pressure measurement scheme using semiconductor diaphragm.

2.4 INDUCTIVE SENSORS

Although specific cases of inductive technique of sensing and/or transducing have been dealt in detail in Chapter 4 on magnetic sensors, a generalized discussion on inductive sensing is given in this section.

The inductive transducer utilizes the simple principle that the physical quantity, such as motion, to be measured can be made to vary the inductance of a coil, maintaining a relation between the two. This variation of inductance can often be measured by ac bridge circuits, or can be made to produce a voltage if it is magnetically coupled to another coil carrying a flux or voltage. If a magnetostrictive core material is used, force or pressure can change the permeability which can be measured as a change in inductance of a coil around the core.

The two most common methods of achieving variation in inductance are (i) by changing the reluctance of the magnetic path and (ii) by coupling two or more elements. The latter technique works by (a) change of mutual inductance, (b) change of eddy current when one element is just a short-circuited sleeve, and (c) transformer action. These are shown schematically in Figs. 2.17(a), (b), (c), and (d) respectively.

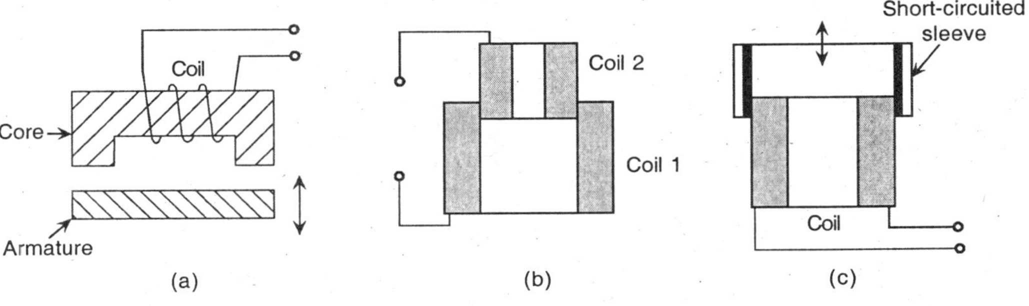

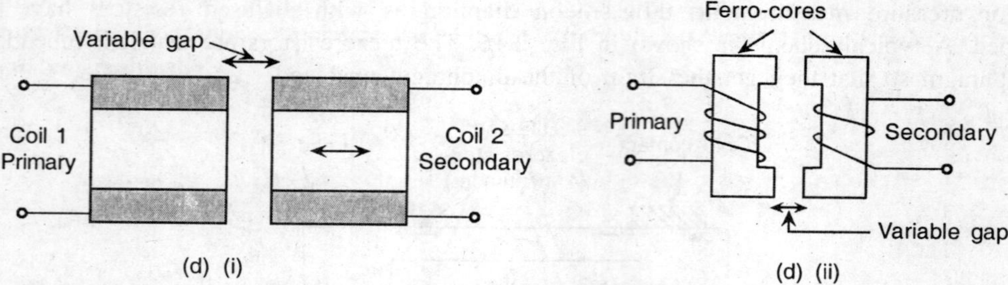

Fig. 2.17 Inductive sensors using (a) change of reluctance of magnetic path, (b) change of mutual inductance between two coils, (c) change of mutual inductance between a coil and a sleeve, and (d) (i) and (ii) transformer action.

Then there are inductive sensors of (i) the electromagnetic type which are bilateral in operation with electrical and mechanical input/output relationship and (ii) the magnetostrictive type. A sensor that uses a magnetostrictive core material is shown in Fig. 2.18.

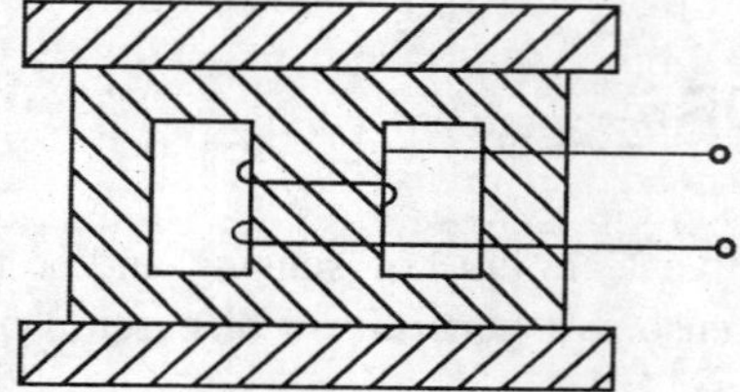

Fig. 2.18 Sensor using a magnetostrictive effect.

Inductance variation can also be achieved by variation of coil geometry such as coil length but such a procedure is not very convenient to be adopted in practice.

It is observed that a coil is an essential part of inductive transducers and the coil may be wound on a metal (iron) core or an air core. In the variable reluctance type, the core is a ferromagnetic material as also the armature. This type of sensors are, perhaps, the most extensively used because it (i) is the most sensitive one, (ii) is least affected by external fields as the air gap is least, and (iii) requires less number of turns than in air core design for same value of inductance so that interwinding or self-capacitance and stray effects are less. The copper coil on a ferromagnetic core has an equivalent circuit that consists of an inductance L in series with copper loss resistance R_c and a resistance R_e, representing eddy loss resistance in the core in parallel with L. Interwinding or self-capacitance, important specially at high frequencies, is in parallel to the coil resistance R_c and inductance L. The equivalent circuit is shown in Fig. 2.19.

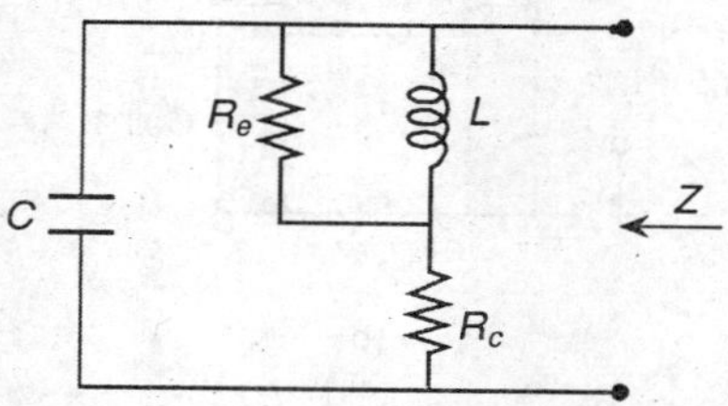

Fig. 2.19 Equivalent circuit of a ferromagnetic coil.

If a coil has n turns, a current I, and the core length l, the field strength H is given by

$$H = \frac{nI}{l} \quad \text{(A/m)} \tag{2.16}$$

For a core material of permeability μ, which often is expressed as the product of its relative permeability and the permeability of the free space or vacuum ($\mu_0 = 4\pi \times 10^{-7}$ H/m), and core cross-section area a, the self inductance L of the coil is the flux linkage per unit current so that

$$L = \frac{n\phi}{I} = n\frac{Ba}{I} = n\frac{\mu Ha}{I} \tag{2.17}$$

where B is in Tesla or Wb/m^2 and ϕ is in Wb.

Using Eq. (2.16), one derives

$$L = \frac{\mu n^2 a}{l} \quad \text{(Henries)} \tag{2.18}$$

The copper resistance R_c is also easily calculated if the coil wire diameter d and the copper resistivity ρ are known, so that

$$R_c = \frac{4\rho n l_t}{\pi d^2} \tag{2.19}$$

where l_t is the average length per turn of the coil. The coil dissipation factor D_c is usually defined as

$$D_c = \frac{R_c}{\omega L} \tag{2.20}$$

which decreases with increasing frequency.

For reducing eddy loss or core loss as it is called (the core is usually made of laminations of certain thickness, say t_l), the depth of penetration of eddy current, d_p is given by

$$d_p = \sqrt{\frac{\rho_e}{\pi \mu f}} \tag{2.21}$$

where ρ_e is the resistivity of the core material and $f = \omega/(2\pi)$ is the frequency. The eddy loss resistance is then given by

$$R_e = \left(\frac{2d_p \omega L}{t_l}\right)\left[\frac{\cosh\left(\dfrac{t_l}{d_p}\right) - \cos\left(\dfrac{t_l}{d_p}\right)}{\sinh\left(\dfrac{t_l}{d_p}\right) - \sin\left(\dfrac{t_l}{d_p}\right)}\right] \tag{2.22}$$

Equations (2.21) and (2.22) are valid only for low frequencies when $\rho_t = (t_l/d_p) \leq 2$. The frequency range, however, varies depending on the core material as well as lamination thickness. Figure 2.20 shows the plots of f versus t for different materials of commercial importance for $\rho_t \approx 2$, so that within this range of frequency Eq. (2.22) can be simplified using Eqs. (2.18) and (2.21) as

$$R_e \approx \frac{6\omega L}{(t_l/d_p)^2} = \frac{12\rho_e a n^2}{(l t_l)^2} \tag{2.23}$$

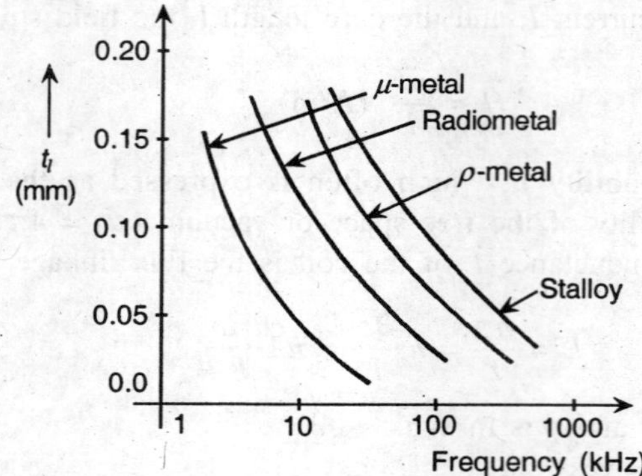

Fig. 2.20 Sheet thickness versus frequency plots for different magnetic materials.

This figure (Fig. 2.20) shows what frequency range can be covered by a specific material with specified thicknesses.

The eddy loss dissipation factor is defined by

$$D_e = \frac{\omega L}{R_e} \tag{2.24}$$

and is directly proportional to frequency.

Magnetic material undergoes hysteresis and this causes dissipation or loss. The area within the hysteresis curve is given by

$$A_h = \int B \cdot dH \tag{2.25}$$

where H is the magnetic field strength and B is the magnetic induction.

The B–H loop for a ferromagnetic material is schematically shown in Fig. 2.21. Following Rayleigh's procedure, the area A_h has been computed and hence, the energy dissipated per unit volume. For a core of cross-sectional area a, and length l, total hysteresis loss, in this way, is obtained as

$$P_h = \left(\frac{16\pi}{3}\right) a l \alpha_r H_l^3 f \times 10^{-7} \quad \text{(watts)} \tag{2.26}$$

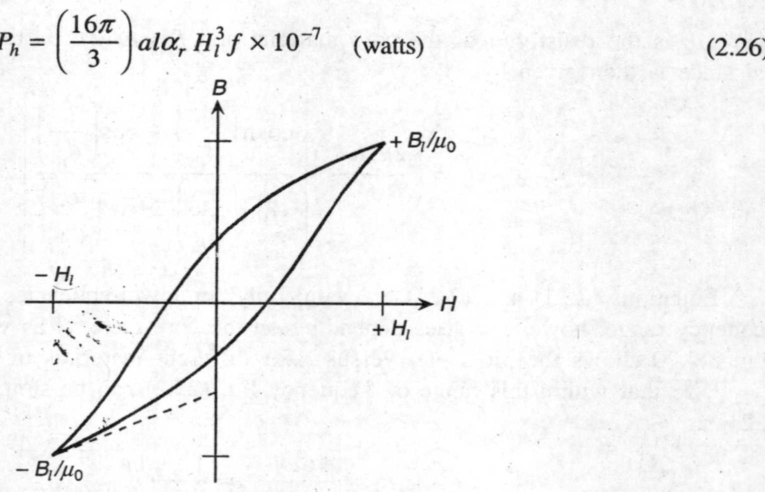

Fig. 2.21 The B–H loop for a magnetic material.

where α_r is the Rayleigh's constant which may be defined by the equation

$$\alpha_r = 2\frac{\left(\dfrac{\Delta B}{\mu_0} - \mu_i H\right)}{(\Delta H)^2} \tag{2.27}$$

where μ_i is the initial permeability, that is, permeability at $H = 0$. With change from zero values of B and H, Eq. (2.27) is written as

$$\alpha_r = \frac{2(B/\mu_0 - \mu_i H)}{H^2} \tag{2.28}$$

Using $P_h = E^2/R_h$, R_h, being the equivalent hysteresis loss resistance, is

$$R_h = \frac{\omega^2 L^2 I^2}{P_h} \tag{2.29}$$

which is proportional to the square of the frequency. However, the hysteresis dissipation factor D_h is given by

$$D_h = \frac{\omega L}{R_h} = \frac{2\alpha_r H_l}{(3\pi \mu_i)} \tag{2.30}$$

which is independent of frequency.

A sensor or a transducer involves the movement of an armature, that is, the situation demands that the core has an air gap, the length of which varies with the value of the measured quantity such as a motion. This is taken into consideration by determining the effective permeability of the core when the sample permeability μ_s is known and a relation between L and the gap length l_g can be found. Thus, for a torroidal ring sample of total path length l, gap length l_g, cross-sectional area a, the effective permeability μ, we obtain

$$\frac{\left(\dfrac{(l - l_g)}{\mu_s} + l_g\right)}{a} = \frac{l}{\mu a} \tag{2.31}$$

yielding

$$\mu = \frac{\mu_s}{\left\{1 + \left(\dfrac{l_g}{l}\right)(\mu_s - 1)\right\}} \tag{2.32a}$$

Since $\mu_s \gg 1$,

$$\mu \approx \frac{\mu_s}{\left\{1 + \left(\dfrac{l_g}{l}\right)\mu_s\right\}} \tag{2.32b}$$

Substituting this in Eq. (2.18),

$$L = \left[\frac{\mu_s}{\left\{1 + \left(\dfrac{l_g}{l}\right)\mu_s\right\}}\right]\left(\frac{n^2 a}{l}\right) \quad \text{(Henries)} \tag{2.33}$$

Before moving on to the analysis of change of inductance with air gap and its nature, the effect of the capacitor C of Fig. 2.19 is considered. This capacitance arises, as already mentioned, due to the coil self-capacitance, that is, interwinding capacitance as also due to the connecting cable capacitance. The effect of parallel resistance R_e can be considered in series with the inductance so that the total series resistance R, is then used to calculate the impedance Z as

$$Z = \frac{R + j\omega L}{(1 - \omega^2 LC) + j\omega RC} \tag{2.34}$$

which, on rationalization, can be written as

$$Z = \frac{R}{(1 - \omega^2 LC)^2 + (\omega^2 LC/Q)^2} + j\omega L \frac{(1 - \omega^2 LC) - (\omega^2 LC/Q^2)}{(1 - \omega^2 LC)^2 + (\omega^2 LC/Q)^2} \tag{2.35}$$

where $Q = L/R$.

For a good inductor with $Q^2 \gg 1$, we get

$$Z = \frac{R}{(1 - \omega^2 LC)^2} + \frac{j\omega L}{(1 - \omega^2 LC)} = R_{eq} + j\omega L_{eq} \tag{2.36}$$

indicating that both R_{eq} and L_{eq} increase but the effective Q, Q_{eq} decreases

$$Q_{eq} = \frac{\omega L \, (1 - \omega^2 LC)}{R} \tag{2.37}$$

2.4.1   Sensitivity and Linearity of the Sensor

For a small air gap l_g and effective permeability of the core μ, the inductance is given by Eq. (2.33). Now since n and a are constants, using

$$K_l = 4\pi \times 10^{-7} n^2 a \tag{2.38}$$

Equation (2.33) can be written as

$$L = \frac{K_l}{\left(l_g + \dfrac{l}{\mu_s} \right)} \tag{2.39}$$

from which assuming $l \gg l_g$, for small increase or decrease in gap l_g and ∂l_g,

$$\frac{\partial L}{L} = \frac{\partial l_g}{\left(l_g \pm \partial l_g + \dfrac{l}{\mu_s} \right)}$$

$$= \frac{\partial l_g / l_g}{1 + \dfrac{l}{l_g \mu_s}} \cdot \frac{1}{1 \pm \dfrac{(\partial l_g / l_g)}{\left\{ 1 + l/(l_g \mu_s) \right\}}} \tag{2.40a}$$

and for $(\partial l_g / l_g)/(1 + l/(l_g \mu_s)) \ll 1$.

$$\frac{\partial L}{L} = \frac{\partial l_g/l_g}{1 + \dfrac{l}{l_g\,\mu_s}}\left[1 \mp \frac{\partial l_g/l_g}{1 + \dfrac{l}{l_g\,\mu_s}} + \left(\frac{\partial l_g/l_g}{1 + \dfrac{l}{l_g\,\mu_s}}\right)^2 \mp \cdots\right] \tag{2.40b}$$

If only the first term is accepted for ∂l_g being very small, there appears to be a linear variation between L and l_g, and the sensitivity $S_{l_g}^L = (\partial L/L)/(\partial l_g/l_g)$ is given as

$$S_{l_g}^L = \frac{1}{1 + l/(l_g\,\mu_s)} \tag{2.41}$$

However, presence of higher order term increases nonlinearity. Figure 2.22 shows the nature of L versus l_g curve. It is possible to have two coils in the variable inductance transducer such that inductance in one coil increases and that in the other decreases. This can be adapted in the plunger type design, discussed later, where a push–pull arrangement of the coils and their connections would produce an output which is the sum of the fractional changes in the values of inductances in the two coils. This would make the even order terms in Eq. (2.40b) disappear and result in improvement of linearity over a wider gap range as shown in Figs. 2.23(a) and (b).

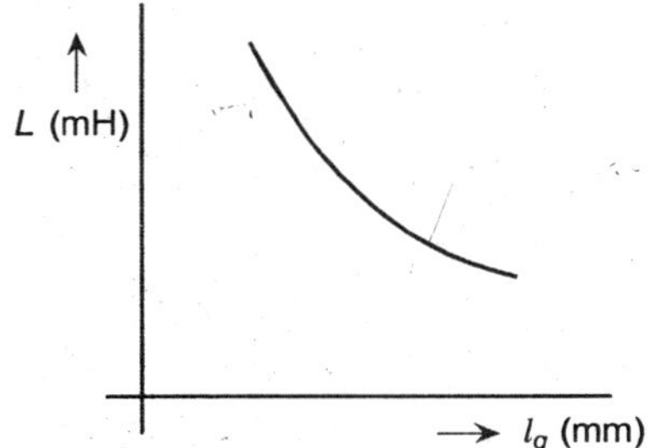

Fig. 2.22   Variation of inductance with air gap.

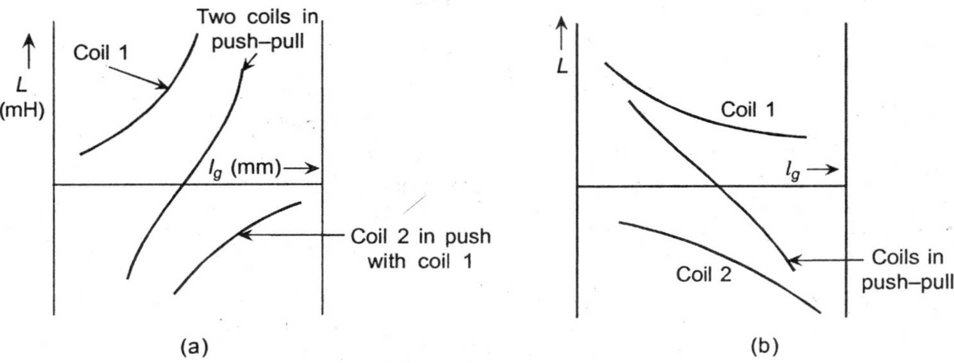

Fig. 2.23   Linearity improvement by using two coils in push–pull.

It must be remembered that air gap is likely to vary because of the eddy loss effect that is assumed parallel to the inductance which when transferred to series path, is given by

$$R_{es} = R_e(1 + Q^2)^{-1} \approx \frac{R_e}{Q^2} \tag{2.42}$$

As Q contains L as well as frequency f and L being a functions of μ, the resistance R_{es} changes with change in μ or in air gap. The values can actually be computed by using the equations that have been discussed here.

2.4.2 Ferromagnetic Plunger Type Transducers

A variation of the variable air gap core type design is the one in which usually a solid ferromagnetic plunger moves inside a helical coil wound on a ferromagnetic sleeve, such that the inductance of the coil depends on the partial core length inside the coil. Such a design can also be analyzed with an equivalent air gap which is now large. The coil itself can be split up into two equal parts such that in their push–pull operation, the field is constant and the inductance has a linear range over wider range of movement of the core. In this case, the sensitivity is also larger.

Figures 2.24(a) and (b) show a single coil with its magnetic field distribution and Figs. 2.25(a) and (b) show a double coil design with field distribution on both sides with respect to the central point.

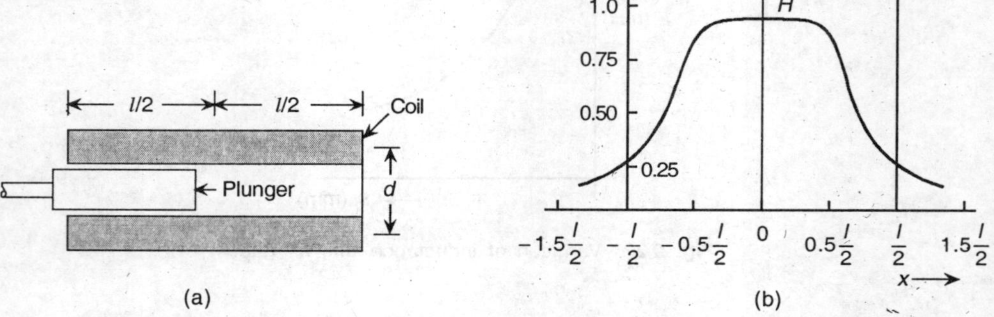

(a)   (b)

Fig. 2.24   (a) Single coil plunger type transducer design, (b) field versus length plot for the system.

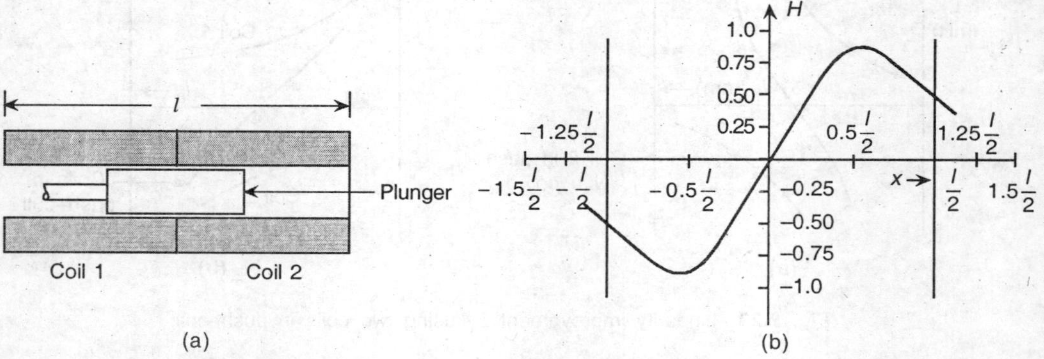

(a)   (b)

Fig. 2.25   (a) Double coil transducer, (b) the response plot.

For a current I in the coil of Fig. 2.24(a) with the coil length l, diameter d, and number of turns n, the field strength along the axis is given by

$$H = \frac{nI}{2l}\left[\frac{l + 2x}{\sqrt{d^2 + (l + 2x)^2}} + \frac{l - 2x}{\sqrt{d^2 + (l - 2x)^2}}\right] \tag{2.43a}$$

while for push–pull coils 1 and 2 design of Fig. 2.25(a), the field strength is given by the relation

$$H = \frac{nI}{2l}\left[\frac{l - 2x}{\sqrt{d^2 + (l - 2x)^2}} - \frac{l + 2x}{\sqrt{d^2 + (l + 2x)^2}} + \frac{2x}{\sqrt{(d/2)^2 + x^2}}\right] \tag{2.43b}$$

The corresponding field plot is shown in Fig. 2.25(b).

A coil, theoretically of a very large length l, having n number of single layer turns and radius $d/2$ has an inductance [see Eq. (2.18)]

$$L = \left(\frac{\pi^2 n^2 d^2}{l}\right) \times 10^{-7} \quad \text{(H)} \tag{2.44}$$

A ferromagnetic plunger of diameter d_p, length l_p with $l_p < l$, covering the middle part of the coils-length and having effective permeability μ_p, the inductance increases to

$$L_p = \left(\frac{\pi^2 n^2}{l^2}\right)\left[ld^2 + (\mu_p - 1)l_p\, d_p^2\right] \times 10^{-7} \quad \text{(H)} \tag{2.45}$$

With the movement of the plunger, there occurs a change in l_p, say ∂l_p, and correspondingly L_p changes by ∂L_p, and if in coil 1 there is increase in a parameter, coil 2 observes a decrease in the same. The change ∂L_p in a coil is given by

$$\partial L_p = \frac{\pi^2 n^2 d_p^2 (\mu_p - 1)\, \partial l_p}{l} \times 10^{-7} \quad \text{(H)} \tag{2.46}$$

so that per unit change in ∂L_p is

$$\frac{\partial L_p}{L_p} = \frac{\partial l_p}{l_p}\left(\frac{1}{1 + \left(\dfrac{l}{l_p}\right)\left(\dfrac{d}{d_p}\right)^2\left(\dfrac{1}{\mu_p - 1}\right)}\right) \tag{2.47}$$

In the other coil, same change occurs with opposite sign. If the length of the plunger is the same as that of the coil sleeve, this change is given by

$$\frac{\partial L_p}{L_p} = \frac{\partial l_p}{l_p}\left(\frac{1}{1 + \left(\dfrac{d}{d_p'}\right)^2\left(\dfrac{1}{\mu_p - 1}\right)}\right) \tag{2.48}$$

For very large μ_p, $d_p \to d$ and $l_p \to l$. Therefore,

$$S_{l_p}^{L_p} \approx 1 \tag{2.49}$$

Compared to the performances of the transverse type of design (refer Fig. 2.17(a)), this one has a number of disadvantages, mainly because of large air paths and leakage for which more number of turns per coil would be necessary to achieve same inductance value resulting in larger capacitance and higher rise time. Also plunger gives larger core loss and less Q-factor, external pick up increases and finally, design with split-coil tends to introduce asymmetry and a mutual coupling between the two coils. The mutual coupling increases with the two coils sharing a common magnetic path. The directions of the magnetic field of the coils with respect to the core fields are also important.

Considering ideal coils, the equivalent circuit of two coupled coils is given in Fig. 2.26 where Z_a represents the impedance of the individual coils considered identical, and Z_b represents the coupled impedance which is considered positive for opposing fields and negative when fields are in the same direction. The coupling coefficient k is given by

$$k = \frac{Z_b}{(Z_a + Z_b)} = 1 - \frac{2Z_a}{2(Z_a + Z_b)}$$

$$= 1 - \frac{Z_{11}}{2Z_{12}} \tag{2.50}$$

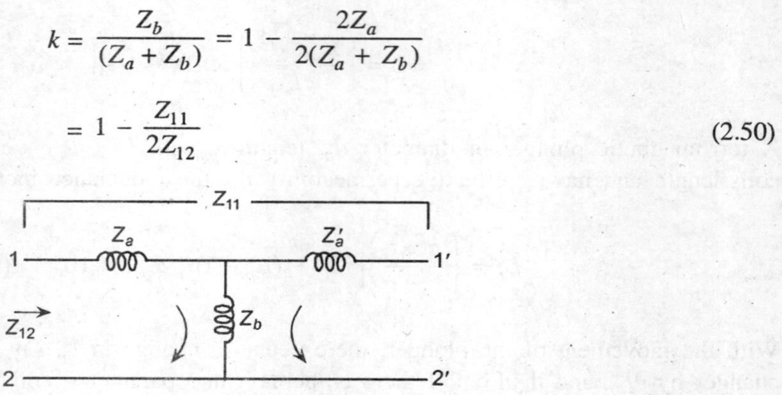

Fig. 2.26 Circuit of a pair of ideal-coupled coils.

As stated already, $k > 0$ for opposing fields and $k < 0$ for fields in the same direction.

Coupling can be eliminated or at least reduced to a greater extent if the magnetic paths for the two coils are independent which is realized to a certain extent by having an E-shaped core as shown in Fig. 2.27.

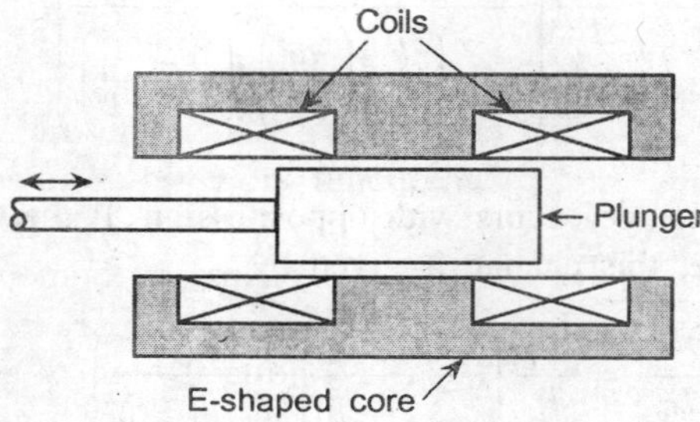

Fig. 2.27 Double coil design with E-shaped coil cores.

2.4.3   Inductance with a Short-circuited Sleeve

A schematic of such a sensor has been given in Fig. 2.17(c) where only a single coil arrangement is given. A double-coil with push–pull arrangement for better sensitivity and linearity is also available in this kind. Figure 2.28(a) shows such a scheme with Fig. 2.28(b) showing the variation of magnetic field with the sleeve position.

Considering, however, the single coil design first with the short-circuited sleeve covering only a part of the 'coil 1' of length, say l_2, (Fig. 2.29) and with coil and sleeve diameters d_1 and d_2, the voltage across the terminals 1 and 2 for a current i_1 in coil 1 would be changed due to presence of the sleeve around the coil (1) as this shorted sleeve acts a secondary of the transformer. If the self-inductance of the sleeve is L_s (its resistance is ignored), the coil inductance and resistance are L_1 and R_1 respectively, and the mutual inductance between the two is M, then the voltage v_1 would be given by

$$v_1 = i_1 \left[R_1 + j\omega L_1 + \frac{\omega^2 M^2}{j\omega L_s} \right] \tag{2.51}$$

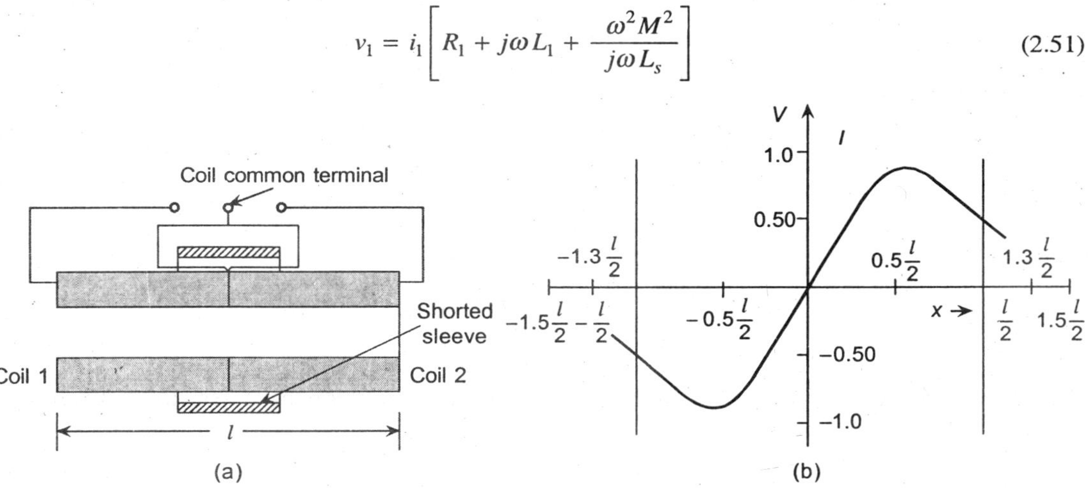

Fig. 2.28   (a) Structure of a double coil shorted sleeve **transducer**, (b) response characteristics of the transducer.

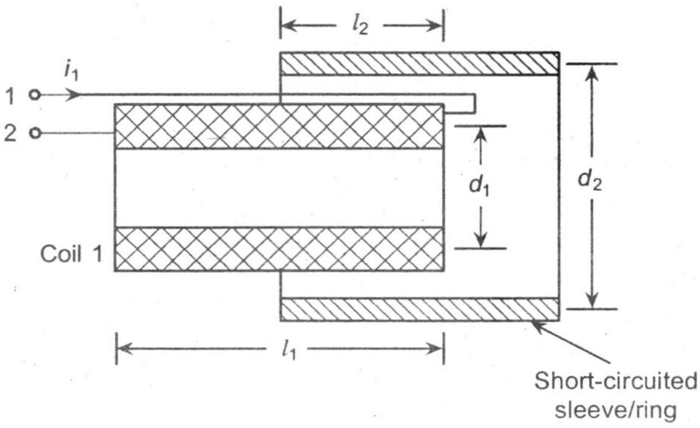

Fig. 2.29   The single coil sleeve type design.

where mutual inductance is produced due the coupling factor

$$k = \frac{M}{\sqrt{(L_1 L_s)}} \tag{2.52}$$

so that

$$\frac{v_1}{i_1} = Z_{\text{coil1}} = R_1 + j\omega L_1(1 - k^2) \tag{2.53}$$

or in other words, the inductance L_1 is changed by a factor $(1 - k^2)$. In fact, the coupling coefficient can be calculated for a pair of long coils one covering the other. Thus,

$$k = \left(\frac{d_1}{d_2}\right)\left(\frac{l_2}{l_1}\right)^{1/2} \tag{2.54}$$

so that the changed inductance of coil 1 is

$$L_{1c} = L_1\left[1 - \left(\frac{d_1}{d_2}\right)^2\left(\frac{l_2}{l_1}\right)\right] \tag{2.55}$$

From this, the change L_{1c} for the sleeve movement is obtained (sign change ignored) as

$$\partial L_{1c} = \frac{L_1 k^2 \partial l_2}{l_2} \tag{2.56}$$

so that

$$\frac{\partial L_{1c}}{L_{1c}} = \frac{\partial l_2}{l_2}\left(\frac{k^2}{1 - k^2}\right) \tag{2.57}$$

where $k \neq 1$ and sensitivity is never infinitely large. In fact, the normalized sensitivity, given by,

$$S_{l_2}^{L_{1c}} = \frac{k^2}{1 - k^2} \tag{2.58}$$

which becomes unity for $k = \pm 1/\sqrt{2}$ and more than unity for $k > 1/\sqrt{2}$. For shorted coils, it is very difficult to make $k \geq 1/\sqrt{2}$, mainly because of fringing effect and non-ideal coupling, so that

$$\frac{\partial L_{1c}}{L_{1c}} < \frac{\partial l_2}{l_2} \tag{2.59}$$

For shorted sleeve or ring type design, the magnetic field changes as shown in Fig. 2.28(b). On the other hand, for a single coil it changes similar to that already shown in Fig. 2.24(b) and the induced voltage and hence, current variation would be similar to variation of H. This, then, is the calibration curve of the transducer.

It would be seen that the nature of this transducer and that of the plunger type are same except that the plunger type transducer has higher iron loss and the shorted sleeve type has lower sensitivity, lower by about 35–40%.

2.4.4 The Transformer Type Transducer

The transformer type transducer can be formed like a transformer with a variable iron core coupling between a pair of coils or more. Figures 2.17(d)(i) and (ii) show two such kinds, of which the latter one can be considered as a typical case where one coil acts as a primary (in which an ac voltage is impressed) and the other acts as the secondary. But for this type, there often occurs a 'no signal' output and this can be compensated by another coil or a compensating current. In this group, the transducer used most is the linear variable differential transformer (LVDT) whose operation has been described in detail later in the book. In LVDT, a plunger type armature moves into a pair of secondary coils and a primary coil, the secondaries being connected in differential mode. The simple plunger type sensor has also been thoroughly discussed in Chapter 4.

While LVDT in the Chapter 4 on Magnetic sensors has been analyzed using an equivalent circuit of a transformer, basic equations are quoted here without the complex deduction process. It takes help of the properties of the magnetic circuit and flux leakages. It has been assumed that the mmf in ferromagnetic/iron is negligible in comparison with that in air paths of the leakage flux. Figure 2.30 is a schematic representation of the differential transformer. For magnetic circuit-based deduction, the gaps and material dimensions are very important.

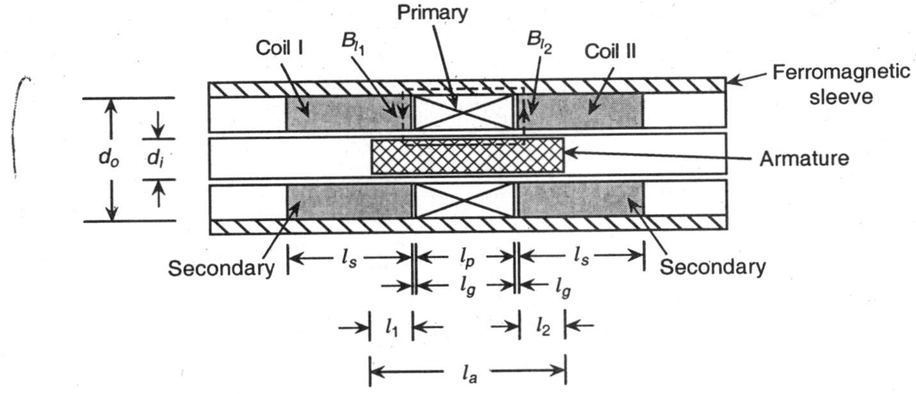

Fig. 2.30 The linear variable differential transformer.

Assuming current in the primary as I_p (rms) and number of turns n_p, if the number of turns in each secondary is n_s, it can be shown that the flux densities around the primary coil linking the secondaries are given as

$$\frac{B_{l_1}}{B_{l_2}} = -\frac{2l_2 + l_g}{2l_1 + l_g} \tag{2.60}$$

The negative sign comes because of direction (see Fig. 2.30).

For a supply frequency ω, the induced emf's in coils I and II are given respectively by

$$e_1 = \frac{2\pi^2 \omega I_p n_p n_s}{\ln(d_o/d_i)} \cdot \frac{2l_2 + l_g}{l_s l_a} \cdot x_1^2 \times 10^{-7} \tag{2.61a}$$

and

$$e_2 = \frac{2\pi^2 \omega I_p n_p n_s}{\ln(d_o/d_i)} \cdot \frac{2l_1 + l_g}{l_s l_a} \cdot x_2^2 \times 10^{-7} \tag{2.61b}$$

where x_1 and x_2 represent penetration of armature from nominal position beyond the primary coil length including the air gap.

Thus, the differential voltage

$$e_o = e_1 \sim e_2 = \left[\frac{2\pi^2 \omega I_p n_p n_s}{\ln(d_o/d_i)} \times 10^{-7}\right] \frac{l_g}{l_s l_a} \left[\left(\frac{2l_2}{l_g} + 1\right) x_1^2 - \left(\frac{2l_1}{l_g} + 1\right) x_2^2\right] \tag{2.62a}$$

$$= K_1 \left[\left(\frac{2l_2}{l_g} + 1\right) x_1^2 - \left(\frac{2l_1}{l_g} + 1\right) x_2^2\right] \tag{2.62b}$$

$$= K_1 \left[x_1^2 - x_2^2 + \left(\frac{2}{l_g}\right)(l_2 x_1^2 - l_1 x_2^2)\right] \tag{2.62c}$$

where

$$K_1 = \left(\frac{2\pi^2 \omega I_p n_p n_s}{\ln(d_o/d_i)}\right) \frac{l_g}{l_s l_a} \times 10^{-7}$$

In normal condition, if $l_1 = l_2 = l$, then

$$e_o = K_1 \left(1 + \frac{2l}{l_g}\right)\left(x_1^2 - x_2^2\right) \tag{2.63}$$

Approximate linearization is done by making $(1/2)(x_1 + x_2) = x_0 = $ constant, and $(1/2)(x_1 - x_2) = x$, the weighted differential movement, then

$$e_o = \left[4K_1\left(1 + \frac{2l}{l_g}\right) x_o\right] x$$

$$= K_2 x \tag{2.64}$$

A rearrangement of Eq. (2.61) converts e_o in the form

$$e_o = K_3 x \left(1 - K_4 x^2\right) \tag{2.65}$$

where

$$K_3 = \frac{8\omega I_p n_p n_s (l_p + 2l_g + x_o) x_o \times 10^{-7}}{\ln(d_o/d_i) \cdot l_s l_a} \tag{2.66a}$$

and

$$K_4 = \frac{1}{(l_p + 2l_g + x_o) x_o} \tag{2.66b}$$

In fact, there is a nonlinearity in the output which is given by the relation

$$\eta_l = K_4 x^2 \tag{2.67}$$

Assuming that $2l_g \ll l_p$ and that even at maximum movement the armature remains within the secondary coils, one can simplify the output relation as

$$e_o = \left(\frac{8\pi^2 \omega I_p n_p n_s}{\ln(d_o/d_i)} \right) \frac{2l_p}{3l_s} \left(1 - \frac{x^2}{2l_p^2} \right) \times 10^{-7} \tag{2.68}$$

The maximum movement of x, $x_{\max}$, l_p, and l_s can now be given for a given nonlinearity.

With iron core, power frequency is usually preferred although a frequency of upto about 5000 Hz can be used with sufficient accuracy. Above this, the core loss rises enormously. This core loss even at lower frequencies creates a different problem—a non-zero output at balance condition mainly because of dissimilar losses due to harmonic contents and varying capacitive effects.

2.4.5 Electromagnetic Transducer

It is a bilateral double-function type transducer, as has already been mentioned, that has 'mechanical input–electrical output' and 'electrical input–mechanical output' construction.

A general name of such systems is *electromechanical energy converters* which are governed simultaneously by (i) Faraday's law of electrodynamics and (ii) piezoelectric effect as postulated by Curie. Such transducers can be used both as 'generators' and 'sensors' often termed as 'senders' and 'receivers' respectively. Only the latter usage is of relevance here and is discussed.

Similar to the reluctance type transducer, such a system consists of an inductance coil wound on a ferromagnetic core and a variable gap provides the variation in the output. For producing unidirectional flux, a magnetizing coil with a bias current may be provided or the core can itself be a permanent magnet. Generally, a permanent magnet is used as a core.

If a coil of n turns wound on the core has a coil flux ϕ and coil inductance L, then as shown earlier

$$L = \frac{\mu_0 a n^2}{d} \quad \text{(H)} \tag{2.69}$$

for coil cross-section a, the effective gap d is given by (refer Fig. 2.31)

$$d = x_1 + \left(\frac{\mu_1}{\mu_2} \right) l_1 \tag{2.70}$$

μ_1 and μ_2 being relative permeabilities of air and core material respectively. Usually μ_1 is unity. Also the magnetic energy stored in the coil is

$$E_m = \frac{1}{2} \frac{(n\phi)^2}{L} \tag{2.71}$$

If a current

$$I = \frac{n\phi}{L} \tag{2.72}$$

flows in the coil, the stored energy is obtained by combining Eqs. (2.69), (2.71) and (2.72) as

$$E_m = \frac{1}{2}\frac{\mu_0 I^2 a n^2}{d} \tag{2.73a}$$

$$= \frac{1}{2}LI^2 \tag{2.73b}$$

This energy leads to development of a force f across the gap d (sign ignored) as

$$f = \frac{\partial E_m}{\partial d} = \frac{1}{2}\frac{LI^2}{d} \quad \text{(N)} \tag{2.74}$$

which would consist of a number of components depending on the 'magnetic condition' of the core. If the bias magnetizing current or its equivalent is given by I_o and a sinusoidal current of amplitude of i and frequency ω with $i \ll I_o$ energizes the core, then (ignoring the i^2 terms)

$$f = \left(\frac{L}{2d}\right)\left(I_o^2 + 2I_o i\right) \tag{2.75}$$

If the coil resistance is negligible, then

$$i = \frac{V}{j\omega L} \tag{2.76}$$

and the varying force term is

$$f_v = \frac{LI_o}{d}\cdot i \tag{2.77a}$$

and using Eq. (2.76)

$$f_v = \frac{I_o V}{j\omega d} \tag{2.77b}$$

The force f is associated with a velocity v as shown in Fig. 2.31 so that with analogy of electrical parameters, we can write

$$f_m = Z_m v \tag{2.78}$$

where Z_m is the mechanical impedance.

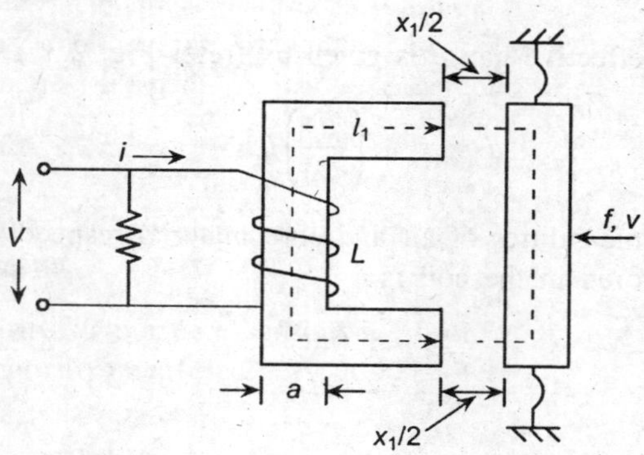

Fig. 2.31 Double-function electromagnetic transducer.

If m, k, and δ represent the mass, stiffness, and damping of the transducer (mechanical), then

$$Z_m = \delta + j\left(\omega_m - \frac{k}{\omega}\right) \qquad (2.79)$$

Equations (2.77b) and (2.78) may be combined to give

$$f_t = Z_m v \frac{I_o V}{j\omega d} \qquad (2.80)$$

A relation between voltage V, velocity v, and current i can be written as

$$V = \alpha_{vV}\, v + \alpha_{iV}\, i \qquad (2.81)$$

where α_{iV} is the electrical impedance and α_{vV} is complex transducer coefficient. Equation (2.81) can also be used to write, v in terms of V and i.

It can be shown that 'receiver' has a voltage to force ratio

$$\frac{V}{f} = \frac{I_o/(j\omega d)}{\left(\dfrac{I_o^2}{\omega^2 d^2}\right) + \left(\delta + j\left(\omega_m - \dfrac{k}{\omega}\right)\right)\left(\dfrac{1}{R} + \dfrac{1}{R_o} + \dfrac{1}{j\omega L}\right)} \qquad (2.82)$$

Where R_o is the load (indicator) resistance and is large, and R is the coil resistance (considered parallel to inductance). The characteristic transfer matrix equation of the transducer can be written as

$$\begin{bmatrix} f \\ v \end{bmatrix} = \begin{bmatrix} \dfrac{I_o}{j\omega d} & 0 \\ \dfrac{-d}{LI_o} & \dfrac{j\omega d}{I_o} \end{bmatrix} \begin{bmatrix} V \\ i \end{bmatrix} \qquad (2.83)$$

To this, the mechanical impedance and electrical impedances are superposed so that we get V/f as given by Eq. (2.82).

2.4.6 Magnetostrictive Transducer

Magnetostrictive transducer is not popular as a transducer mainly because of its limitations with respect to materials. Besides the input quantity, its output depends on some other variables also. It is of two different types, namely (i) the variable permeability type and (ii) the variable remanence type.

In general, a magnetostrictive material such as pure nickel has a slope of the hysteresis curve that decreases with increasing tension σ, as shown in Fig. 2.32. This change alters the value of the permeability μ, which also decreases with stress and hence, inductance of a coil wound on it. Also, with increasing tension, the remanence magnetism B_0 decreases. Ni is seen to be a material with negative magnetostriction.

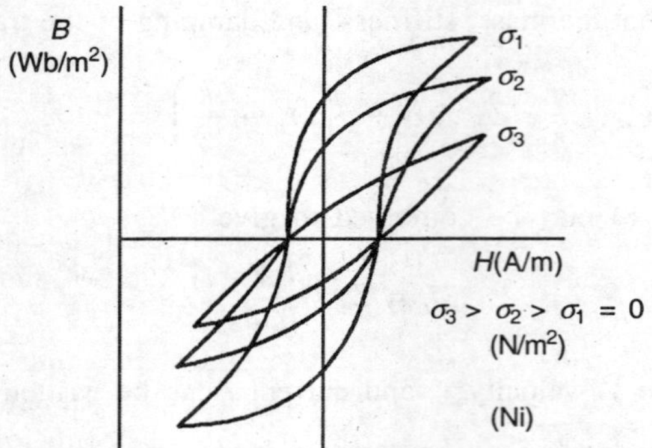

Fig. 2.32 *B–H* loops of magnetostrictive material with changing tension.

However, in case of Ni–Fe alloy known as permalloy such as 68 permalloy (Ni 68), 45 permalloy (Ni 45), the picture is reversed. Increasing tension increases B_0 as also permeability. The shapes of the *B–H* curves for such a situation are depicted in Fig. 2.33.

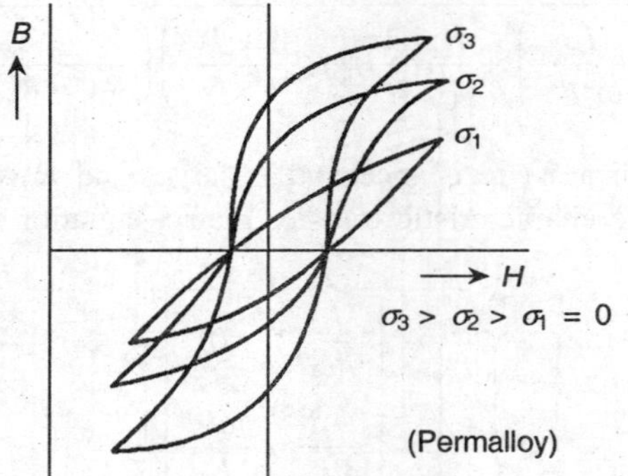

Fig. 2.33 *B–H* loops of another type of material with varying tension.

A typical scheme of the transducer using variable permeability is shown in Fig. 2.34(a). The coil inductance changes with change of force as the latter changes the core permeability. The coil inductance is measured through a bridge with the current and frequency, the coil also changes the inductance. These quantities and temperature have to be kept under strict regulation.

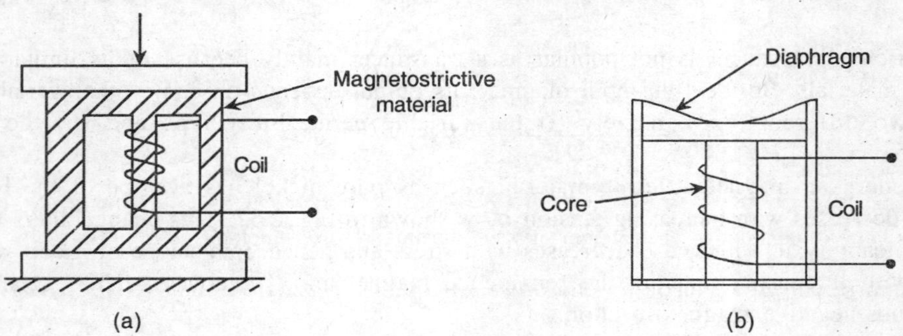

Fig. 2.34 (a) Scheme of a sensor with magnetostrictive material, (b) transducer operated by a diaphragm usually used in accelerometers.

The variable remanence type transducer is used for specific applications such as an accelerometer where the transducer is designed to receive the stress through a metal diaphragm as shown in Fig. 2.34(b). The open circuit voltage is proportional to the rate of change of the remanence magnetism. In fact, a relation is given as

$$B_0 - B_{0i} = k_1 \sigma \tag{2.84}$$

and for n turns of coil, the output voltage V is

$$V = nk_2 \frac{dB_0}{dt} \tag{2.85}$$

The k_i's in these equations are constants.

2.4.7   Materials—Some Comments

The core and armature material is essentially ferromagnetic that has high permeability, low loss, high Curie temperature, and low cost. Soft magnetic Ni–Fe alloy is good for the purpose in which there are a few commercial variety such as (i) Mu-metal and (ii) Radiometal (radiometal can further be subdivided into a few types). The permeability in the two cases varies as 60×10^3 to 240×10^3 and 4×10^3 to 65×10^3 respectively. Hysteresis losses are 4 and $40 \, \text{J/m}^3/\text{cycle}$ respectively while Curie temperatures are $350°$ and $540°C$ respectively.

Magnetically soft ferrites consisting of mixed crystals of cubic ferrites are good alternatives, which again have a number of varieties represented by the general formula MFe_2O_4 where M is a divalent metal such as manganese–zinc, magnesium–zinc, nickel–zinc, and so on. Such materials have initial permeabilities varying from 0.7×10^3 to 1.8×10^3. One special feature is that ferrites have resistivities about 10^6 times higher than ferromagnetics such that the eddy losses are negligible. Some of such ferrites can be used in high frequency ranges, for example, the Ni–Zn ferrite is particularly suitable for the purpose.

2.5   CAPACITIVE SENSORS

Three types of capacitive sensors can be listed under this category, namely

1. variable capacitance type with varying distance between two or more parallel electrodes (Fig. 2.35(a)).
2. variable capacitance obtained by variable area between the electrodes. An interesting variation of this is obtained by making serrated electrodes or electrodes with teeth, one of which moves (Fig. 2.35(b)), and
3. variable capacitance obtained by having variable dielectric constant of the intervening material. For this the material has to move between the pair of electrodes, and the change in capacitance is obtained and measured (Fig. 2.35(c)).

A fourth variety, the piezoelectric type, depends on the piezoelectric properties of specific kinds of dielectric materials and would be considered later. The movement of the moving electrode of the type shown in Fig. 2.35(b) is restricted to be short, while that of the dielectric material such as an insulation tape is not restricted.

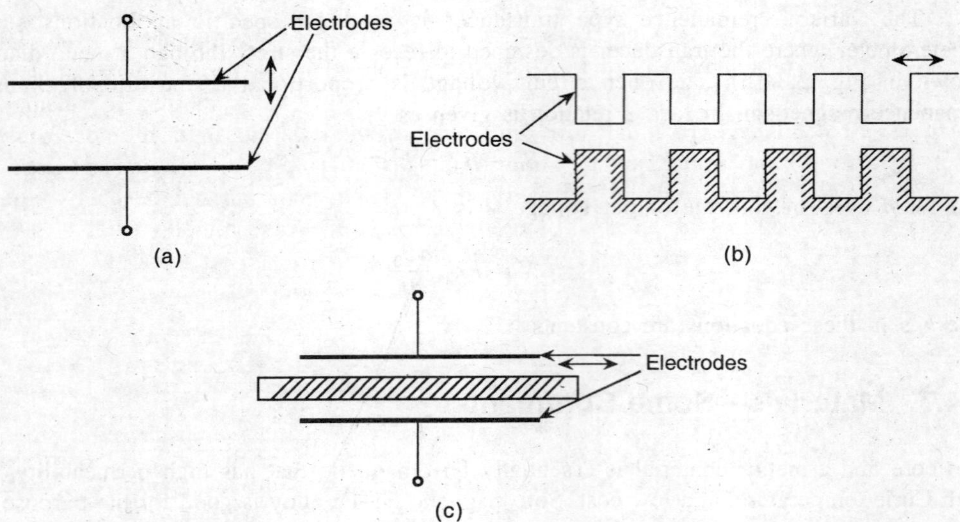

Fig. 2.35   (a) Parallel plate capacitance type, (b) capacitance type with serrated electrodes, and (c) capacitance type with varying dielectric type material.

A variation in parallel type design is the cylindrical design. Besides, the parallel plate capacitive sensor is often used in a differential form with three plates as shown in a Fig. 2.36(a). For a parallel plate capacitor with dielectric constant or permittivity ε, which is the product of its relative permittivity and the permittivity of the free space (vacuum, often taken as air) of value 8.85×10^{-12} F/m and plate area α, each separated by a distance x from the other, the capacitance is

$$C_p = \frac{\varepsilon\alpha}{x} \tag{2.86}$$

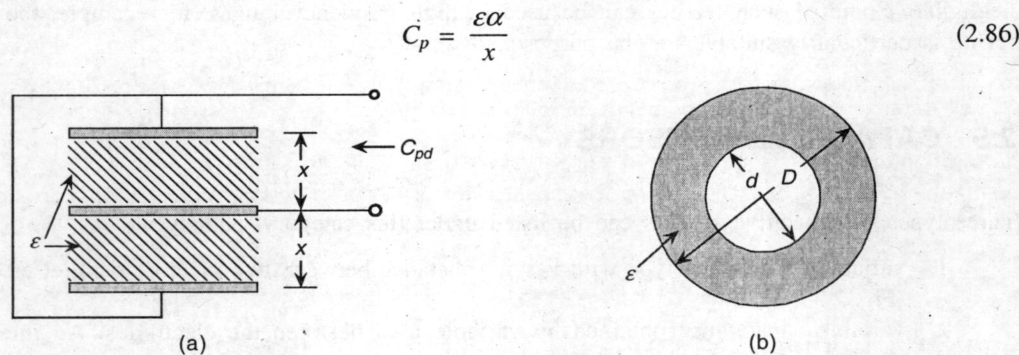

Fig. 2.36   (a) Parallel plate capacitance sensor, using three plates, (b) cylindrical type capacitance sensor.

A typical three plate capacitor arrangement is shown in Fig. 2.36(a). The capacitance C_{pd} is then given as

$$C_{pd} = \frac{2\varepsilon\alpha}{x} \tag{2.87}$$

For the cylidrical sensor with the electrode thickness negligible as compared to dielectric thickness (Fig. 2.36(b)), the capacitance is

$$C_c = \frac{2\pi \varepsilon l}{\ln(D/d)} \tag{2.88}$$

where l is the cylinder length.

For very thin layer of dielectric material, Eq. (2.88) can be approximated to

$$C_{ca} = \frac{\pi \varepsilon l(D + d)}{(D - d)} \tag{2.89}$$

If in a parallel plate pair the dielectric has a number of layers of dielectric constants with corresponding permittivity ε_i for thickness x_i, the relation (2.86) can be modified to

$$C_{pi} = \frac{\alpha}{\sum x_i/\varepsilon_i} \tag{2.90}$$

The capacitance is, in general, associated with a high resistance, called *leakage*, because the dielectric materials do not have infinite permittivity. This leakage is represented by a parallel resistance R_p, particularly at lower frequencies of measurement. This loss consists of dc conductance, dielectric loss of insulators supporting the electrodes, and the actual dielectric loss. With increasing frequency, the load resistances R_l contribute to loss factors and the complete equivalent circuit is given by the circuit of Fig. 2.37, where the inductance L represents the inductance between the terminals as also the cable inductance whenever such cable is used. Such an equivalent circuit would be taken up at a later stage.

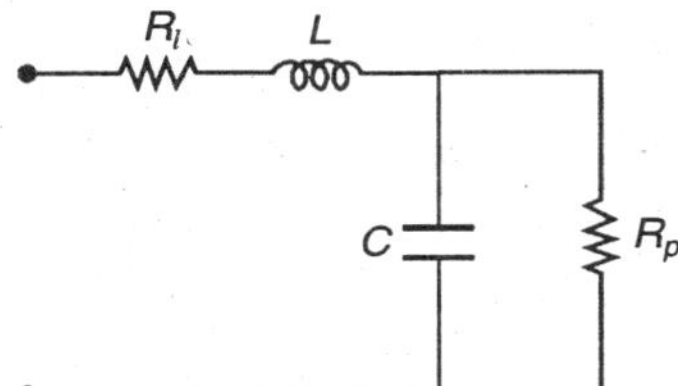

Fig. 2.37 Equivalent circuit of the capacitance transducer.

2.5.1 The Parallel Plate Capacitive Sensor

Considering now a general case of a pair of parallel plates with a solid dielectric of a certain thickness x_s and an air gap x_a as shown in Fig. 2.38, the capacitance C is given by

$$C = \frac{\alpha}{\left(\dfrac{x_a}{\varepsilon_a}\right) + \left(\dfrac{x_s}{\varepsilon_s}\right)} \tag{2.91}$$

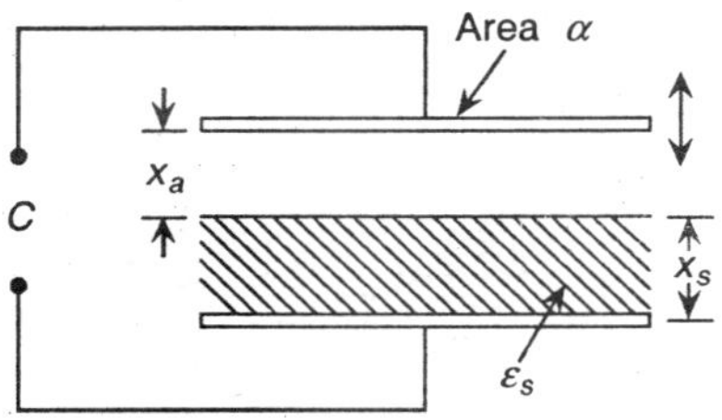

Fig. 2.38 Parallel plate sensor with different dielectric materials.

With the plate moving, a decrease in x_a increases C and vice versa. Thus,

$$C \pm \partial C = \frac{\alpha}{\left(\dfrac{x_a \mp \partial x_a}{\varepsilon_a} + \dfrac{x_s}{\varepsilon_s} \right)} \tag{2.92}$$

Considering, however, $\varepsilon_a \approx 1$, for simplicity, we obtain

$$\mp \frac{\partial C}{C} = \pm \left(\frac{\partial x_a}{x_a + x_s} \right) \left(\frac{1}{\dfrac{1 + \dfrac{x_s}{x_a \varepsilon_s}}{1 + \dfrac{x_s}{x_a}} \mp \dfrac{\partial x_a}{x_a + x_s}} \right) \tag{2.93}$$

In Eq. (2.93), the quantity $(1 + x_s/(x_a\varepsilon_s))/(1 + x_s/x_a)$ is an important factor in determining the value of $+\partial C/C$ as well as its nature. This quantity is represented as $1/\beta$, where β is often referred to as the *sensitivity factor*, but it also is responsible for the nonlinearity. Writing $(\partial x_a/x_a)/(1 + x_s/x_a) = (\partial x_a/x_a)/(1 + \lambda)$, $\pm\partial C/C$ can be expanded as

$$\mp \frac{\partial C}{C} = \pm \left(\frac{\partial x_a}{x_a} \right) \left(\frac{\beta}{1 + \lambda} \right) \left[1 \pm \left(\frac{\partial x_a}{x_a} \frac{\beta}{1 + \lambda} \right) + \left(\frac{\partial x_a}{x_a} \frac{\beta}{1 + \lambda} \right)^2 \pm \cdots \right] \tag{2.94}$$

As β is a function of x_a, x_s, and ε_s, the plots of β versus λ with ε_s as a parameter show that with increasing λ, β increases with ε_s, its minimum value being 1 for $\varepsilon_s = 1$.

It must be stressed here that capacitors have fringing effects which are usually taken care of by providing guard ring which is a ring surrounding a plate of the capacitor, the ring and the plate both being at the same potential.

2.5.2 Serrated Plate Capacitive Sensor

As has been discussed earlier, a pair of flat serrated plates, one of which is fixed in position, the other with a small relative movement show change in capacitance and this principle is utilized in some cases to measure small angular variations. For the measurement to be of any significance, the relative movement has to be small. Figure 2.39 shows the active tooth length (on the fixed plate) as l, air gap as x, tooth width as w; if number of teeth-pair is n and air permittivity is ε_a, the capacitance C is given as

$$C = \frac{\varepsilon_a l w n}{x} \tag{2.95}$$

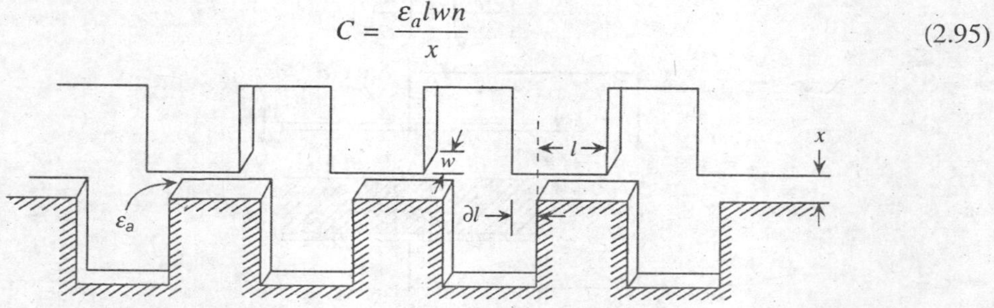

Fig. 2.39 Serrated electrode capacitance sensor with changing active tooth length.

so that for a small relative movement ∂l of the moving plate, we obtain

$$\frac{\partial C}{C} = \frac{\partial l}{l} \tag{2.96}$$

This simplified relation assumes no fringing effect. However, by drawing actual equipotential lines and parallel flux lines between the pair of teeth, the leakage can be allowed in the relation. Therefore,

$$\frac{\partial C}{C} = \frac{\partial l}{l}\left(\frac{1}{1 + \dfrac{kx}{l}}\right) \tag{2.97}$$

where the expression within the brackets can be termed as the sensitivity factor, β_s, which decreases with increasing x/l as shown in Fig. 2.40. This factor β_s is actually the ratio of nonleakage to total flux.

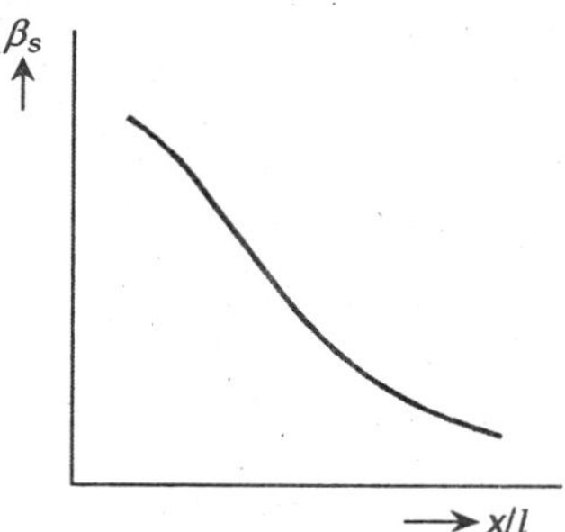

Fig. 2.40 Sensitivity versus normalized gap curve.

2.5.3 Variable Permittivity or Variable Thickness Dielectric Capacitive Sensor

This type of capacitive sensors can be represented as shown in Fig. 2.41. With plate effective area α and other dimensions as shown in the figure, the capacitance C is given by

$$C = \frac{\alpha}{l - x + \dfrac{x}{\varepsilon_d}} \tag{2.98}$$

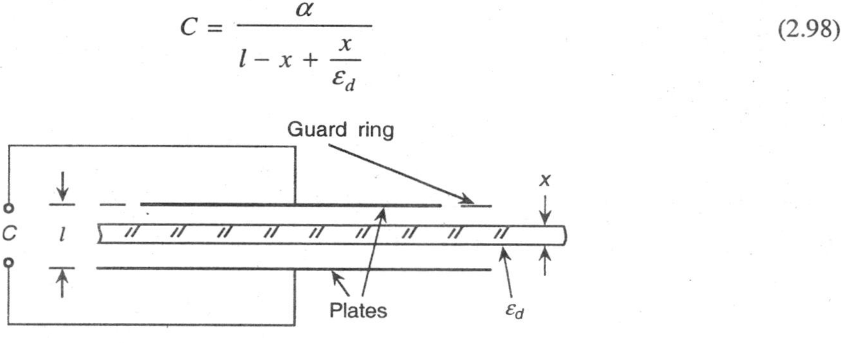

Fig. 2.41   Scheme of a variable permittivity (or thickness) dielectric type sensor.

where ε_d is the permittivity of the dielectric material. Following the development in Section 2.5.1, one obtains the normalized change in capacitance as

$$\left(\frac{\partial C}{C}\right)_{\varepsilon_d} = \pm \frac{\partial \varepsilon_d}{\varepsilon_d} \frac{1/[1 + \varepsilon_d(l - x)/x]}{1 \pm \dfrac{1}{1 + x/(\varepsilon_d(l - x))} \cdot \dfrac{\partial \varepsilon_d}{\varepsilon_d}} \tag{2.99}$$

Here, $1/(1 + \varepsilon_d(l - x)/x)$ is the sensitivity factor β_s and the nonlinearity factor is $\eta_n = 1/(1 + x/(\varepsilon_d(l - x)))$. If $\eta_n \partial \varepsilon_d/\varepsilon_d$ is small, we obtain, with first order approximation,

$$\left(\frac{\partial C}{C}\right)_{\varepsilon_d} = \frac{\partial \varepsilon_d}{\varepsilon_d} \cdot \frac{1}{1 + \varepsilon_d(l - x)/x} \left[1 \mp \frac{\partial \varepsilon_d/\varepsilon_d}{1 + x/(\varepsilon_d(l - x))}\right] \tag{2.100}$$

Obviously, with $x/(l - x)$ high, β_s is high and η_n is low which must be a good choice.

Instead of variation in ε_d, there may be variation in x, so that we have

$$\left(\frac{\partial C}{C}\right)_x = \frac{\partial x}{x} \frac{\dfrac{\varepsilon_d - 1}{1 + \varepsilon_d(l - x)/x}}{1 \mp \dfrac{\varepsilon_d - 1}{1 + \varepsilon_d(l - x)/x} \dfrac{\partial x}{x}} \tag{2.101}$$

and if $[(\varepsilon_d - 1)/(1 + \varepsilon_d(l - x)/x)]\partial x/x \ll 1$, taking the first order term only, the expression for $(\partial C/C)_x$ is obtained as

$$\left(\frac{\partial C}{C}\right)_x = \frac{\partial x}{x} \frac{\varepsilon_d - 1}{1 + \varepsilon_d(l - x)/x} \left[1 + \frac{\varepsilon_d - 1}{1 + \varepsilon_d(l - x)/x} \frac{\partial x}{x}\right] \tag{2.102}$$

In this case, the sensitivity factor and the nonlinearity factor are identical and given by $(\varepsilon_d - 1)/(1 + \varepsilon_d (l - x)/x)$. It means that the sensitivity is good with high $x/(l - x)$ as also ε_d, but the nonlinearity also increases.

2.5.4   Stretched Diaphragm Variable Capacitance Transducer

Such transducers with a very thin diaphragm are used in differential arrangement in transmitters these days where the stiffness in bending is ignored. Other kinds are used for direct pressure measurement which are thicker though thin with respect to diameter. Their stiffness in bending cannot be ignored like ones with thin diaphragms.

A typical scheme of the former type (non-differential mode) is shown in Fig. 2.42. With the dimensions shown after a maximum central deflection d of the diaphragm with a pressure p has occurred, the depression x at a radial distance r from the centre of the diaphragm under tension t_d is given by

$$x = \frac{2t_d}{p} \left[\sqrt{1 - \left(\frac{rp}{2t_d}\right)^2} - \sqrt{1 - \left(\frac{Rp}{2t_d}\right)^2}\right] \tag{2.103}$$

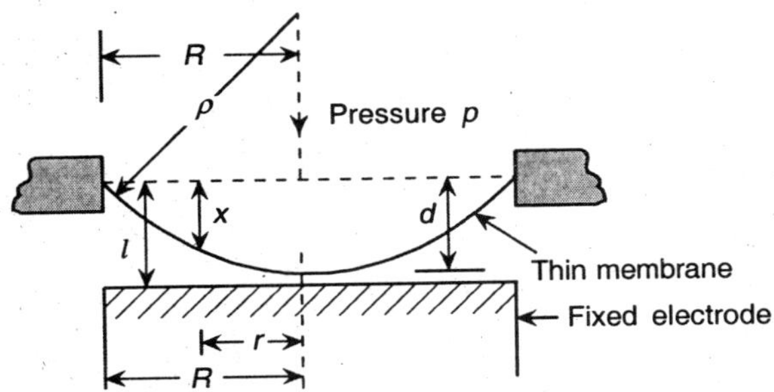

Fig. 2.42 Stretched diaphragm type capacitance sensor.

Expanding the right hand side in series form and accepting only the first order term for small deflection, that is, $(d/R)^2 \ll 1$,

$$x = \frac{p}{4t_d}(R^2 - r^2) \tag{2.104}$$

The small capacitance formed by the fixed electrode and a narrow annular area of the depressed membrane of width dr at a radius r, as shown in the figure, is given by

$$dC = \frac{2\pi r\, dr \varepsilon_a}{l - x} \approx \left\{\frac{2\pi r\, dr\, \varepsilon_a}{l}\right\}\left(1 + \frac{x}{l}\right) \tag{2.105}$$

for $x/l \ll 1$, so that the total capacitance, using Eq. (2.104), is

$$\int_0^R dC = \frac{2\pi \varepsilon_a}{l} \int_0^R \left(1 + \frac{x}{l}\right) r\, dr$$

$$= \frac{2\pi \varepsilon_a}{l} \int_0^R \left(1 + \frac{p}{4t_d l}(R^2 - r^2)\right) r\, dr \tag{2.106}$$

The first part $\pi \varepsilon_a R^2/l$ is the capacitance of the pair when the diaphragm is undeflected, while the incremental capacitance ∂C is computed as

$$\partial C = \left(\frac{\pi \varepsilon_a R^4}{8l^2 t_d}\right) p \tag{2.107}$$

so that

$$\frac{\partial C}{C} = \frac{R^2}{8t_d l} p \tag{2.108}$$

represents the sensitivity of the capacitance system.

Diaphragm system is frequently used in microphones where the vibrating diaphragm is backed by a thin layer of air which also tends to vibrate with it giving a cushioning effect, and by reducing the dynamic sensitivity, it increases dynamic stiffness. The effect is reduced to a greater extent by providing perforation in the fixed electrode. The vibrating layer of air also increases the diaphragm inertia affecting the frequency response.

A diaphragm formed by machining from the solid to avoid large hysteresis losses is said to be clamped type and although it may be made to have its thickness small enough with respect to its diameter, it does provide a stiffness to bending. If the diaphragm thickness is τ and material Poisson's ratio v, Young's modulus Y, for other dimensions as shown in Fig. 2.42, the deflection x is given by

$$x = \frac{3p}{16} \cdot \frac{1 - v^2}{Y\tau^3} (R^2 - r^2)^2 \tag{2.109}$$

Following the similar procedure as above, the sensitivity may be derived as

$$\frac{\partial C}{C} = \frac{(1 - v^2)R^4}{16Ylt^3} p \tag{2.110}$$

2.5.5 Electrostatic Transducer

Similar to the electromagnetic transducer discussed in Section 2.4.5, capacitive type transducer can also be developed with bilateral characteristics, where it is used with dc polarization. Such a transducer is also referred to as an *electrostatic transducer*. A typical scheme of such a system is shown in Fig. 2.43. A capacitor is formed with a 'flexible' diaphragm which can move due to application of force and a rigid plate p_1. There is bias voltage V_s which is sufficiently large. When the system acts as a transducer, the gap x between the plates changes as by some pressure in case of an 'electrostatic microphone'. This pressure may be considered sinusoidal in nature for analysis purpose. A circuit consisting of a resistance R and capacitance C 'varying sinusoidally' allows V_s to send a sinusoidal current i to flow in it and hence, a sinusoidal output V_o across resistance R is obtained.

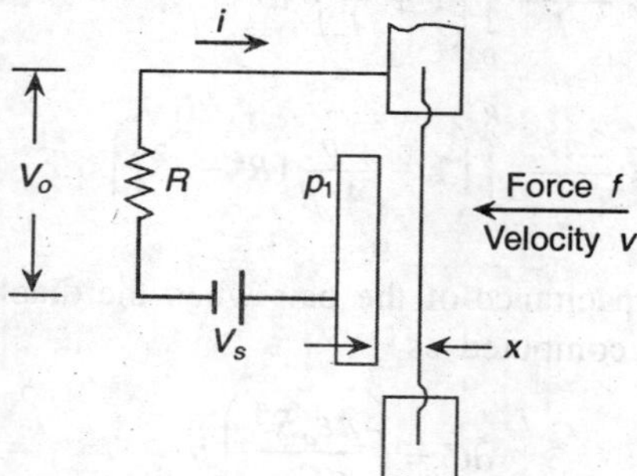

Fig. 2.43 Electrostatic transducer.

Analyzing as in the case of electromagnetic transducer, V_o corresponding to a force f can be obtained in terms of the parameters V_s, x, R, C, ω, mass m, stiffness k, and damping (ζ) of the system. In fact, the dynamic transfer function is given by

$$\frac{V_o(s)}{f(s)} = \frac{sx_oRC_oV_s}{s^3x_o^2mRC_o + s^2(mx_o^2 + x_o^2 RC_o\zeta) + s(x_o^2\zeta + x_o^2RC_ok) + (x_o^2k + V_s^2C_o)} \tag{2.111}$$

where, C_o and x_o are the initial values of x and C, and s may be replaced by $j\omega$ where ω is the input circular frequency.

Frequency response analysis of this shows a flat response upto a frequency $\omega_o = (k/m)^{1/2}$, at which a resonance occurs and range is obviously specified by the same. Also below $\omega_b = (C_oR)^{-1}$, the response is not constant. Hence, the frequency range is $(\omega_o - \omega_b)$.

In case of generating action, alongwith bias V_s, a sinusoidal input voltage is also applied so that the diaphragm undergoes electrostatic vibration.

2.5.6　Piezoelectric Elements

Crystals of certain classes are said to show piezoelectric effect which essentially means electric polarization produced by mechanical strain in the crystals. Such a polarization is believed to occur because of asymmetric crystal structure. The effect is reversible in the sense that a strain may be produced in the crystal by electrically polarizing it using an external source. While the mechanical input to electrical output form is used in developing transducers extensively, the reverse effect is used in many modern gadgets such as sonar systems, ultrasonic non-destructive test equipment, ultrasonic flowmeters, pump for inkjet printers, and so on.

Also a piezoelectric crystal is represented by a set of three Cartesian coordinates so that the polarization P can be represented in the vector form as

$$\mathbf{P} = P_{xx} + P_{yy} + P_{zz} \tag{2.112}$$

However, P_{xx}, P_{yy}, and P_{zz} are again related to the stresses, axial and shear, σ, and χ, in terms of a set of axes-dependent coefficients called d-constants of the crystal. With the axial and shear axes as shown in Fig. 2.44 with reference to the crystal axes X-Y-Z, we obtain

$$\begin{bmatrix} P_{xx} \\ P_{yy} \\ P_{zz} \end{bmatrix} = \begin{bmatrix} d_{11} & d_{12} & d_{13} & d_{14} & d_{15} & d_{16} \\ d_{21} & d_{22} & d_{23} & d_{24} & d_{25} & d_{26} \\ d_{31} & d_{32} & d_{33} & d_{34} & d_{35} & d_{36} \end{bmatrix} \begin{bmatrix} \sigma_{xx} \\ \sigma_{yy} \\ \sigma_{zz} \\ \chi_{yz} \\ \chi_{zx} \\ \chi_{xy} \end{bmatrix} \tag{2.113}$$

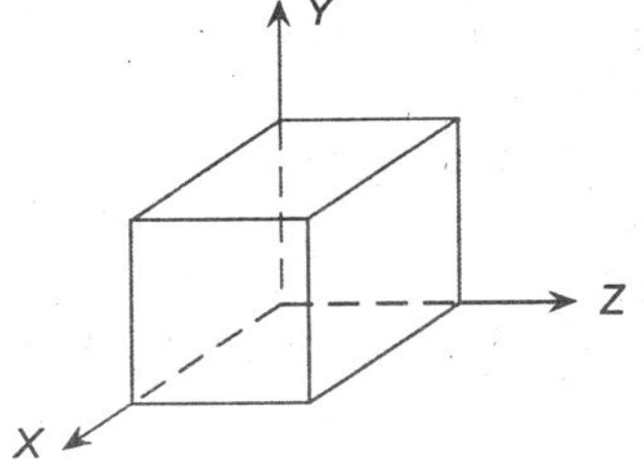
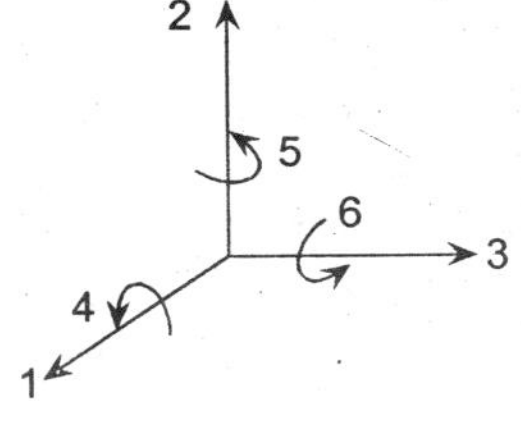

Fig. 2.44　The piezoelectric crystal defined in *X-Y-Z* axes.

The d-constants are defined as

$$d_{ij} = \frac{\text{charge generated in direction } i}{\text{force applied in direction } j} = \frac{Q_i}{f_j} \tag{2.114}$$

expressed as coulomb per Newton usually. The reverse effect d-coefficients are similarly defined as

$$d_{ij} = \frac{\text{strain in direction } i}{\text{field applied in direction } j} = \frac{\varepsilon_i}{E_j} \qquad (2.115)$$

expressed usually in (m/m)/(V/m).

One other coefficient which is of importance in practical design is the g-coefficient and is related to the d-coefficient by the dielectric constant of the material. It is defined as the voltage gradient or field in the crystal per unit pressure imparted to it. Maintaining the direction as before, it can be shown that

$$g_{ij} = \frac{Q_i}{\varepsilon_d f_j} = \frac{d_{ij}}{\varepsilon_d} \qquad (2.116)$$

A third coefficient, the h-coefficient, is defined as the voltage gradient per unit strain which also appears to be the reciprocal of d_{ij} given by Eq. (2.115). The h-coefficient is easily obtained from the g-coefficient by multiplying it with the Young's modulus in the appropriate direction.

Crystals, for various uses, are characterized by coupling coefficient which actually is a measure of the efficiency of the crystal as energy converter. Its application in transducer engineering is limited but it is a necessary parameter when used in generators.
The numerical value of coupling coefficient is given by

$$K_{ij} = (d_{ij} \, h_{ij})^{1/2} \qquad (2.117)$$

The value of d_{11} for quartz is 2.3×10^{-12} coulombs/N and its dielectric constant is 4.06×10^{-11} F/m. Hence, its g_{11} value is 56×10^{-3} (V/m)/(N/m^2).

Piezoelectric materials

Materials for piezoelectric sensors have been divided into two groups: (i) those occurring naturally such as quartz, rochelle salt $NaKC_4H_4O_6$, $4H_2O$, tourmaline and so on, (ii) those produced synthetically such as lithium sulphate (LS), $NH_4H_2PO_4$ or ammonium dihydrogen phosphate (ADP), and $BaTiO_3$ or barium titanate (BT). Barium titanate is actually a ferroelectric ceramic and requires to be polarized before use. Besides, there are certain polymer films which also exhibit the piezoelectric property.

Crystals like quartz have natural asymmetric structure which is responsible for this property. Quartz is representable as a helix along which one silicon and two oxygen atoms interlace. The planar view of the crystal cell, perpendicular to the z-axis also called the optic axis, shows a hexagonal shape with one Si and two oxygen occupying the vertices alternately as shown in Fig. 2.45(a). The chemical structure gives the formula SiO_2. In the normal unstressed condition, the positive charges of silicon and the negative charges of oxygen compensate each other without showing any electrical output.

However, with application of a force (compression) in the direction of x-axis, the crystal is deformed to the extent of being polarized so that positive and negative charges are generated as shown in Fig. 2.45(b). If this force is applied in the Y-direction, the deformation produced is such that opposite charges are now generated on the two faces as shown in Fig. 2.45(c). These two cases are known as longitudinal and transverse effects respectively. Changing the type of force, that is, from compression to tension, reversal in the polarity of charge generation occurs.

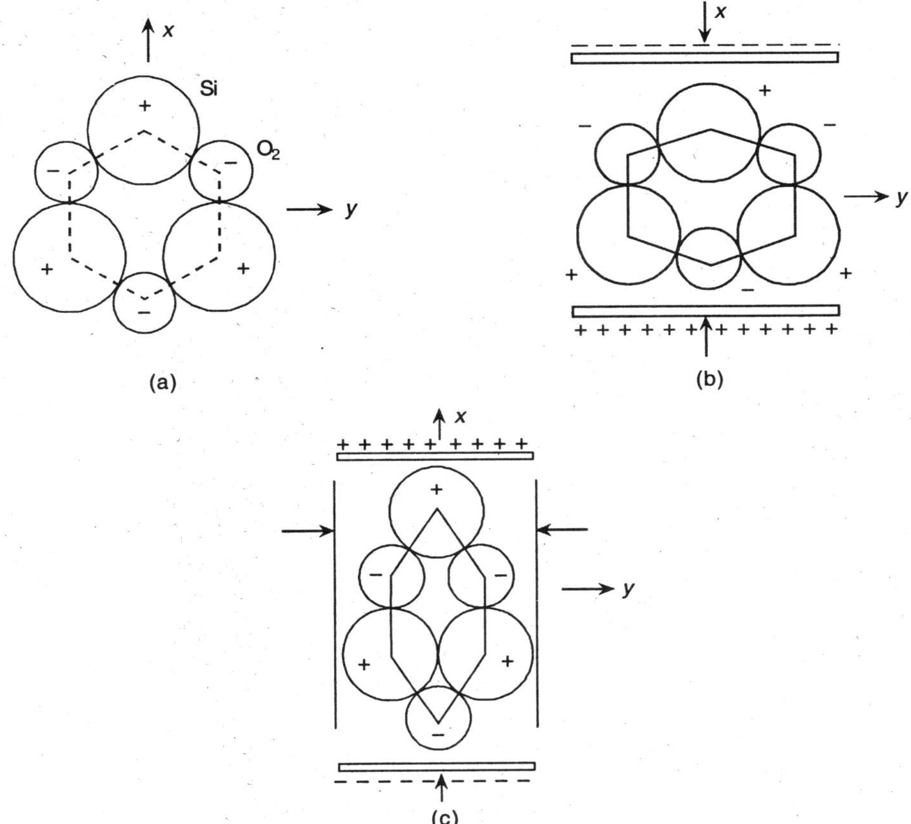

Fig. 2.45   (a) The quartz crystal model, (b) charge generation with force applied in the direction of the electrodes, and (c) charge generation with the force applied perpendicular to the position of electrodes.

As a result of the symmetry of the crystal structure in the z-direction, there does not occur any charge 'discrepancy' or polarization when force is applied in this direction and this axis is, therefore, termed as the *optic axis*.

It may be noticed that polarization deep inside the crystal is cancelled out and only the surface layers are affected to produce the free charges. The degree of distribution is thus, important for the amount of charges on the two faces which means that the force applied is the main criterion. This is true specially for the case of Fig. 2.45(b). However, in case of Fig. 2.45(c), the transverse charge 'size' in the x-direction has a multiplying factor α_x/α_y where α is the face area.

The material properties that are relevant to the piezoelectric sensors are (i) dielectric constant, (ii) d-coefficients (xx, say), (iii) resistivity (specifically, volume resistivity is considered), (iv) Young's modulus, (v) humidity range (since above or below this range large absorption of moisture occurs changing volume resistivity and performance characteristics), (vi) temperature range, and (vii) density. A comparative study of these properties is made in Table 2.5.

Table 2.5 Properties of piezoelectric materials

Material	d (relative)	$d_{xx}(\times 10^{-12})$ (cou/N)	ρ_v ($\Omega - $m)	$Y(\times 10^9)$ (N/m^2)	H_R (%)	T_R (°C(max))	Density ($\times 10^3$) (kg/m^3)
Quartz	4.5	2.3	10^{12}	80	0–100	550	2.65
Rochelle Salt	350	550	10^{10}	10–20	40–70	45	1.77
Tourmaline	6.7	2–2.25	10^{11}	160	0–100	1000	3.10
LS	10.3	13–16	10^{10}	46	0–95	75	2.05
ADP	15.3	25–45	10^8	19.5	0–94	125	6.8
Titanates	500–1750	80–500	10^9–10^{13}	47–80	—	200–400	5.8–7.8

Inspite of some deficiencies such as low mechanical strength, limited humidity and temperature range, large hysteresis, and fatigue, Rochelle salt is often used in microphones and also in gramophone pickups because of high shear sensitivity and permittivity. Although available as naturally occurring, it is industrially grown now for bigger requirements.

Tourmaline has poor sensitivity ($d_{xx} \approx 2$–2.5) and is costly. It is, therefore, rarely used as a sensor of this type. But it has two specific advantages—(i) it has a long, perhaps the longest, temperature range, and (ii) it is the only naturally occurring variety that shows large volume-expander mode capability, that is, with high force in all three directions it gives a large d-value in x-x direction.

Lithium Sulphate is good in volume-expander mode but ammonium dihydrogen phosphate is used quite extensively for acceleration and pressure sensing purposes although it has low permittivity. It can also be used in twisting applications.

Among the titanates, barium titanate (BaTiO$_3$)—a polycrystalline ceramic has high ε_d and with induced polarization is very conveniently used in many transducers. Ferrroelectric materials can be analyzed analogous to the ferromagnetic ones and its polarization is effectively explained with the help of the 'domain' structure. The material is assumed to consist of 'zones' with spontaneous polarization (for example, Weiss zones in ferromagnetics) which can be partially oriented by the application of external electric field. A barium titanate crystal is modelled as shown in Fig. 2.46. The crystal cells are tetragonal with the axes ratio 1.01 and the central Ti atom has a preferred direction of movement between the oxygen atoms in which polarization can occur. Above Curie point (120°C), the structure reduces to a cubic form and the polarizability is lost.

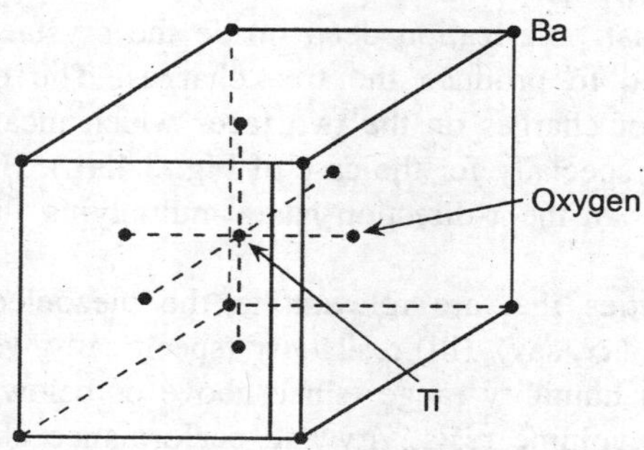

Fig. 2.46 The model of a BaTiO$_3$ crystal.

As in the case of soft magnetic material, ferroelectric material also loses polarization with time as the remanent polarization depends on the coercive force of the dipoles. This is understood from the hysteresis loop. This loss can be prevented and stability increased, by introducing polarization impurities such as lead, calcium, yittrium and so on. However, for transducers, lead zirconate titanate has been found to be more suitable than the simple ones suggested previously. Lead zirconate titanate is a solid solution of lead titanate and lead zirconate which is only 10–60 mole percent of the former. Depending on the amount of lead zirconate and also on processing techniques, values of d-coefficients differ greatly, the Curie point being pushed up in almost all the cases from 200° to 300°–350°C. Another composition consists of lead actaniobate which has the highest Curie point.

The dielectric constant, d-coefficients, and dissipation in a ferroelectric ceramic change with temperature. The nature of such changes are shown in relative response plots in Fig. 2.47. These can be compared with those of quartz, specially the variation of d-coefficients and ε_d. Figure 2.48 shows the plots.

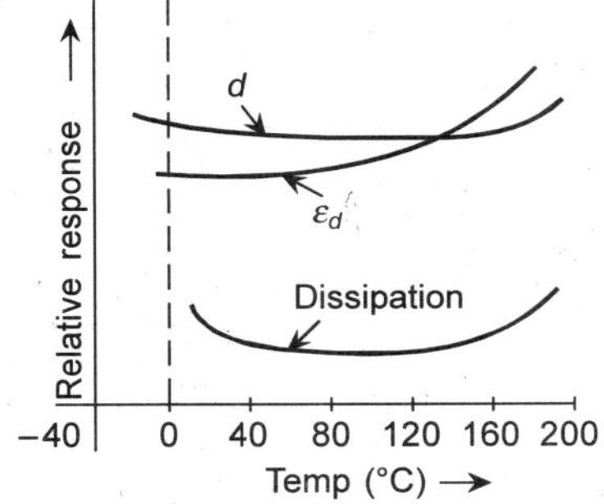

Fig. 2.47   Relative response-temperature curves for *d*-coefficients, dielectric constant and dissipation of BaTiO₃.

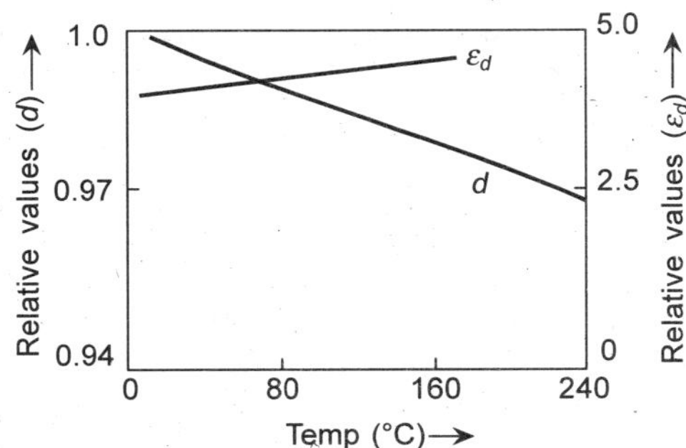

Fig. 2.48   Variation of *d*-coefficients, dielectric constant with temperature for quartz.

Titanates are synthetically produced by pressure, film-casting or extrusion, and finally sintering—the ohmic contacts are obtained by silver or palladium coating on which soldering of lead-wires can be done before polarization. Polarization is usually affected at a voltage of 2 KV/mm and is kept for a few minutes depending on the material.

Considering a quartz sensor of thickness t obtained by cutting perpendicular to its x-axis, two faces which have same areas (α each) and are perpendicular to this axis are metallized; if now, a force f_x is applied to it along the x-direction, the charge Q_x generated would be

$$Q_x = d_{11} f_x \tag{2.118}$$

The capacitance C_x of the sensor is then given by

$$C_x = \frac{\varepsilon_d \alpha}{t} \tag{2.119}$$

so that voltage V_x is

$$V_x = \frac{Q_x}{C_x} = \frac{d_{11} f_x t}{\varepsilon_d \alpha} \tag{2.120}$$

For a crystal of dimensions as shown in Fig. 2.49 with a force f_y in the y-direction, the charge on the plates perpendicular to x-direction is given by (as already mentioned)

$$Q_x = d_{12} \left(\frac{l_y}{l_x} \right) f_y \tag{2.121}$$

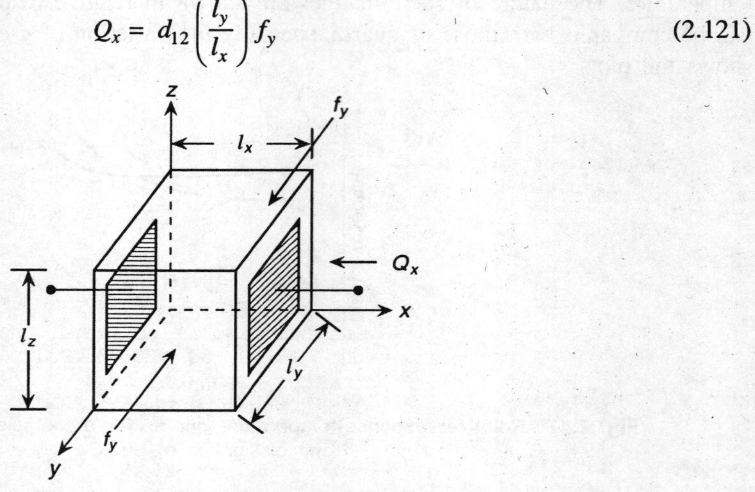

Fig. 2.49 A crystal with electrodes and marked dimensions.

However, for quartz, all the d-coefficients given in Eq. (2.113) are not finite nonzero values. In fact, the d-matrix for quartz is given as

$$[d] = \begin{bmatrix} d_{11} & -d_{11} & 0 & d_{14} & 0 & 0 \\ 0 & 0 & 0 & 0 & -d_{14} & -2d_{11} \\ 0 & 0 & 0 & 0 & 0 & 0 \end{bmatrix} \tag{2.122}$$

so that Eq. (2.121) is modified as

$$Q_x = -d_{11} \left(\frac{l_y}{l_x} \right) f_y \tag{2.123}$$

and a voltage V_x is given by

$$V_x = \frac{-d_{11} f_y}{\varepsilon_d l_z} \tag{2.124}$$

Deformation modes and multimorphs

Piezoelectric sensors can produce outputs in the form of charge or voltage with force, acceleration, velocity, as (displacement) inputs and then occurs 'deformation' (in the crystals). This deformation is of different types depending on the application of inputs in it. Accordingly, a number of modes are listed. In the preceding subsection, it was the thickness that changed, and accordingly the mode is named 'thickness expander mode' (TEM). The others of consequence are shown in Fig. 2.50. Other modes are length expander mode (LEM), thickness shear mode (TSM), face shear mode (FSM) and volume expander mode (VEM).

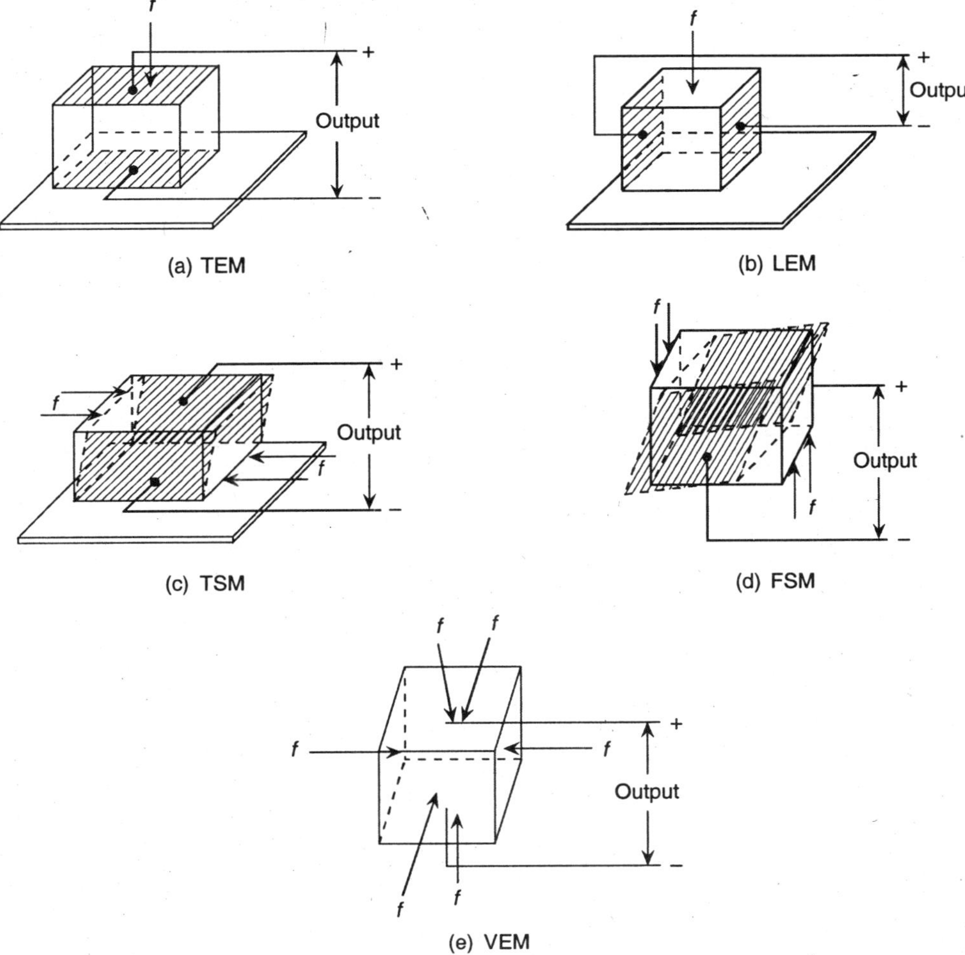

Fig. 2.50   Representation of different deformation modes: (a) thickness expander mode, (b) length expander mode, (c) thickness shear mode, (d) force shear mode, and (e) volume expander mode.

Instead of a single element sensor, it is possible to cement together two such elements as in a sandwich to obtain larger (ideally double) output. Such elements are often termed as 'bimorphs'. Proceeding in a similar way, multimorphs may be obtained for more than two elements. Bimorphs may be obtained by series sandwiching or by parallel arrangement. Figure 2.51(a) and (b) show the two cases. In these cases, the polarization of the two plates with respect to each other, is different so that the series or parallel arrangement may be achieved. Typical bimorph cantilevers for bending (strain) and torque are shown in Figs. 2.52(a) and (b) respectively.

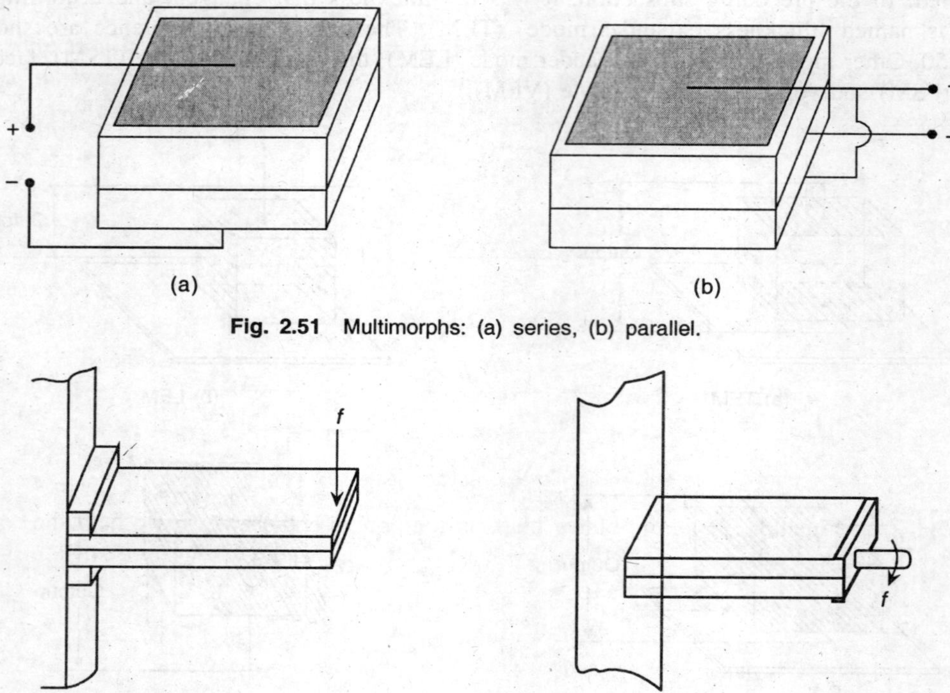

(a)

Fig. 2.51   Multimorphs: (a) series, (b) parallel.

(b)

Fig. 2.52   Multimorphs applied in (a) bending, (b) torque.

2.5.7   The PZT Family

The compounds of the solid-solution system $PbZrO_3$–$PbTiO_3$, called PZT, show strong piezoelectric effect and they have same structures as pervoskites. The piezoelectric properties depend on the Ti/Zr ratio. In the phase diagram, it is seen that there are both rhombohedral and tetragonal ferroelectric phases and the best composition for piezoelectricity is one where it lies close to the morphotropic boundary between the rhombohedral and tetrahedral phase when Ti : Zr : : 1 : 1. Most piezoelectric ceramics are based on this PZT group. In fact, attempts have been made to develop 'better' materials by replacing Pb^{2+} with bivalents like Ba, Ca, Sr, Cd, and Ti^{4+} and Zr^{4+} with tetravalents such as Sn^{4+} (the results are available in literature and patents).

A very important material is obtained by incorporating lanthanum into PZT that shows both piezoelectric as well as electro-optic effects such as change in refractive index with application of external fields. This variety produces a new group of materials called PLZT, usually Pb replaced by La. A generalized structure may be written as $Pb_{1-x}La_x(Zr_yTi_{1-y})_{1-x/4}O_3$. For piezoelectric applications, PLZT has less than 5% lanthanum whereas for electro-optic applications, PLZT contains about 6% lanthanum.

Piezoelectric ceramics are used as capacitors, pressure sensors, resonators, electroacoustic transformers, and so on. In fact, PZT as piezoelectrics are applied in developing ultrasonic motors and ultraprecision grinders whereas the electro-optic variety is used in optical shutters and modulators, displays, optical waveguides, holographic recording, image storage, and so forth.

2.6　FORCE/STRESS SENSORS USING QUARTZ RESONATORS

When stress is applied to a flexurally vibrating quartz beam through its mountings (the stress producing a tension along the axis), the beam has a fundamental mode of flexural resonance frequency f given by

$$f = f_o\sqrt{1 + k_1\left(\frac{S}{Y}\right)\left(\frac{l}{t}\right)^2} \tag{2.125}$$

with f_o as the frequency in absence of stress and is given by

$$f_o = k_2\left(\frac{t}{l^2}\right)\left(\frac{Y}{\rho}\right)^{1/2} \tag{2.126}$$

where　S = stress,
　　　　Y = Young's modulus,
　　　　l = beam length,
　　　　t = beam thickness,
　　　　ρ = quartz density, and
　　　　k_1 and k_2 are constants.

The beam is generally machined out from a stock of quartz with semiflexible mounting for minimizing the mounting misalignments that can affect the vibrating frequency mode. One such scheme is shown in Fig. 2.53(a). The vibrational bending moments that might induce distortion are sought to be cancelled by a double beam design as shown in Fig. 2.53(b). The vibration in the

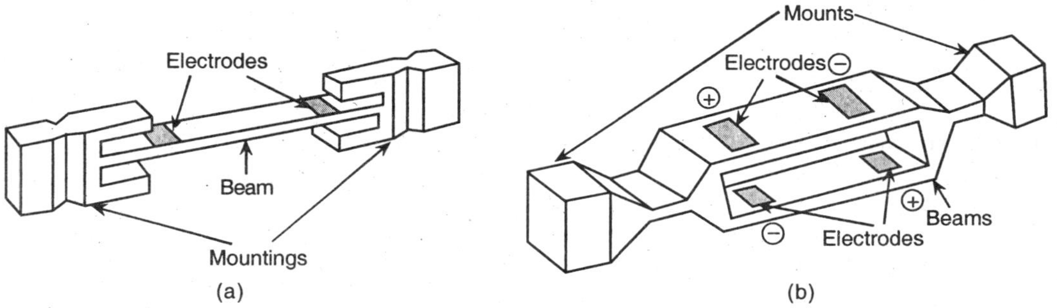

Fig. 2.53　(a) A quartz beam with semiflexible mounting, (b) Double-beam design of the cantilever.

two beams are opposite to each other. The Q-factor in such a structure can be made as high as 10^5. For sustenance of the fundamental mode and frequency measurement, the beams are excited by applying a voltage to the electrode pairs on the beam, when a shearing force is set up in the quartz crystal in consequence of which a small component of flexural deformation is produced in it. The electrodes are produced on the faces of the beams by evaporation technique. As shown in Fig. 2.53(b), the electrodes are so supplied that opposite polarization is set up in the beams that can initiate the desired stress pattern for obtaining the fundamental flexural mode. Electronic oscillators are used for supplying the electrodes.

Tuning fork type design of the beams has also been developed using photolithographic techniques, two schemes of which are shown in Figs. 2.54(a) and (b). Beam depicted in Fig. 2.54(a) vibrates in the plane of the 'plate' while the other vibrates perpendicular to it.

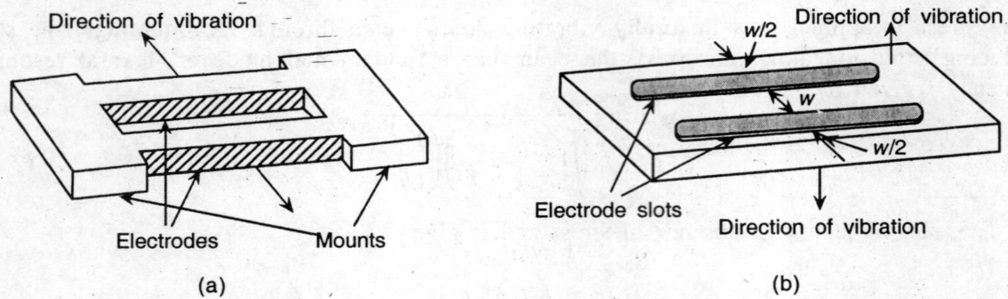

Fig. 2.54　Tuning fork type design of the beams vibrating in opposite phase (a) in the plane of the paper, (b) at right angles to the plane of the paper.

In the scheme of Fig. 2.54(a), the two beams are polarized by oscillator input in such a way that the beams vibrate in opposite phase. Polarization in the other case is so arranged that the central beam, which has a width double that of the side ones, vibrates in opposition to the side beams. More number of beams can be produced in a single assembly if desired.

During production by etching or machining, the crystal axes are properly oriented such that thickness-shear deformation is produced with the application of field. AT and NT cuts are generally recommended.

The stress sensitivity of frequency $\Sigma_S^f = (df/dS)/(f/S)$ is obtained from Eq. (2.125) as

$$\sum{}_S^f = \frac{k_1\left(\dfrac{S}{Y}\right)\left(\dfrac{l}{t}\right)^2}{1 + k_1\left(\dfrac{S}{Y}\right)\left(\dfrac{l}{t}\right)^2} \tag{2.127}$$

indicating nonlinearity in f–S relationship. Σ_S^f, however, increases with (l/t), but increase of (l/t) reduces mechanical strength.

2.7　ULTRASONIC SENSORS

Piezoelectric effect of certain crystalline materials has been successfully utilized in ultrasound production and sensing. This is described in detail in Section 2.5.6. Basically, it is the converse piezo-effect, that is, when an electrical field is applied to the crystal it changes its shape. This property is utilized in generating acoustic or ultrasound wave. It is to be noted that for transmitting the wave through a medium, it is necessary that an appropriate interfacing is provided. Special types of grease are available for the purpose. Good contact is established by this interfacing.

Of the synthetic piezoelectric crystals, barium titanate ($BaTiO_3$) stands out as the major material which, however, requires prior polarization. It consists of randomly oriented tiny piezoelectric crystallites which are properly oriented mostly by DC polling field of several thousand volts per cm, and the material is cooled through Curie temperature. A strong piezoelectric effect has been observed in compounds such as $PbZrO_3$–$PbTiO_3$ called PZT materials (Section 2.5.7). This also has perovskite structure like $BaTiO_3$ (shown in Fig. 2.46).

Piezoelectric transducers can generate continuous wave ultrasound or pulsed ultrasound—latter being used in SONAR or other similar systems. Ultrasonic piezocrystals operate in the range of 0.5–10 MHz. They are directly attached to the transmitting medium or are separated by a small distance which is filled with coupling materials of suitable acoustic properties. Typical couplants at low temperatures are water, grease, and petrojelly and for higher temperatures special polymer couplants may be used.

For continuous wave operation, the sensor is energized by a tuned oscillator while for pulsed application 'relaxation' oscillators are used to charge a capacitor which is discharged across the sensor.

Analytical models describing the interactions of electrical and mechanical phenomena in piezoelectric media have been proposed but found to be inadequate for the design of piezoelectric transducers with realistic geometries and parameters of the material. Numerical solutions in three dimensions of the fundamental equations of the system, coupling the electrical and mechanical phenomena in the piezo element, are found to be necessary for the purpose. A finite element scheme is often adopted because of its inherent flexibility in handling arbitrary device geometries and anisotropies in the materials. Besides, one has to take account of the interactions of the transducer with the ambient media solutions to the wave equations which govern the propagation if acoustic waves in the ambient media flourish.

REVIEW QUESTIONS

1. (a) How is the output of a potentiometric sensor affected due to shorting of windings by jockey?

   For a 100 turn potentiometer, if once the 50th wire is only contacted while at the next instant, 50th and the 51st wires are shorted by the jockey, what would be the percent loss in resolution in the second case if the supply voltage is 10 V?

   [*Hint:* Actual resolution in percentage is

   $$100 \times \frac{\Delta V - \Delta V_k}{\Delta V} = \{1 - nk[1/(n-1) - 1/n]\}100$$

   $$= \{1 - 100 \times 50[1/99 - 1/100]\}100$$

   $$= 49.49\%]$$

   (b) What are the different principles or schemes adopted to eliminate or at least reduce this effect? Explain with diagrams.

2. What are the different materials of the wire and corresponding ones for jockey in potentiometric sensors? What happens for arbitrary choice of materials?

3. What is longitudinal piezoresistance coefficient in connection with strain gauges? How does the resistivity of a gauge material having Poisson's ratio 0.4 and gauge factor 2 change with strain? If the Young's modulus of the material is 2×10^6 kg/cm^2, what is its longitudinal piezoresistance coefficient?

[*Hint:* $\Delta\rho/\rho = (\lambda - 1 - 2\mu)(\Delta l/l) = (2 - 1 - 0.8)\,\varepsilon$

or, percent change in resistivity is proportional to 20% of strain. Longitudinal piezoresistance coefficient is

$$(\Delta\rho/\rho)/(\Delta l/l)/E = 0.2/2 \times 10^6 = 10^{-7} \text{cm}^2/\text{kg}]$$

4. (a) How are bonded strain gauges produced now without requiring bonding materials? What is the difference between the vacuum deposition and sputtering deposition processes?

   (b) What are the important properties necessary for bonding materials? How are they realized in practice?

   (c) Explain the ceramic spray technique of bonding.

5. (a) How does the gauge factor of a semiconductor strain gauge vary with doping level? Discuss with the help of diagrams.

   (b) Describe a piezoresistive type strain gauge sensor appending appropriate diagrams.

6. What are the different types of inductive sensors? How is displacement measured by such sensors? How does the inductance change in such a system?

   A core having 20 single-turn coil and a magnetic path length, including an armature of 10 cm, has a cross-section area of 1.2 sq cm. For a displacement to be measured, the gap length changes from 0.1 cm to 0.15 cm. By what percentage does the inductance of the coil change and for what original inductance value? Assume a core permittivity of 10.5.

   [*Hint:* Percent change in $L = 100\{(\mu_s/l)(l_{g2} - l_{g1})/(1 + l_{g2}\,\mu_s/l)\}$

   $$= (10.5/10)\,\{0.05/(1 + 0.15 \times 0.5/10)\}100 = 4.5\%$$

   Also, $L_1 = \mu_s/(1 + l_{g1}\,\mu_s/l)(n^2 a/l) = \dfrac{10.5 \times 400 \times 1.2 \times 10^{-4}}{(1 + 0.1 \times 0.5/10) \times 10 \times 10^{-2}} = 4.56$ H]

7. Does the sensitivity of an inductive transducer vary? If yes, how does it vary?

   Calculate the normalized sensitivity of an inductance type transducer formed with a core material of permittivity 10.5 and the 'closed' length of path of 10 cm for a gap length of (i) 0.1 cm and (ii) 0.05 cm.

   [*Hint:* $S_{l_g}^L = 1/(1 + l/(l_g\mu_s)) = 1/(1 + 0.952/l_g)$;

   case (i) $l_g = 0.1$ cm, $S_{l_g}^L = 0.095$

   case (ii) $l_g = 0.05$ cm, $S_{l_g}^L = 0.0499$

8. How is linearity of input–output relation improved in case of a ferromagnetic plunger type transducer? What material is usually the plunger made of?

   How does the inductance of the system change for a push–pull type design with a change in position of the plunger, for a plunger permittivity 12.2, coil average diameter 1.1 cm, and plunger diameter 0.86 cm.

   [*Hint:* $S_{l_p}^L = 1/[1 + (d/d_p)^2 (1/(\mu_p - 1))] = 1/[1 + (1.1/0.86)^2 (1/(12.2 - 1))] = 0.875$]

   How does the situation change if the plunger is replaced by short-circuited sleeve over the pair of coils, the sleeve being movable? Discuss with diagram.

9. Describe the operation of an LVDT for measuring displacement. How is its operation dependent on the position of the core?

10. Sketch a magnetostrictive type transducer for measurement of force. On what principle does this transducer work? Explain with diagrams.

11. What are the different types of capacitive sensors used for displacement measurement? How do they differ in operating principles?

    If the air gap between the teeth of the two electrodes of a serrated type capacitive transducer is 0.1 cm and the 'active' tooth length is 1 cm, what is the sensitivity factor of the sensor? Assume constant term as 4.

    [*Hint:* $\beta_s = 1/(1 + K(x/l)) = 1/(1 + 4 \times 0.1/1) = 0.96$]

12. What type of capacitive sensors are used in pressure transmitters? Explain its operation with appropriate diagrams.

    If the diaphragm diameter is 2.8 cm (in a diaphragm type capacitive sensor), separation between the fixed and the movable plate is 0.4 cm in normal condition, and the diaphragm is kept taut with a tension of 2 kg/cm, what is the change in the capacitance for an input differential pressure of 1 kg/cm^2?

    [*Hint:* $\partial C/C = \{(2.8/2)^2/(8 \times 4 \times 0.4)\} \times 1 = 0.15$]

13. Explain the phenomenon, how charges develop on two plates placed across a piezoelectric crystal with a force applied on it?

    Define *d*-constants and *g*-constants of a piezoelectric crystal. How are they useful in the design of a transducer with a piezoelectric crystal?

14. (a) Explain how piezoelectric crystals are used as bimorphs and multimorphs. Where are they used?

    (b) What are PZT and PLZT? Why are they gaining importance in sensor technology?

15. Describe how quartz resonators are used as stress sensors? On what factors does the resonance frequency of a resonator depend? Discuss.

Chapter 3

Thermal Sensors

3.1 INTRODUCTION

Thermal sensors are primarily temperature sensors and their operating principles have long been established, specially those of the primary sensors, also called thermodynamic sensors, and which are the subjects of discussion in this chapter. Any physical quantity, say Q, is usually expressed as its magnitude in number N and in unit U so that

$$Q = NU \tag{3.1}$$

If it is possible to relate temperature T directly in the form of Eq. (3.1), from the first principles, in a sensing system, then it is called a *primary sensor*.

Even though the principles of thermal sensing are well established, newer innovations are added to the stock of sensors dependent on these principles with improved quality and better practical approaches. Many of the commonly used practical 'thermometers' are, however, not primary in that sense and may be called *secondary* as the relationship between Q and T used by them is largely empirical.

A brief classification of primary and secondary temperature sensors is presented in Table 3.1.

Table 3.1 Classification of sensors

Primary sensors	Secondary sensors
1. Gas thermometer	1. Thermal expansion types: solid, liquid and gas
2. Vapour pressure type	2. Resistance thermometer
3. Acoustic type	3. Thermoemf type
4. Refractive index thermometer	4. Diodes, transistors, or junction semiconductor types
5. Dielectric constant type	5. Adapted radiation type
6. He low temperature thermometer	6. Quartz crystal thermometer
7. Total radiation and spectral radiation type	7. NQR thermometer
8. Magnetic type	8. Ultrasonic type
9. Nuclear orientation type	
10. Spectroscopic techniques (not sensors in that sense)	
11. Noise type	

There are different kinds of heat flux sensors which measure heat flux in terms of temperature difference. Even in temperature measurement, there are special types of sensors such as pneumatic type, pyroelectric type and so on.

The following discussion describes the primary sensors, perhaps, in principles alone, mainly because of their limited applications in industry. Some of these are transformed to or adapted in commercial applications with minor changes. The overall measuring systems are not discussed but the basic sensing mechanisms are dealt with greater emphasis.

3.2 GAS THERMOMETRIC SENSORS

Gas thermometric sensors are based on the gas law

$$PV = nRT \tag{3.2}$$

where P = the pressure,
V = volume of the gas,
R = the gas constant,
T = the temperature in K-scale, and
n = the number of moles

The relation is, however, true for all ideal gases and is approximately true for real gases at low pressures. For a real gas, a series relation is usually considered which is given by

$$PV = nRT\left[1 + \beta_1(T)\left(\frac{n}{V}\right) + \beta_2(T)\left(\frac{n}{V}\right)^2 + \dots\right] \tag{3.3}$$

where β_i's are virial coefficients which are different for different gases and are functions of temperature. Contributions of the higher order terms become larger at lower temperatures and higher pressures since the gases depart from the ideal nature under such conditions. Even if the first order term is to be retained, knowledge of $\beta_1(T)$ is required.

Gas thermometers can be of two different types, both based on the same basic law. These are: (i) constant volume thermometers where P is proportional to T and (ii) constant pressure thermometers where volume V is proportional to T. However, constant volume thermometer is more easily realized in practice. Usually, the term higher than first order term in Eq. (3.3) is negligible so that, on approximation, we obtain

$$PV = nRT\left[1 + \beta_1(T)\left(\frac{P}{RT}\right)\right] \tag{3.4}$$

If the reference conditions are known *a priori* and also the corresponding pressure and temperature P_r and T_r, then, for constant volume

$$T = T_r\left(\frac{P}{P_r}\right)\left[\frac{1 + \beta_1(T_r)\dfrac{P_r}{RT_r}}{1 + \beta_1(T)\dfrac{P}{RT}}\right] \tag{3.5a}$$

Obviously, $\beta_1(T)$ and P/T in the denominator of the right hand side of Eq. (3.5a) pose problem even if $\beta_1(T_r)$ is known. Extrapolation techniques are known for determining the virial coefficients but obviously, these are not applicable in day-to-day activities. The secondary means is adopted where Eq. (3.2) is good enough for the purpose of simplifying Eq. (3.5a) to

$$T = T_r\left(\frac{P}{P_r}\right) \tag{3.5b}$$

For commercial use, direct application of Eq. (3.2) is also suitable when, by enlarging the volume of the gas, the output may be made large as the scale multiplier is $V/(nR)$ between T and P. The schematic of such a transducer is shown in Fig. 3.1.

In measurement, often a reference system is coupled and the pointer movement may be made differential. The gas used is an inert one though nitrogen is a good choice. The bulb volume is made at least 100 times larger than the combined volume of the capillary and Bourdon. Any change in the temperature of the bulb which is immersed in the process causes change in gas pressure which is transmitted to the Bourdon.

Care must be taken to compensate for the expansion of the bulb volume at the process temperature which is varying. If the expansion coefficient of the material is known, it can be easily corrected. Also, the gas in the capillary and Bourdon does not expand and correction for this must be made knowing the temperature distribution along the capillary line and the Bourdon tube. Besides, for such a system some other errors are there which need be taken into consideration.

A consequence of the gas pressure thermometer is the vapour pressure thermometer where the temperature scale is obtained from calibration with respect to reference points. A feasible practical form of it is similar to that of Fig. 3.1 except that a suitable liquid now fills the bulb partially, keeping enough space above the surface of the liquid for vapour pressure to form and even be saturated at all temperatures of interest for the particular liquid. With temperature, the above pressure increases according to Claussius–Clapeyron equation

$$T\frac{dP_s}{dT} = \frac{H_v}{V_g - V_l} \tag{3.6}$$

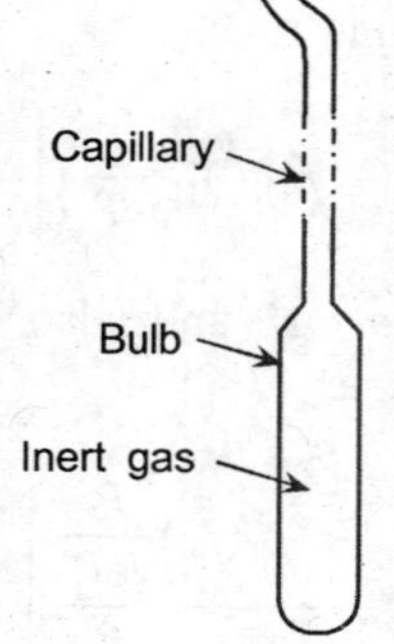

Fig. 3.1 The schematic of a gas pressure thermometer.

where P_s = saturated vapour pressure
   H_v = molar heat of vaporization
   V_g = molar volume in gaseous state
   V_l = molar volume in liquid state.

The solution of Eq. (3.6) for deriving a relation between P_s and T for practical utility is quite involving. Instead, a series equation of the form

$$\ln\left(\frac{P_s}{P_{os}}\right) = \sum_{j=-k}^{k} \alpha_j T^j \tag{3.7}$$

where P_{os} is the saturation pressure at reference temperature (0°K) and k is an arbitrary order determined by the 'characteristic vaporization' of the gas.

The important aspect of the sensor is that it is the temperature of the liquid–gas interface that determines the pressure P_s and hence, the indication made by the Bourdon element would be independent of the volumes of the bulb, capillary, and Bourdon as well as ambient temperature. As is apparent in Eq. (3.7), the P_s versus T curves for different liquids are of the type shown in Fig. 3.2. Wide variety of ranges can be expected with different liquids and the scale range of each liquid is limited by its boiling point and a critical temperature at which the vapour disintegrates. Large factors of safety are, however, kept on both sides of the scale. Table 3.2 shows the ranges for some commonly used liquids.

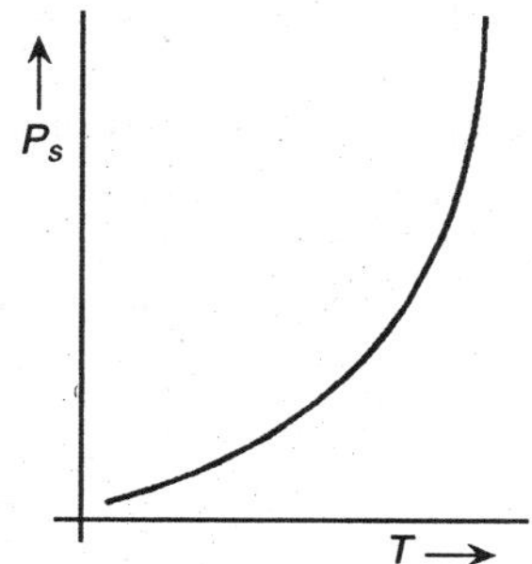

Fig. 3.2 Vapour pressure variation with temperature.

Table 3.2 Temperature ranges in vapour pressure thermometers

Liquid	Range (°C)
Methyl alcohol	0–50
n-Butane	20–80
Methyl Bromide	30–85
Ethyl chloride	30–100
Ethyl ether	60–160
Ethyl alcohol	30–180
Toluene	150–250

For very low temperature, Argon is used covering a range of –250°C to –100°C while water can be used between 120°C and 200°C. Rapid heating often produces bubbles which pass on to the capillary causing errors in indication in which cases it is recommended that a nonvolatile liquid immiscible with the operating liquid be used to fill the Bourdon and the capillary keeping the vaporizing surface as usual as shown in Fig. 3.3

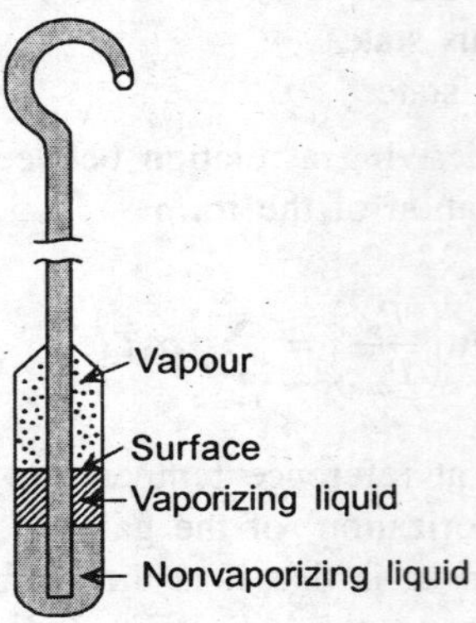

Fig. 3.3 Scheme of a vapour pressure thermometer.

3.3 THERMAL EXPANSION TYPE THERMOMETRIC SENSORS

The thermal expansion types thermometric sensors including the ones specified in Section 3.2 are, perhaps, the oldest varieties still used commercially to a certain extent.

Earliest of this kind is the solid expansion type bimetallic sensor which uses the difference in thermal expansion coefficients of different metals. Two metal strips A and B of thickness t_A and t_B and thermal expansion coefficients α_A and α_B are firmly bonded together at a temperature, usually the lowest or the reference temperature, to form a cantilever or a helix with one end fixed as shown in Figs. 3.4(a) and 3.4(b) respectively. When the temperature of the cantilever or the

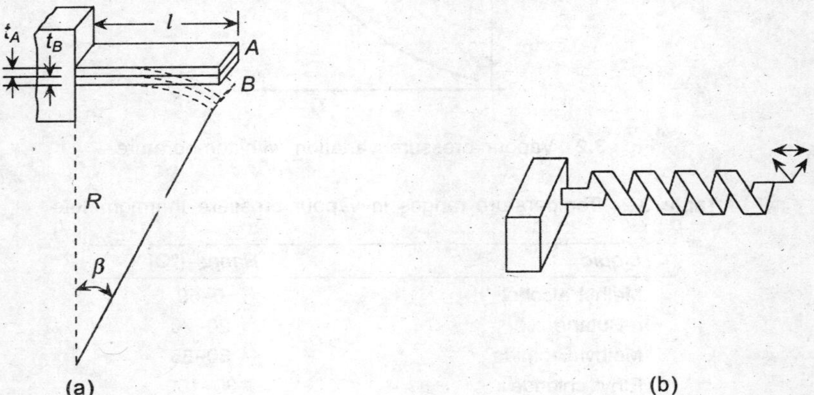

Fig. 3.4 (a) Cantilever type bimetal thermometer, (b) helix type bimetal thermometer.

helix is raised by heating or lowered by cooling, one strip expands or contracts more and free end of either of the two moves as shown. The cantilever, in fact, bends into a circular arc with radius of curvature R given by the relation

$$R = \frac{(t_A + t_B)\left[3\left(1 + \frac{t_B}{t_A}\right)^2 + \left(1 + \left(\frac{t_B}{t_A}\right)\left(\frac{Y_B}{Y_A}\right)\right)\left\{\left(\frac{t_B}{t_A}\right)^2 + \frac{t_A Y_A}{t_B Y_B}\right\}\right]}{6(\alpha_A - \alpha_B)(T_h - T_b)\left(1 + \frac{t_B}{t_A}\right)^2} \qquad (3.8)$$

where Y is the Young's modulus,
   T_h is the raised temperature, and
   T_b is the bonding temperature.

Equation (3.8) is simplified using $t_A = t_B = t$ and $Y_A \approx Y_B$. This gives

$$R = \frac{4t}{3(\alpha_A - \alpha_B)(T_h - T_b)} \qquad (3.9)$$

The angular deflection, β, per unit temperature change, that is, sensitivity (for small β) is given by

$$S_T^\beta = \frac{\beta}{(T_h - T_b)} = 3l\frac{\alpha_A - \alpha_B}{4t} \qquad (3.10)$$

where l is the length of the cantilever.

S_T^β increases linearly with length and inversely with strip thickness for a given pair of metal elements. Usually element B is made of invar (a Ni–Fe alloy) of $\alpha_B \approx 1.7 \times 10^{-6}/°C$ which is quite low and element A is brass or steel of different alloying compositions. Such sensors can work precisely but not very accurately in a range –50–400°C. Besides cantilever and helix forms, they are also made in spiral and disc forms in different control applications.

Next in line is the liquid-in-glass thermometer—the liquid in majority of the cases being mercury. With mercury, this thermometer is almost the basic temperature measuring unit in home (as clinical thermometer), in laboratories and even in industries. It utilizes the expansion property of the liquid kept in a bulb to which a capillary, closed at the far end, is attached through which the expanded liquid rises and an indication in mm, calibrated directly in temperature scale, is obtained. The schematic is shown in Fig. 3.5. The range of mercury thermometer is normally

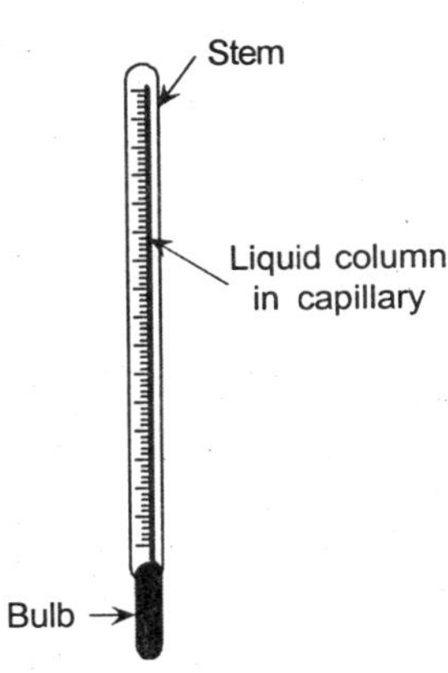

Fig. 3.5   Liquid-in-glass thermometer.

–35–300°C and the upper limit is 357°C, its boiling point. The range can be extended upto 600°C by filling the volume above mercury with pressurized dry nitrogen. The volume of the bulb is made 100 to 400 times larger than the capillary volume. Other liquids used as expansion media are given in Table 3.3 with their corresponding ranges.

Table 3.3 Thermometric liquids and their ranges

Liquid	Range (°C)
Pentane	–200–30
Alcohol	–80–70
Toluene	–80–100
Creosote	–5–200

When the measurement is made, the thermometer should be immersed upto the meniscus in the capillary which means that the thermometer is to be moved for varying temperatures. Alternatively, the entire thermometer is immersed, or, only the bulb is immersed. The last alternative is the most common one and for this purpose, a correction has to be applied for the mercury column above the immersion line because the column is at a different temperature t_c than the measured value t_m. The correction term is

$$\Delta t = \gamma_d n \, (t_m - t_c) \tag{3.11}$$

where γ_d is the differential thermal expansion coefficient of volume between mercury and glass and

n is the number of degrees indicated by the column, that is, exposed degrees.

γ_d has a value of about $1.6 \times 10^{-5}/°C$.

An extension of this is the industrial type liquid filled-in system which consists of a metallic bulb attached to a metallic capillary. The other end of capillary is fitted with a Bourdon. The expansion of the liquid in the bulb is transmitted to the Bourdon which uncurls in the usual manner. The basic scheme is similar to that presented in Fig. 3.1. A number of compensations are necessary to obtain correct indication by the measurement system using such a sensor. The correction methods are available in standard texts on industrial instrumentation.

3.4 ACOUSTIC TEMPERATURE SENSOR

When a longitudinal (acoustic) wave propagates through an ideal gas, it has a speed C_i given by

$$C_i = \left(\frac{\gamma RT}{M}\right)^{1/2} \tag{3.12}$$

where M is the molecular weight of the gas and $\gamma = C_p/C_v$ is the ratio of specific heats ($\gamma = 5/3$ for monoatomic gases).

Knowing the gas and measuring velocity C_i, temperature T can be given by

$$T = \frac{MC_i^2}{\gamma R} \tag{3.13}$$

The realization of this technique is made in acoustic helium interferometer whose working is explained through Fig. 3.6. A quartz crystal excited to its resonance frequency is used to transmit this wave through a gas (He) column, to be faced by a piston. The wave is reflected at the piston surface to form a pattern as shown. When the path length l has a multiple number of half-wavelengths and correspondingly the gas column is set to resonate at each such half-wavelength gap, with the piston moving away from the crystal at each resonant peak, the crystal gives out maximum energy and hence the voltage V_Q across the crystal defines peaks as shown in Fig. 3.6(c). If the piston moves by a distance d to give n such peaks, $d = n\lambda/2$ from which C_i is determined and thence temperature T. The piston movement must be accurately monitored to within, say, 1 µm.

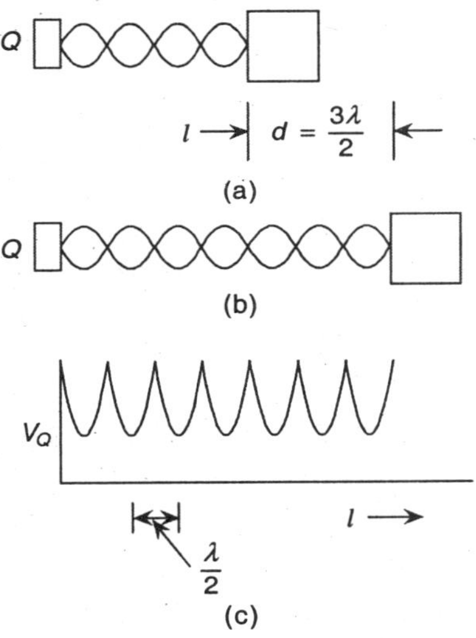

Fig. 3.6 Principles of acoustic temperature sensor: (a) the system, (b) the system with changed position of piston for maintaining resonance, and (c) the crystal output peak positions.

In non-ideal gas, correction as per Van der Waal's equation

$$V - b\left(P + \frac{a}{V}\right) = MRT \tag{3.14}$$

has to be applied, where M is the molecular weight of the gas, and a and b are functions of 'molecular' constants. The corrected velocity C_c is then given by

$$C_c = \sqrt{\frac{\gamma RT}{M}\left[1 + \frac{\alpha P}{RT}\right]} \tag{3.15}$$

where α is a function of a, b, T, and V.

There is a nonresonant acoustic sensor that utilizes the pulse-echo transit time difference which changes with temperature. Figure 3.7 is a schematic representation of the sensory parts of the measurement system. An ultrasonic pulse is transmitted through the sensor, a part of which is reflected at the entrance (a discontinuity) and a part at the end, as shown. The reflected pulses are received by the transreceiver coil at an interval of t_{tr} called the transit time. The pulse that travels the entire length of the sensor is delayed more/less depending on the change in the sensor temperature. This temperature dependence is a function of the path length l, sensor material, temperature range, and vibration mode even if the first echo is considered. The materials which show distinctive t_{tr} are listed in Table 3.4 with their temperature ranges.

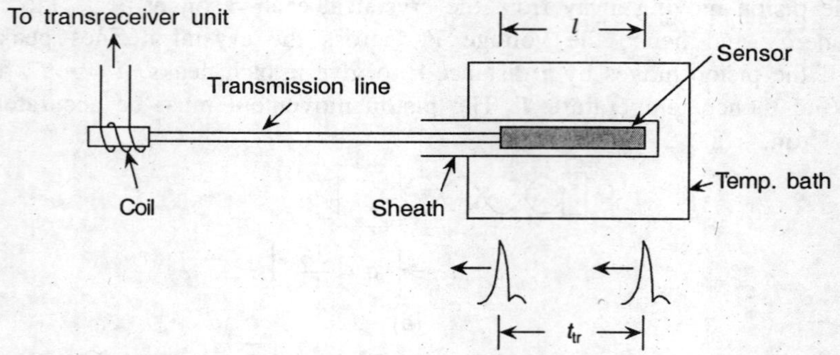

Fig. 3.7 Pulse-echo transit time difference technique of temperature measurement.

Table 3.4 Materials versus temperature range

Material	Temperature range (°C)
Aluminium	≤500
Stainless steel	≤1100
Sapphire	≤1600
Molybdenum, Ruthenium	≤2100
Wolfrum, Rhenium, ThO_2–W(2%)	≤2700

The nature of the plot between t_{tr} and temperature T is shown in Fig. 3.8.

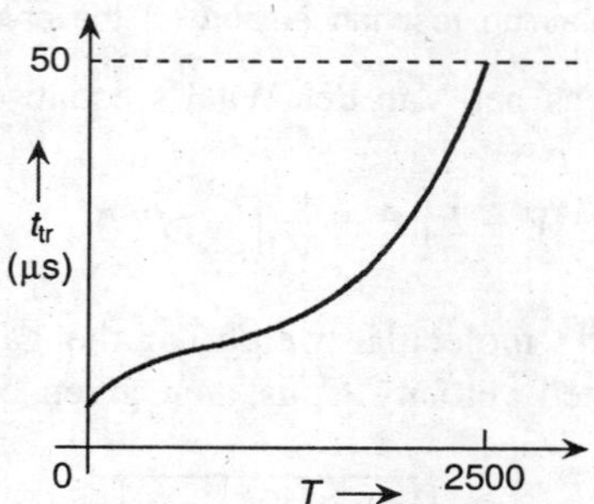

Fig. 3.8 Transit time versus temperature plot.

The sensor may be made in the form of a thin wire with restrictions or constrictions at intervals of space where the reflections would occur. The wire diameter varies from 0.03–3 mm and spacing between restrictions varies from 5–10 mm in a sensor length of 15–50 mm, and a number of echos can then be produced. There should not be any inhomogeneity in the material faced by the wave except for the restrictions.

3.5   DIELECTRIC CONSTANT AND REFRACTIVE INDEX THERMOSENSORS

The above thermosensors are developed basing on two well known relations: (i) Clausius–Mossotti relation and (ii) the relation between refractive index μ and dielectric constant χ of a gas. The sensors are useful for gas temperature measurement.

The Clausius-Mossotti relation is valid for an ideal gas and is given by

$$\frac{\chi-1}{\chi+2} = M_\chi \frac{n}{V} \tag{3.16}$$

where χ = dielectric constant and
M_χ = the molecular polarizability of the gas.

Using Eq. (3.2), Eq. (3.16) transforms to

$$T = \frac{(\chi+2)M_\chi}{(\chi-1)R} P \tag{3.17}$$

Thus, knowing χ, M_χ, and P, we can measure T. M_χ is available by calculation or in a table or by measurement and χ is measured by measuring capacitance which is given by the relation

$$\chi = \frac{C(P)}{C(0)} \cdot \frac{1}{1+\phi_c P} \tag{3.18}$$

where $C(P)$ denotes capacitance at pressure P and ϕ_c is the capacitance compressibility factor that is considered to account for the change in the capacitor dimensions with pressure.

For real gases, however, virial expansion of the dielectric constant has to be taken into account and 'extrapolation' technique is adopted. For practical purposes, an empirical calibration is the best solution.

Refractive index thermometer uses the relation

$$\frac{\mu^2-1}{\mu^2+2} = \frac{M_\chi n}{V} = M_\chi \frac{P}{RT} \tag{3.19}$$

since $\mu^2 = \chi$, Eq. (3.19) is basically the transformation of the Clausius–Mossotti equation. A practical technique is used to measure $\mu - 1$ by passing a laser beam through a Michelson interferometer with one of its arms containing the gas sample. With pressure, the optical path length L increases and a relation

$$\frac{\Delta L}{L} = \mu - 1$$

$$\propto \rho$$

$$\propto P/T \tag{3.20}$$

holds good. Here, ρ is the gas density. Equation (3.19) can be used for finding T with the knowledge of μ, M_χ, and P.

For non-ideal gases again, a virial expansion of $(\mu^2 - 1)/(\mu^2 + 2)$ is accommodated. The technique is yet to gain commercial importance.

3.6 HELIUM LOW TEMPERATURE THERMOMETER

This is a thermometer developed on the basis of pressure–temperature equilibrium relationship for ^{3}He enclosed in a constant volume chamber provided on one side with a Be–Cu diaphragm on which a strain gauge is bonded for measuring pressure of the expanded He at low temperature where it can exist in liquid–solid equilibrium state. The relation is parabolic in nature with a minimum at a temperature of 0.32°K.

3.7 NUCLEAR THERMOMETER

This type of thermometer can be used at very low temperatures, below 0.1 K, and can be designed with and without a magnetic field. The basic operating principle is a temperature dependent nuclear characteristic which states that the directional distribution of emission of β or γ radiations from a radioactive nucleus is dependent on the degree of ordering of a system of nuclear spins and this ordering is dependent on temperature with Boltzmann distribution. In fact, the relative population p_m of nuclear spin system of the substrate m is given by

$$p_m = \frac{\exp(-E_m/kT)}{\sum_n \exp(-E_n/kT)} \tag{3.21}$$

where E_n are the energies of the nuclear hyperfine states, n taking values in terms of quantum numbers, I, such as $n = I, I - 1, \ldots, - I$.

In thermodynamic equilibrium, the spin system temperature T equals the lattice temperature. Obviously, if T is very large, all p_m's are equal and the radiation is said to be emitted isotropically. However, if T is very small, $T \leq E_n/k$, p_m's have different values and an anisotropic emission occurs resulting in the difference in the nuclear orientation about the axis of quantization of the nuclear spin system. The pattern of distribution is, to some extent, like the radiation emission pattern from an antenna. When this radiation is counted in different directions at different temperatures, a pattern is generated but only in a three dimensional 'field' with intensity, direction and temperature as coordinates. The relation that expresses such a distribution with temperature is actually an implicit relation and for actual measurement, prior calibration is often a better alternative.

In practical sensors, suitable radioactive nuclei are incorporated in a 'host lattice' which usually is ferromagnetic in nature. The nuclei used are ^{54}Mn in Fe, Ni, Al, Cu, Zn or ^{60}Co in Fe, Co, and Ni with known decay schemes. The normalized direction distribution (NDD) with specified angles (α_s) are measured and it is known that NDD (α_s) is a unique function of temperature T. Often the distribution is measured as a relative value with respect to its counts at isotropic condition, that is, when T is large.

3.8 MAGNETIC THERMOMETER

Another low temperature thermal sensor is based on the change of magnetic susceptibility χ of a paramagnetic substance (salt). The materials used for different temperature ranges are as listed in Table 3.5.

Table 3.5 Materials and temperature ranges

Materials	Temperature ranges (°K)
Cerium Magnesium Nitrate (CMN)	0.01–2.5
Chromic Methylammonium Alum (CMA)	0.3–30
Manganous Ammonium Sulphate (MAS) and Godolinium Sulphate (GS)	0.9–80

Proper materials have to be chosen because, for the sensors to be able to measure temperature using this technique, it is required that the temperature dependent susceptibility should be solely an outcome of electron spin magnetism and not of orbital angular momentum. In case of the materials being not as stipulated, ranges becomes limited due to the effect of orbital angular momentum which is predominant at 'higher' temperatures. When an external magnetic field is applied, the magnetic dipoles of the paramagnetic salts are ordered.

According to the modified Curie's law, the susceptibility χ, is given by the relation

$$\chi = \chi_o + \frac{C}{T + \bar{\theta} + \dfrac{\alpha}{T}} \qquad (3.22)$$

where χ_o is the temperature independent susceptibility,

$\bar{\theta}$ accounts for the dipole coupling of the local field and ion-exchange interaction, and,

α/T accounts for the Stark splitting of the ground state with field.

Both $\bar{\theta}$ and α/T are dependent on sample geometry and on the angle between the applied field and the sample crystal axis. For a spherical sample, for example, dipole coupling of the local field vanishes. Measurement is made by putting the sample crystal in a pair of mutually coupled coils (Fig. 3.9) which are connected to one arm of a measuring bridge for measuring the change in the mutual inductance M as given by the relation

$$M = M_0 + \frac{B}{T + \bar{\theta} + \alpha/T} \qquad (3.23)$$

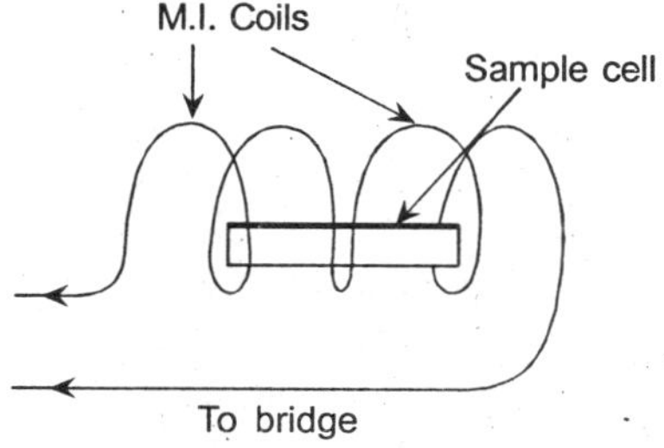

Fig. 3.9 Principle of a magnetic thermometer.

In practice, M_0 and B depend upon χ_0, C, and the coil design, $\bar{\theta}$ is a salt property but also depends on the sample shape, and α is dependent only on the property of the salt. The four constants M_0, B, $\bar{\theta}$, and α are evaluated by calibrating at four temperatures and this calibration is necessary for different samples individually.

For CMN, both $\bar{\theta}$ and α are very small and we can use the simple relation

$$M = M_0 + \frac{B}{T} \tag{3.24}$$

3.9 RESISTANCE CHANGE TYPE THERMOMETRIC SENSORS

The principle of temperature dependence of electrical conduction in conductors and semiconductors is the basis of resistance type thermometric sensors. It is a complex phenomenon. Basically, there is a change in resistance ΔR with change in temperature ΔT, and there is a specific relation between ΔR and ΔT that is used to evaluate ΔT by measuring ΔR mostly by electrical circuits and indicating systems.

If mass and charge of an electron are given respectively by m and e, then postulating an electron gas model of a metal as per Drude's suggestion, its conductivity σ is given by

$$\sigma = n_e \frac{e^2}{m} \tau_r \tag{3.25}$$

where n_e = number of electrons per unit volume, and
τ_r = relaxation time. τ_r is a function of temperature which at room temperature is of the order of 10^{-13}s.

Conductivity of metals is therefore, believed to be due to the travel of free electrons through them but modified by a function that has the periodicity of the lattice. Conductivity arises because of the imperfections of this lattice such as presence of foreign atoms, point defects, irregular grain boundaries which lead also to scattering and decrease in conductivity. Besides, lattice vibration by emission and absorption of phonon leads to additional scattering and further decrease in conductivity. Each scattering mechanism has its own relaxation time τ_{ri} and the rate of change of drift velocity of electrons. The overall relaxation time is given by τ_{ro}, as

$$\tau_{ro} = \left(\sum_{i=1}^{n} \left(\frac{1}{\tau_{ri}} \right) \right)^{-1} \tag{3.26}$$

However, resistivity, because of scattering due to phonon absorption/emission, is only temperature-dependent and the resistivity of a sample conductor remains constant otherwise and reduced considerably by small concentrations of point defects and foreign atoms.

Since, in practice, a few metals and certain semiconductors are considered as viable sensor materials, their temperature-dependent resistance is of importance for the purpose.

Resistance variation in metals depends on temperature in various ways depending, in turn, on the type of metal and range of temperature.

For simple metals such as sodium and potassium, resistance is a simple function of T as long as $T > T_c$ where T_c is a characteristic temperature of the metal related to Debye temperature θ_d, given by

$$\theta_d = \frac{h}{k}\left(\frac{9N}{4\pi V}\right)^{1/3}\sqrt{d}\left[\frac{1}{(\beta + 4C/3)^{3/2}} + \frac{2}{C^{3/2}}\right]^{1/3} \tag{3.27}$$

where h = Planck's constant,
        k = Boltzmann constant
        N = number of molecules in volume V,
        d = density,
        C = modulus of rigidity, and
        β = bulk modulus.

For the same types of materials, the relation varies proportionally to T^5 at very low temperatures.

For transition metals such as iron, copper, nickel, platinum, and so on the scattering mechanism is quite complex and the simple proportional relation does not hold good over a wide range of temperatures. At low temperatures, the relation with T may be positive for some metals (Ni) and negative for some metals (Pt). At temperatures below 10 K, electron-electron scattering becomes predominant and resistance becomes a function of T^2. At high temperatures ($T > T_c$), the lattice vibration occurs with larger amplitude, thermal expansion is considerable, and resistance also becomes a function of T^2.

Increasing temperature creates lattice vacancies and if E_v is the energy of formation of lattice vacancy, the resistance becomes another function of T given as

$$\partial \rho \propto \exp\left(\frac{-E_v}{kT}\right) \tag{3.28}$$

when magnetic metals such as Fe, Co and so forth are dissolved in nonmagnetic metals, resistance attains a minimum at a very low temperature (*Kondo effect*). However, Fe–Rh or Co–Pt show a certain kind of anomaly such that resistance decreases monotonically with lowering of temperature without a minimum at a specific value and they are, therefore, very conveniently used in low temperature sensing. The nature of the $R–T$ curve of a Fe–Rh alloy is shown in Fig. 3.10.

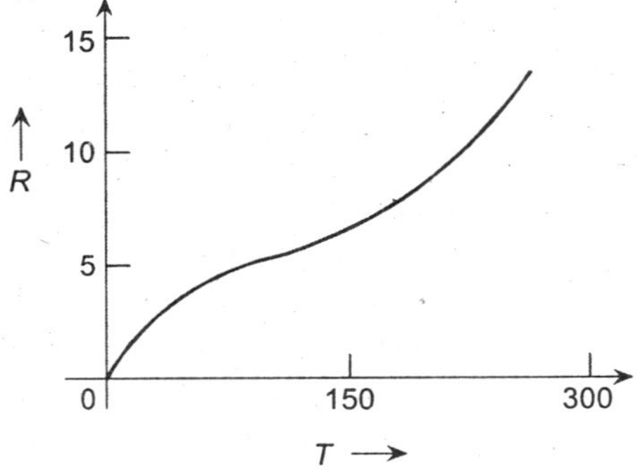

Fig. 3.10 $R–T$ characteristics of Fe–Rh alloy.

In semiconductors, the resistance, in general, has a negative temperature coefficient. This is true for carbon resistors as well as for semiconductor crystals and sintered metal oxides (called thermistors). Conduction in semiconductors occurs due to movement of electrons in conduction band. The energy gap E_g, between the conduction and valence bands is large at ordinary room temperature, much larger than kT and only a limited conduction occurs; but with increasing T, conduction also increases. In fact, if n_e is the conducting electron density, n_p the conducting hole density, m_e and m_p are electron and hole masses respectively, and F_μ is the Fermi level, then the relation shown by Eqs. (3.29a) and (b), as per Fermi–Dirac statistics, hold good

$$n_e = 2\left(2\pi m_e \frac{kT}{h^2}\right)^{3/2} \exp\left(\frac{-(E_g - F_\mu)}{kT}\right) \tag{3.29a}$$

and

$$n_p = 2\left(2\pi m_p \frac{kT}{h^2}\right)^{3/2} \exp\left(\frac{-(E_g - F_\mu)}{kT}\right) \tag{3.29b}$$

For intrinsic semiconductors such as Si, Ge, and diamond (C), $n_e = n_p$ so that expression for conductivity σ, is given by

$$\sigma = 2\left(\frac{2\pi kT}{h^2}\right)^{3/2} (m_e m_p)^{3/4} (\mu_e + \mu_p) \exp\left(\frac{-E_g}{kT}\right) \tag{3.30}$$

where μ_e and μ_p are the mobilities of electrons and holes respectively which show temperature dependence as well. Equation (3.30) can be written in terms of $\rho = 1/\sigma$ as

$$\rho = \alpha \exp\left(\frac{\beta}{T}\right) \tag{3.31}$$

where α is also temperature dependent as

$$\alpha = \frac{T^{-3/2}}{2}\left(\frac{h^2}{(2\pi k)^{-3/2}} \frac{(m_e m_p)^{-3/4}}{(\mu_e + \mu_p)}\right) \tag{3.32}$$

Doped crystals have varying performance characteristics in different zones. They are either n-type or p-type and have lower resistivity and lower temperature coefficients because doping with foreign atoms extends donor and acceptor properties and generates extra electrons or holes for better conduction.

Ranges of such sensors are generally limited upto about 300–400°C. Consider n_d and n_a donor and acceptor atoms per unit volume, produced by doping, depending on the ranges of temperature. Then ionization level of n_d and n_a atoms changes so as to bring about a change in the conduction level. At relatively higher temperatures, greater than 150°C, all these n_d and n_a atoms are ionized and conduction occurs by thermally excited electrons from valence band. Hence, the behaviour of the material is like an intrinsic semiconductor. For temperature range between –150–150°C, $n_d - n_a$ atoms are ionized and in the conduction band electrons from this ionization are available so that their mobility determines the conductivity. For temperatures below –150°C but not too low, ionization of the effective donor atoms is not complete and conductivity obviously depends exponentially on temperature as is resistivity. At very low temperatures, below

–260°C conduction is by electron jumps but still the resistance variation with temperature remains exponential. The values of α and β in these cases differ. Choice of material plays a dominant role in making their nature 'soft' or 'hard' in exponential nonlinearity.

Commercial passive semiconductor transducers, as has been indicated, are known as *thermistors* which are 'mixtures' of metal oxides and/or complex oxidic systems such as pervoskites with varying proportions and sintered into desired shapes. Metals whose oxides are commonly used for the purpose are Fe, Co, Mn, Ni, Cu, Ti, Li, Mg, Cr and so on. The most stable are those that are made from oxides of Mn–Ni and Mn–Ni–Co.

The two types described here, that is, the intrinsic single crystal type and extrinsic sintered metal oxide type are compared in Table 3.6.

Table 3.6 Comparison of semiconductor properties

Count	Type	
	Monocrystalline Ge/Si	*Oxide type*
Conduction	Intrinsic	Extrinsic
Bond	Covalent	Covalent and ionic
Impurities	Chemical	Mostly physical
N/P types	Impurity gives excess valence electrons/produces deficiency of valence electrons (holes) in covalent bonds	Frankel/Schottky defects

3.9.1   Metal Resistance Thermometric Sensors

As has already been discussed, different types of scattering would lead to the resistivity relation

$$\rho = \rho_0 \left[1 + \frac{m}{n_e e^2} \left(\frac{1}{\tau_r} \right) \right] \tag{3.33}$$

which for a specific resistance element as a sensor can be transformed into

$$R = R_0 \left[1 + \sum_{i=1}^{n} \alpha_i (\Delta T)^i \right] \tag{3.34}$$

Actually, the coefficients α_i, for $i > 1$, are quite small for any metal useful for practical purposes and in stipulated ranges, so that a relation that is followed is effectively linear. Per unit resistance change from the initial value R_0 is then given by

$$\frac{R - R_0}{R_0} = \alpha_1 \Delta T \tag{3.35}$$

With appropriate circuit, $(R - R_0)$ is measured, R_0 and α_1 are found and ΔT is thus, evaluated.

Commonly used materials for this application are Pt, Ni and Cu. Of these, Pt has a number of added advantages over the others in that it can be drawn into a thin wire yet maintaining its purity to 99.99%. A comparative study of the properties of these metals as thermal sensing elements is made in Table 3.7 as follows.

Table 3.7 Comparative study of metallic thermal sensors

Property	Pt	Ni	Cu
Range (°C)	−250–650	−100–350	−200–250
Resistivity, ρ, ($\mu\Omega$ cm, 20°C, Pure metal)	10.60	6.34	1.67
α_1 ($\Omega/\Omega/°C$)	0.00397	0.0067	0.0043
Length (m) and mass (g) for 100 Ω (at 20°C) wire of 0.005 cm diameter	11.73, 0.205	2.87, 0.050	1.85, 0.078

The values of ρ and α_1 depend on the purity of the materials. As per International Temperature Scale (ITS–90), the required purity, specifically for platinum, is 99.999%. This, of course, is for the standard platinum resistance thermometer (SPRT) when the ratio of $R_{100}/R_0 \geq$ 1.3925 whereas for Industrial Platinum Resistance Thermometer (IPRT), this ratio normally equals 1.385 which means platinum is not as pure as it is in case of SPRT. Ni and Cu thermometers are used for special cases only. Figure 3.11 illustrates the R–T characteristics of the three metals.

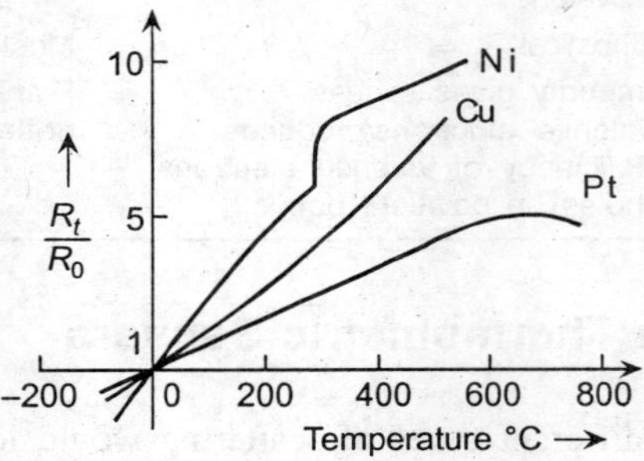

Fig. 3.11 Normalized resistance versus temperature curves of Pt, Cu, and Ni.

The SPRT variety usually is a long-stem variety for relatively higher temperatures while for relatively low temperatures, the capsule type is used. For making the sensor in usable form drawn and annealed platinum wire, diameter varying between 0.001–0.01 cm, is wound bifilarly on an insulating former which also provides freedom to the wire to expand and contract. Mica is usually used in the form of cross strips and borosilicate glass, or silica, or ceramic insulators are used in the form of flats, or arbours, or even crosses. A typical SPRT of the long stem type is shown in Fig. 3.12. For high temperature operations, arrangements are different and are shown in Figs. 3.13(a), (b), and (c). In the bird cage type (Fig. 3.13(a)) a fifth lead in the terminals is provided for insulation testing.

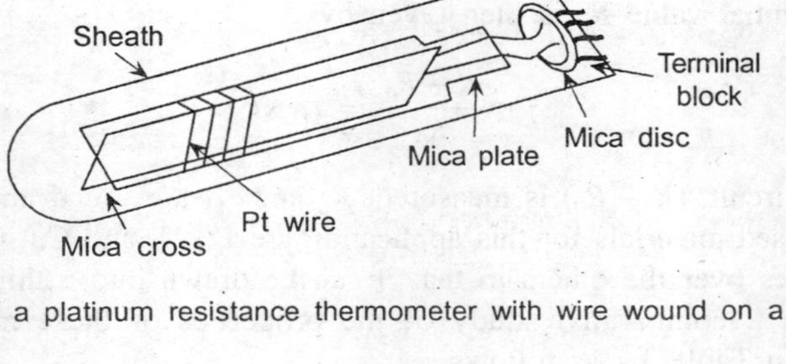

Fig. 3.12 Sketch of a platinum resistance thermometer with wire wound on a mica-cross.

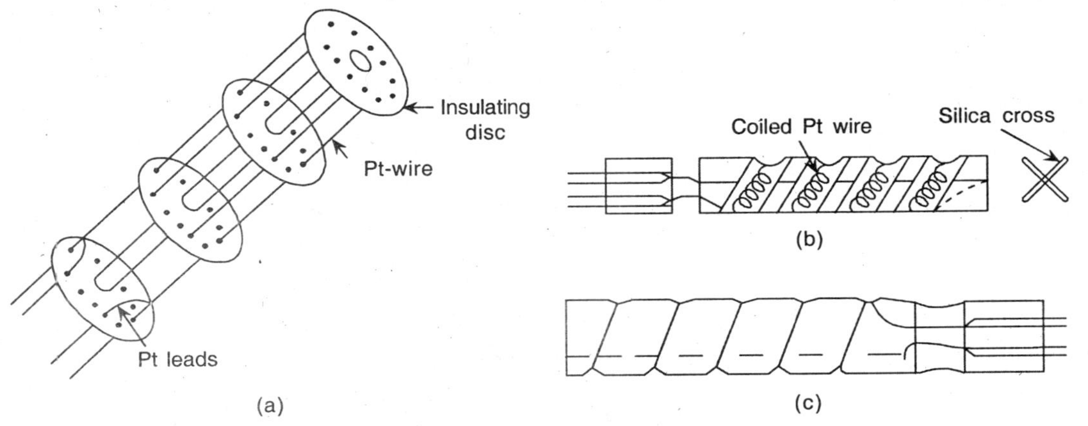

Fig. 3.13 Other designs of resistance thermometers: (a) birdcage type, (b) notched silica cross-coiled type, and (c) single-wire silica cross-coil type.

The thermometer is enclosed in a sheath of fused silica (upto 500°C) and sand blasted glass or alumina is used upto 1000°C. The sheath diameter is about 0.7 cm and a length of upto 400–800 mm is recommended. Capsule types are small-sized ones with platinum sheath and notched mica crosses are often used as insulators over which the wire is wound. Sheath diameter is about 5 mm and overall length is around 60 mm. In both the cases, four lead wires are used. These leads are made of gold or platinum. Gold has better thermal conductivity than platinum, is easier to work with, and does not contaminate platinum. The leads are passed through mica capillaries and alumina discs at the head. All the leads should be identical in length, diameter, and spaced identically.

The sheath is filled with dry air at a pressure of 30 kPa, 1/3 to 1/2 atmosphere or 1 atmosphere for low temperature. This filling with dry gas is necessary to avoid insulation leakage through the release of moisture from mica, for example, which contains about 5% w/w water, specifically at low temperatures. For very low temperatures, filling gas is often replaced by helium. Air is also used to make the atmosphere oxidizing to a certain extent. Platinum does not get oxidized easily even in the presence of oxygen, but in reducing atmosphere silica in the insulation formers is reduced to silicon which then reacts with platinum to form an alloy making it brittle and changing its characteristics.

Annealing is necessary as resistivity depends on internal strains that may be produced during the hard-drawn process. Annealing is done at a temperature higher than the highest measuring value and it is done over a prolonged period. The strains that are produced as a consequence of higher order point defects in the lattice, quenching of excited vacancies and occasional oxidation, are eliminated by proper quenching during the annealing process.

For industrial and commercial applications in varied conditions, IPRTs should be designed to withstand vibrations and shocks and should respond faster and must be designed (in size and shape) to suit specific requirements. The wider field of application requires it to be provided with a calibration graph/table to specified tolerance limits. Accordingly, 'International Electrotechnical Commission' has declared two accuracy classes for IPRTs designated as class A and class B for two ranges –200–650°C and –200–850°C. For class A, the limit is less than that for class B. Standard resistance value considered is again 100 Ω but at 0°C. IPRT, however, can have other values at the manufacturers' and users' prerogative. Figure 3.14 shows the chart of the tolerance limits as prescribed. This is based on the relation

$$R(t) = R_0[1 + \alpha_1 t + \alpha_2 t^2 + \alpha_3 t^3(t - 100°C)] \tag{3.36}$$

where $R_0 = 100 \ \Omega$.

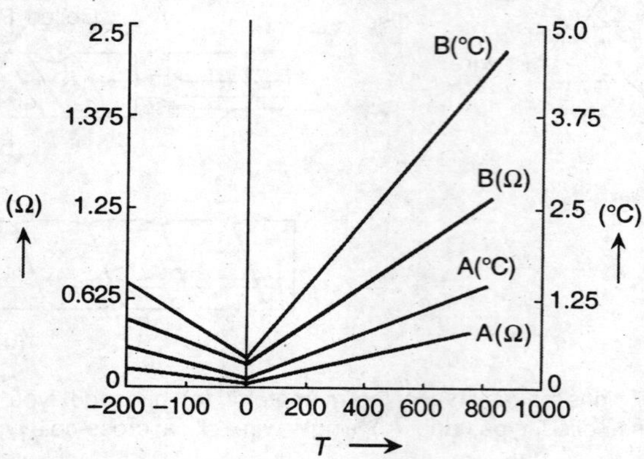

Fig. 3.14 Tolerance chart of the standard platinum resistance values as per IPRT.

α_1, α_2, and α_3 have specified values and are reviewed at intervals of time with advancement in technology. It is kept in mind that the differences should not exceed certain preset values.

IPRTs are made in two different types using (i) wires of diameter between 0.001–0.005 cm and (ii) thin or thick films deposited on ceramic substrates. Because the wire now has a finer diameter, the winding needs better support, at least, partially. One way of doing this is to embed the coil in alumina powder or provide the coil with better supporting cage and guard as shown in Fig. 3.15(a). Also, the coiled elements may be glued with cement/enamel for insulating support or embedded in glass as shown in Fig. 3.15(b) and (c) respectively.

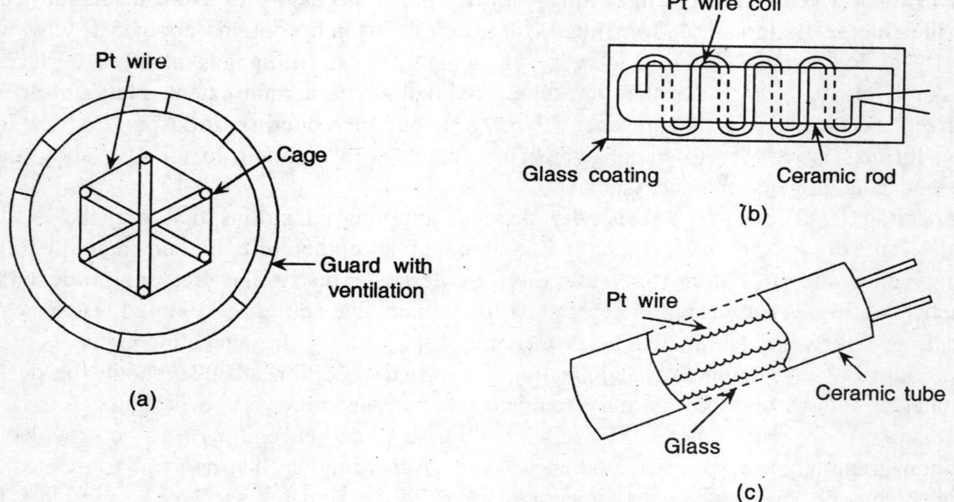

Fig. 3.15 Resistance thermometers for finer wires: (a) design with support cage and guard, (b) wound on ceramic insulating rod and glass coated, and (c) enclosed in ceramic/glass tube.

The thick film is more rugged, low in cost, and is often produced by screen printing process. Platinum ink is printed on an alumina substrate with a thickness of 5–10 μm and then fired. The finishing process includes trimming to obtain the required resistance which is coated with a glaze and fired again. Thin films of thickness 1 μm are produced usually by sputtering.

Nickel, having a very high temperature coefficient of resistance, appears to be more suitable for resistance thermometer although its resistivity is lower than platinum. Besides, it has a higher nonlinearity and it cannot be drawn to as fine a diameter as platinum. Thin film nickel resistance thermometers are, however, a better choice though nickel is easily contaminated and oxidizable. An alloy of nickel (70) and iron (30) has a temperature coefficient of resistance of 0.005/°C but has a better resistance to oxidation and has a resistivity of about 18–19 μΩcm. It has a poorer stability and higher nonlinearity. Ni–Cr alloy has also been used in thick film form for special purposes.

Copper is observed to possess the best linearity in *R–T* response (see Fig. 3.11) over a limited range, 0–100°C, but its resistivity is lowest with almost the same value of α_1 as that for platinum. Copper oxidizes at temperature higher than 90–100°C and the deterioration is faster above 200°C. Response is quick and working is easier with little strain produced in it. Self-heating is also less because of less value of ρ.

With the advent of thin film technology, iridium resistance thermometers have been produced by this process. It is preferred, inspite of its higher cost in the usual coil form, because of its compatible thermal expansion coefficient with alumina substrate. Iridium has an α_1 value comparable to that of copper (see Table 3.7). For surface temperature sensing, such transducers are very useful when flat alumina substrate is used.

Alloyed resistance thermometers such as Rh–Fe and Pt–Co have been considered earlier. In case of the former, the material for the thermometer is obtained by first depositing iron in fine Rh powder and then the powder is dried, sintered, and hot-drawn and finally it is annealed for strain in an atmosphere of hydrogen at about 1100°C. The sensor is designed in the double-capillary form (Fig. 3.16(a)). Figure 3.16(b) shows its sensitivity curve over the stipulated range of 0.5–30 K. The guiding equation is given by

$$T = \sum_{j=1}^{n} \alpha_j (\lambda_1 R + \lambda_2)^j \tag{3.37}$$

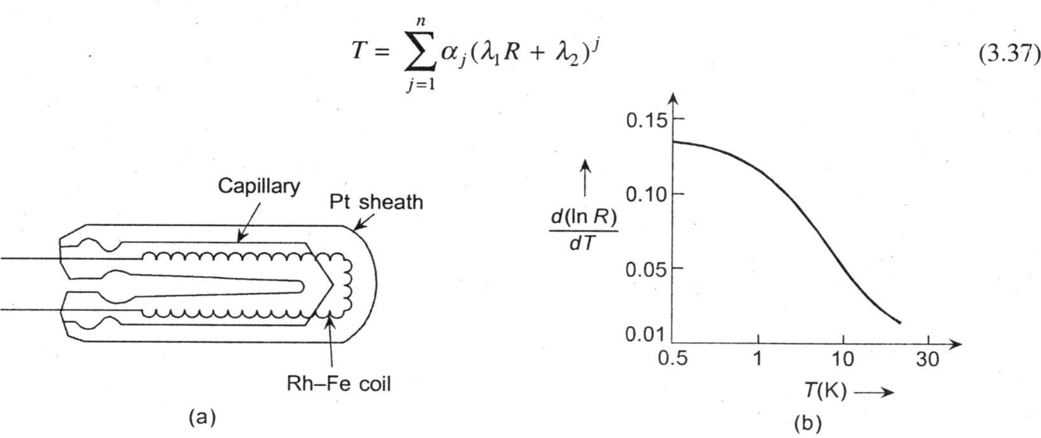

Fig. 3.16   (a) Double capillary housing of Rh–Fe coil of thermometer, (b) sensitivity curve of the same thermometer.

where

> n is chosen for curve-fitting with a specified deviation and
> λ_1 and λ_2 are chosen to keep $(\lambda_1 R + \lambda_2)$ within $[-1, 1]$.

As already mentioned, Pt–Co alloy has similar characteristics as that of Rh–Fe alloy.

3.9.2 Thermistors

Of commercial importance in the category of semiconductor resistance thermometric sensors are the thermistors made from oxides of metals of the transition group. As has been shown earlier, majority members of this group have negative temperature coefficient with the guiding 'empirical' relation given by

$$R_t = R_0 \exp \beta \left(\frac{1}{T} - \frac{1}{T_0} \right) \tag{3.38}$$

Different forms and sizes are given to NTC thermistors and accordingly are named as bead type, rod type, disc type and so on, as shown in Figs. 3.17(a), (b), and (c). Bead type provides the best stability and interchangeability within an operating range of −100–300°C. Two platinum wires are stretched apart to a reasonable distance. Small blobs of the mixed oxides in a suitable binder are applied at appropriate lengths of the pair of wires, so stretched as to form the beads and sintered at about 1300°C. Thus, a series of bead type thermistors are formed, the wires forming the leads. After cutting each part, the individual thermistor is coated with glass for protection. Disc type is made by first forming the disc by pressing the mixture and then heating it to about 1100°C. The two faces of the disc are then deposited with silver by spraying or screen-printing. Finally, wires are soldered for terminals. Sometimes an epoxy coating is provided. In a similar manner, rod types also are produced. The latter two types are less stable, the last being the least stable. The thermistor characteristics such as (i) resistance value, (ii) temperature coefficient of resistance, (iii) response time and so on depend on the ingredients, their mixing, the sintering time, temperature, and some such other factors. Interchangeability is a problem with thermistors. However, Mn–Ni oxides when used with a small amount of binder provide better interchangeability when produced by the same manufacturer.

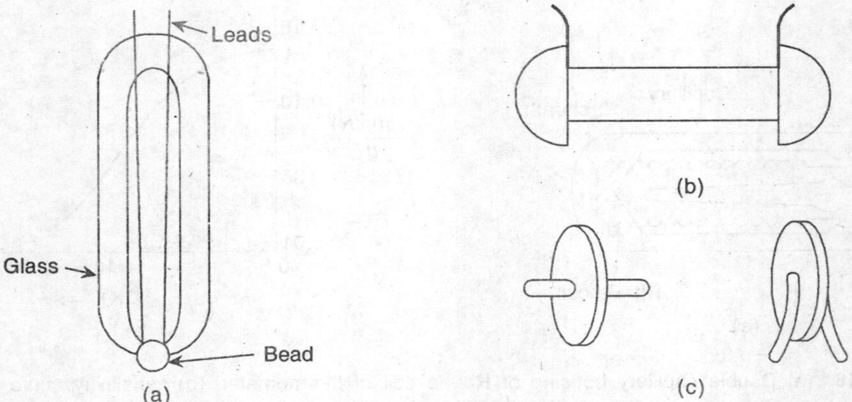

Fig. 3.17 Thermistors: (a) bead type, (b) rod type, and (c) disc type.

Resistivity is usually kept between $100-10^6$ Ωcm and resistance values between $5-50$ Ω. Within working range, a change of R by 30% is preferred. Its coefficient of change in resistance is given by

$$\alpha_{\text{th}} = \frac{1}{R}\frac{dR}{dT} = \frac{-\beta}{T^2} \qquad (3.39)$$

and is, obviously, temperature-dependent, decreasing fast with increasing temperature.

A thermistor has a response time that is dependent, among other things, also on its size. A bead of about 0.2 cm diameter can have a response time as large as 15 s in still condition while in flowing condition this may go down to 30 times depending on the flowing speed of the medium. The time constant of a thermistor in steady condition is calculated as

$$\tau = \frac{mC}{hA} \qquad (3.40)$$

where

 m is the mass of the bead,
 C is the specific heat of the material,
 h is the heat transfer coefficient, and
 A is the area of heat transfer.

Bead diameter is usually kept below 0.2 cm.

Thermistor is characterized by dissipation constant as well which effectively determines its self-heating character. Dissipation constant has a larger value in a flowing medium showing less error in such a condition compared to the steady state medium for the same current flowing through the thermistor. The dissipation factor D is actually given by

$$D = \frac{P}{\Delta T} \qquad (3.41)$$

where

 P is the power dissipated and
 ΔT is the temperature rise.

Investigation on stability has shown that thermistors age but come to stable condition after three to five months. The drift rate for bead type thermistors is, however, not greater than 5×10^{-6} K/day. Often, pre-aging is done by cyclic heating and cooling.

Oxides of rare earth elements are also used to make thermistors in higher ranges but with smaller sensitivity as they are more refractory in nature and possess higher activation energy. They can be used at temperatures upto $750-800°C$. Zr-based thermistors are used even upto $1000°C$. For using thermistors at low temperature (cryogenic level), low activation energy is required. Fe-oxides may be used for the purpose. Below 20 K, the sensitivity rises very sharply— at 20 K it is 15% rising to 300% at 4.5 K. The guiding equation in that stage is better approximated by

$$T = \frac{B}{\log R + \dfrac{C}{\log R} - A} \qquad (3.42)$$

where A, B, and C are constants evaluated by calibration.

Positive temperature coefficient thermistors show large and sudden resistance changes at a temperature called *switching* or *transition temperature* T_s as shown in Fig. 3.18. Such thermistors are made from titanates of barium and/or strontium in pervoskite oxides. The switching temperature is dependent on the Ba/Sr ratio. This is explained by the ferro-electric effect of the material. By proper proportion, T_s may be varied from 15–115°C and such transducers are used as heat switches.

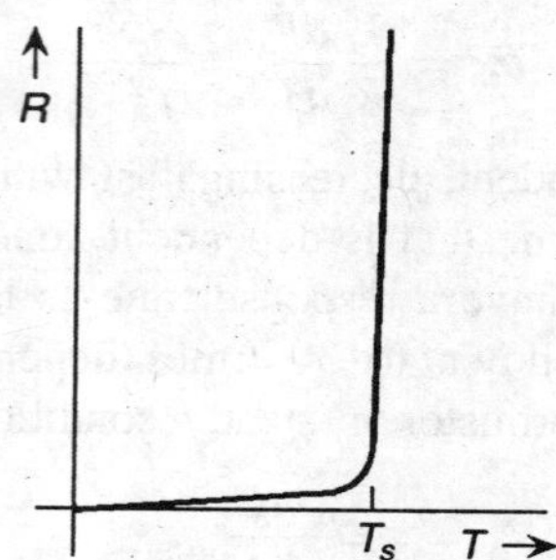

Fig. 3.18 Characteristics of PTC thermistor.

Germanium resistance thermometers have been developed for a range of 1–20 K by doping single crystals with As, Ga, and Sb but these thermometers are to be calibrated at many points and they are sensitive to magnetic fields. However, the calibration curve is reproducible.

Silicon, with boron in various proportions as an impurity, can be used as a sensor both as NTC and PTC with T–R relation showing hard nonlinearity. At lower temperatures, say at –150°C, a typical such thermometer has a –60% resistance change while at +150°C, it may be as high as +150% and at room temperature it may be less than +1% per°C only.

As has been mentioned already, carbon can be temperature sensitive resistor particularly at low temperatures. Ordinary radio carbon resistors can be used for the purpose. They are, however, slightly dependent on pressure as well. Carbon temperature sensors have been developed where a colloidal suspension of carbon in an appropriate organic sample is made or a film of suspension is painted on the sample and cured. It works in a range of 3–60 K with an accuracy of $\pm 3 \times 10^{-2}$ K and follows an interpolating equation

$$\ln R = A_1 T^{-r} + A_2 \tag{3.43}$$

where r, A_1, and A_2 are constants.

For a large range of 1–300 K, carbon-impregnated glass thermometers have been developed which, however, show better performance with respect to stability and sensitivity at lower temperatures.

Resistance thermometers need to have a current passing through them which is likely to cause an error often termed as the *self-heating error*. The heat, produced in the sensor because of this current flows

(i)   towards the zone whose temperature is to be measured through the surrounding walls and sheaths and

(ii) along the leads to a certain extent.

If heat conductances towards the measuring zone and along the leads are χ_m and χ_l respectively and the temperature to be measured and actually measured are t_l and t_M respectively, while a current I flows into the sensing element of resistance R, then

$$I^2 R(t) = (\chi_m + \chi_l)(t_M - t_l) \tag{3.44}$$

The self-heating error, t_h, is however, $(t_M - t_l)$ so that

$$t_h = \frac{I^2 R}{\chi_m + \chi_l} \tag{3.45}$$

If R_1 and R_2 are the resistances when the currents are I_1 and I_2 respectively, and the corresponding measured temperatures are t_1 and t_2, then

$$t_h = t_M - t_l = \frac{(R_2 - R_1)I^2}{S_{th}R(I_2^2 - I_1^2)} \tag{3.46}$$

where S_{th} is the thermal sensitivity of the resistance thermometer.
Or, in terms of temperatures measured, the corrected temperature t_l is given by

$$t_l = t_1 - \frac{(t_1 - t_2)I_1^2}{I_1^2 - I_2^2} \tag{3.47}$$

For NTC, the same condition holds but with opposite signs. Equations (3.46) and (3.47) hold for small currents only with dissipation less than 1 mW or so. There are circuital techniques for minimizing this error.

3.10 THERMOEMF SENSORS

Thermoemf temperature sensors are thermocouples which are most extensively used in industry, over a wide range of temperatures. The range, however, is made wide using different materials. The measurement does not involve separate supply. A resolution of 0.1–0.2°C at ambient condition is obtained which increases at high values to about ±5°C.

It was discovered by J. Seebeck that when two conductors C_1 and C_2 of different compositions are made up into a closed electrical circuit as shown in Fig. 3.19, a small current flows through it if one of the junctions J_1 has a different temperature than the other junction J_2. This current is driven as an emf is generated between these two junctions because of temperature difference. This emf is called the thermoelectric potential or the Seebeck emf which is dependent on the compositions of C_1 and C_2 and the difference of temperatures ΔT with the polarity depending on the sign of ΔT. For measurement of temperature, one junction temperature is held constant.

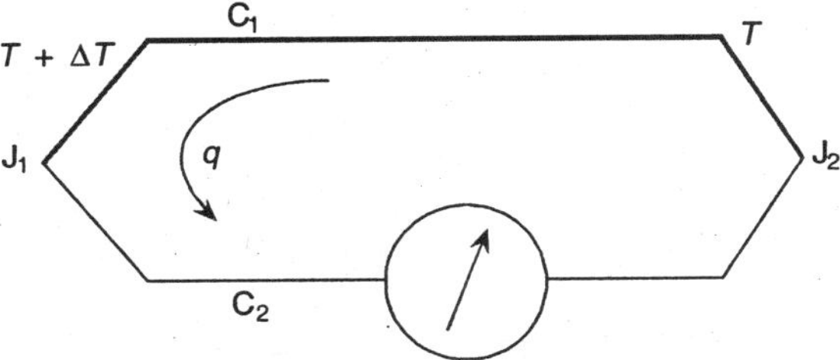

Fig. 3.19 Basic thermocouple.

The Seebeck emf has been found to be the algebraic sum of two potentials named after their discoverers—Peltier and Thomson. The 'Peltier effect' states that *one of the junctions is heated and other cooled if a current is allowed to flow in the circuit, the amount of the temperature rise in one junction and the amount of temperature fall in the other as also which will be heated and which cooled, will depend on the current intensity and direction, besides the compostions of the conductors.* The electrons travelling across the junctions actually do some work or some energy forces them to travel across the junctions, that is, the thermal energy of the electrons is either higher or lower which causes the junctions to get heated or cooled. The heat flow H_f (power) across the circuit is proportional to current I in the circuit so that

$$H_f = \pi I \tag{3.48}$$

where π is a constant called Peltier coefficient and is measured in volts.

Thomson, on the other hand, found that *with a current flowing in a single conductor C_i, its heat content changes and a temperature gradient exists along the length. Accordingly, the heat flow is proportional to current I as well as the temperature gradient ΔT* (see Fig. 3.20). Hence,

$$H_f = \sigma I \Delta T \tag{3.49}$$

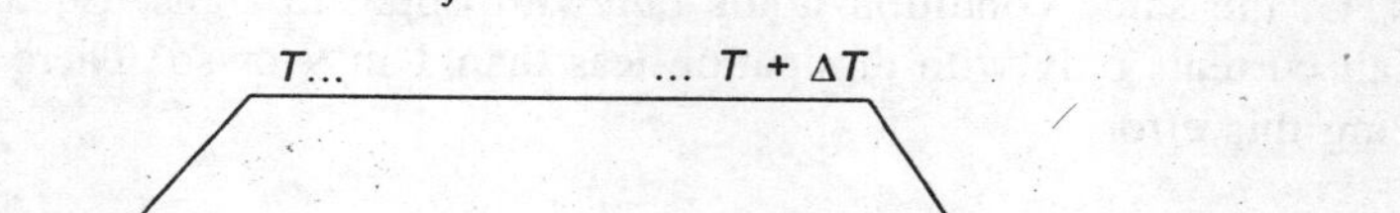
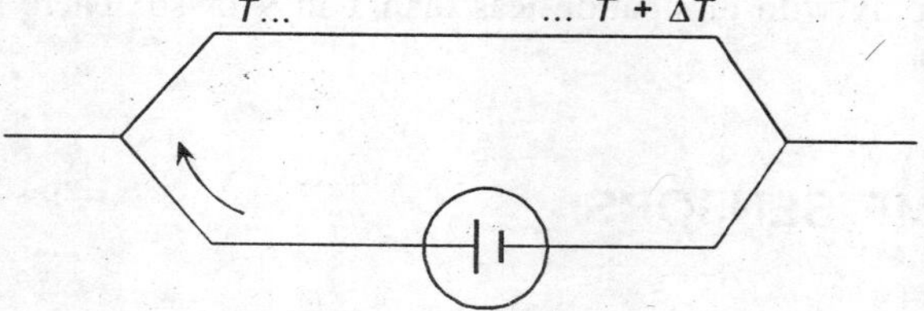

Fig. 3.20 Thermocouple with a source.

Here, σ is a constant called the Thomson coefficient and has the unit V/K.

In Eq. (3.48), π is a potential whereas in Eq. (3.49), σ is the potential per degree temperature. These two equations can be combined to obtain the thermocouple emf. Thus,

$$E = \pi(T + \Delta T) - \pi(T) - (\sigma_{c1}\Delta T - \sigma_{c2}\Delta T) \tag{3.50a}$$

$$= \pi_1 - \pi_2 - \int_1^2 \sigma_{c1}\,dt + \int_1^2 \sigma_{c2}\,dt \tag{3.50b}$$

which can be modified by making ΔT very small to obtain

$$P = \frac{dE}{dT} = \frac{d\pi}{dT} - (\sigma_{c1} - \sigma_{c2}) \tag{3.51}$$

The quantity dE/dt is called the thermoelectric power P, for the two conductors and is defined as the thermal rate of change of emf acting around a couple with change of temperature in one junction.

Now, if a charge q passes around the couple in an anticlockwise direction consisting of metals C_1 and C_2, its junctions J_1 and J_2 at temperatures $T + \Delta T$ and T, then heat (energy) absorbed at $T + \Delta T$ is $q\pi_1$ and heat released at T is $q\pi_2$. Heat released out in metal C_1 at temperature $T + (\Delta T/2)$ is $q\sigma_{c1}\,\Delta T$ and heat absorbed in metal C_2 at temperature $T + (\Delta T/2)$ is

$q\sigma_{c2}\Delta T$. Assuming all these processes are reversible, as is usually the case, the sum total Σ (Heat/Temperature) $= 0$. Hence,

$$\frac{q\pi_1}{T + \Delta T} - \frac{q\pi_2}{T} - \frac{q\sigma_{c1}\Delta T}{T + (\Delta T/2)} + \frac{q\sigma_{c2}\Delta T}{T + (\Delta T/2)} = 0 \tag{3.52}$$

With reference to Fig. 3.21, it is now easily shown that

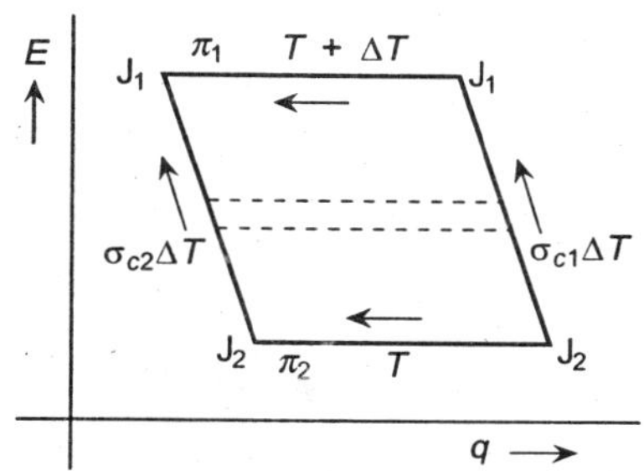

Fig. 3.21   Charge–voltage cycle of a thermocouple.

$$\frac{\pi_1}{T + \Delta T} - \frac{\pi_2}{T} = \int_{T}^{T+\Delta T} \frac{d}{dT}\left(\frac{\pi}{T}\right) dT$$

and

$$\frac{\sigma_{c1}\Delta T}{T + (\Delta T/2)} - \frac{\sigma_{c2}\Delta T}{T + (\Delta T/2)} = \int_{T}^{T+\Delta T} \frac{(\sigma_{c1} - \sigma_{c2})\,dT}{T}$$

so that

$$\int_{J_2}^{J_1} \frac{d}{dT}\left(\frac{\pi}{T}\right) dT - \int_{J_2}^{J_1} \frac{(\sigma_{c1} - \sigma_{c2})\,dT}{T} = 0$$

On differentiating, we get

$$\frac{d}{dT}\left(\frac{\pi}{T}\right) = \frac{\sigma_{c1} - \sigma_{c2}}{T}$$

or,

$$\sigma_{c1} - \sigma_{c2} = T\frac{d}{dT}\left(\frac{\pi}{T}\right) \tag{3.53}$$

Using Eq. (3.51),

$$\frac{dE}{dT} = \frac{d\pi}{dT} - T\frac{d}{dT}\left(\frac{\pi}{T}\right) = \frac{\pi}{T}$$

so that

$$\pi = T\frac{dE}{dT} \tag{3.54}$$

showing that the Peltier coefficient for the junction of a pair of conductors is the product of the absolute temperature of the junction, T, and the thermal rate of change of emf for the whole circuit with that junction temperature change.

The emf values and range of thermocouple can be ascertained from the thermoelectric diagram for different set of conductors. Such a diagram was proposed by Professor Tait in 1871. It is the plot of P with respect to T. For the two conductors forming a couple, the two straight lines for conductors C_1 and C_2 are shown in Fig. 3.22. If we know the equations of the lines, emf is easily obtained for the couple. If they are straight lines, then the equations, from the figure, are

$$P_{c1} = m_1 T + K_1 \tag{3.55a}$$

and

$$P_{c2} = m_2 T + K_2 \tag{3.55b}$$

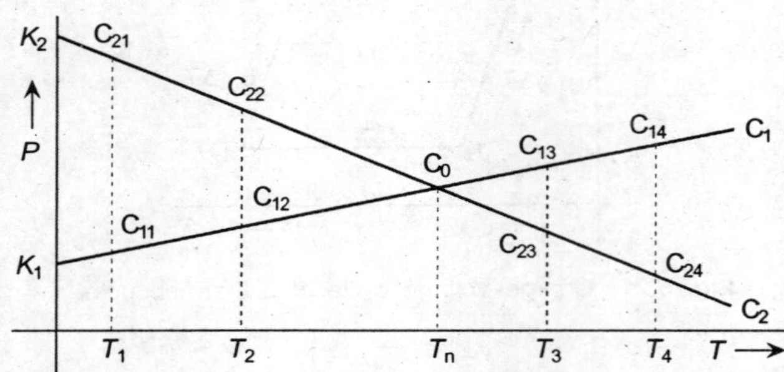

Fig. 3.22   *P–T* plots for the two components of a couple.

so that the emf, for the two conductors with junction temperature T_1 for the cooler and T_2 for the hotter one, is given by

$$[E]_1^2 = \int_1^2 [P_{c1} - P_{c2}] dT = \int_1^2 [(m_1 - m_2)T + (K_1 - K_2)] dT$$

which on integration, yields

$$E = \frac{1}{2}(m_1 - m_2)(T_2^2 - T_1^2) + (K_1 - K_2)(T_2 - T_1) \tag{3.56}$$

Keeping T_1 fixed and T_2 varying, E–T_2 curve is obtained to be a parabola. Equation (3.56) is then transformed to

$$E = (T_2 - T_1)\left[\frac{(T_2 + T_1)}{2}(m_1 - m_2) + (K_1 - K_2)\right] \tag{3.57}$$

It is to be noted that at $T_1 = T_2$, $E = 0$.

Also at,

$$\frac{1}{2}(T_2 + T_1) = \frac{K_2 - K_1}{m_1 - m_2} \tag{3.58}$$

$E = 0$, that is, E is zero when the average temperature of the junction is $(K_2 - K_1)/(m_1 - m_2)$. This temperature is called the *neutral temperature* and occurs when $P_{c1} = P_{c2}$ which occurs at the intersection of the two straight lines. If this temperature is represented by T_n, Eq. (3.56) may be rewritten as

$$E = (m_1 - m_2)(T_2 - T_1)\left(\frac{T_1 + T_2}{2} - T_n\right) \tag{3.59}$$

The plot of E versus T drawn in Fig. 3.23 shows that emf E_n at T_n is maximum after which there is a decrease of E again with difference of temperature increasing. From Fig. 3.22, the emf E for the couple for junction temperatures T_1 and T_2 would be the area $C_{21}C_{22}C_{12}C_{11}$. If one junction temperature is T_1 and the other T_3 beyond T_n, then the effective emf would be

$$E_{\text{eff}} = \text{Area } (C_{21}C_0C_{11}) - \text{Area } (C_{13}C_0C_{23}) \tag{3.60}$$

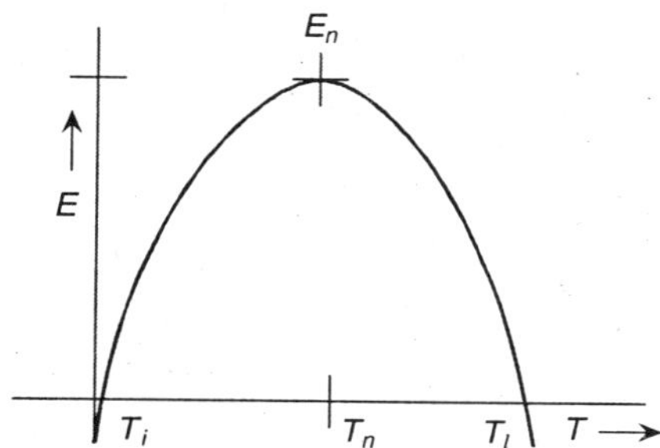

Fig. 3.23 Generalized *E–T* diagram of a thermocouple.

Raising the temperature beyond T_3, say to T_4, it is possible that the emf becomes reversed and hence, T_n is sometimes known as the *temperature of inversion*. The couple is normally to be operated within a hot junction temperature of T_n.

Most metals have emf-T curves as approximate parabolas so that the thermoelectric lines are usually straight. Some exceptions are the cases of nickel and iron which have several points of inflexions. Figure 3.24 shows thermoelectric lines of some common elemental materials in which inflexions of the lines of nickel and iron have been clearly shown. Seebeck himself prepared a table of 25 elemental materials in the order that when any two form a circuit, current flows across the hot junction from the element occurring earlier to that occuring later in the table. The table is reproduced here as Table 3.8.

Table 3.8 Thermoelectric materials

S.No.	Element	S.No.	Element
1	Bi	14	Mo
2	Ni	15	Rh
3	Co	16	Ir
4	Pd	17	Au
5	Pt	18	Ag
6	U	19	Zn
7	Cu	20	W
8	Mn	21	Cd
9	Ti	22	Fe
10	Hg	23	As
11	Pb	24	Sb
12	Sn	25	Te
13	Cr		

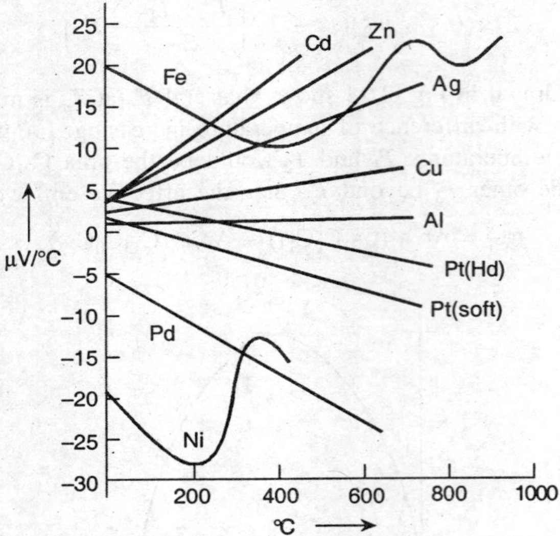

Fig. 3.24 Thermoelectric lines of different elements.

Presently, commercial thermocouple elements are also chosen from alloys for better performance.

While making a measurement with thermocouple sensors, it is necessary to introduce measuring instruments which, therefore, are likely to affect the thermoemf property of the couple. Some new junctions, in effect, are formed because of these insertions. The ideal conditions leading to the thermoemf generation for a couple are rarely met in practice and more often than not empirical situations arise and in-situ calibration of the measuring system with the thermoemf transducer becomes necessary. Some laws of the behaviour of the thermocouple have accordingly been derived. These are:

1. *Law of intermediate temperature:* The emf for a couple, each element of which is homogeneous in constitution, with junctions at temperatures T_1 and T_2 is not affected by temperatures elsewhere in the circuit.
2. *Law of intermediate metals:* If a third homogeneous metal is inserted anywhere in the couple without affecting the junctions J_1 and J_2 and their temperatures T_1 and T_2, and the new junctions of the inserted metal having identical temperature, the thermoemf of the couple remains unaffected.
3. *Law of homogeneous circuit:* If the circuit is made of a single homogeneous metal, no current flows through the application of heat alone and no thermoemf develops.

3.10.1   Materials for Thermoemf Sensors

Material choice is guided by quite a few important factors:

(a)   high thermoemf per unit temperature change, that is, high thermoelectric power,
(b)   low electrical resistance of the couple,
(c)   linearity of *E–T* curve over the range of interest,

(d) high melting point of the couple materials for wider range,

(e) material should be available as pure and homogeneous, workable in desired shapes and should not be easily contaminable,

(f) should be usable over long period of time without getting brittle, or acquiring scales, or change of composition (for alloy type materials),

(g) should be properly annealed to make it free from strains/stresses produced during cold drawing process.

Elemental materials listed by Seebeck are not all suitable for commercial pairing to form thermocouples. Three categories of thermocouple do exist in practice, namely

(i) the base metal type consisting of couple members made of elemental base metals or alloys thereof,

(ii) the noble/precious metal type made from noble metals or alloys thereof, and

(iii) nonmetallic types.

Thermocouples are usually identified by capital letters of the English alphabet. The base metal types are identified by letters E, J, K, N, and T; and the noble metals thermocouples are identified by G, C, D, B, R, and S. Nonmetallic thermocouples are special kind and will be considered separately.

Several countries have included this standardized nomenclature of type letters in specification schedule providing temperature range, tolerance, service, (intermittent or continuous), and quality (standard or special). International Electrotechnical Commission (IEC) publication 584 with various parts (1, 2, 3) is such a standardizing document. Table 3.9 shows a specification sheet of the various types of couples.

Table 3.9 Thermocouple specifications

Type	Materials (Composition in brackets, positive first)	Compensating cable colour	Range (°C) (intermittent in parantheses)	Tolerance	dE/dT μV/°C (range)	Remarks
B	Pt(70) Rh(30) Pt(94) Rh(6)	Grey Red	600–1500 (1750)	±0.0025 $|t|$	5–12	Most stable better life expectance than R, S types at higher T
C	W(95) Re(5) W(74) Re(26)	White, red trace Red	0–2300 (2600)	±1%	5–10	Used for short duration in neutral or reduced atmosphere
D	W(75) Re(25) W(97) Re(3)	White, yellow trace Red	0–2300 (2600)	±1%	5–10	Used for short duration in neutral or reduced atmosphere
E	Chromel (Tophel) Ni(90) Cr(10) Constantan (Cupron) Cu(57) Ni(43) Mn, Fe, C(traces)	Purple Red	−40–800 (1000)	±1.5°C/ ±0.004 $|t|$	15–60 atmosphere	Works in oxidizing
G	W(100) W(74) Re(26)	White, blue traces Red	0–2300 (2600)	±1%	5–10	Used for short duration

(Cont.)

Table 3.9 *Cont.*

Type	Materials (Composition in brackets, positive first)	Compensating cable colour	Range (°C) (intermittent in parantheses)	Tolerance	dE/dT μV/°C (range)	Remarks
J	Fe(100) Constantan	White Red	−200–1000 (1100)	±1.5°C/ ±0.004 $\lvert t \rvert$	45–57	Better in reducing atmosphere, within 600°C in any atmosphere
K	Ni(98) Cr(2)/ Ni (100) Constantan	Yellow Red	−40–1000	±1.5°C/ ±0.004 $\lvert t \rvert$	40–55	Better in oxidizing atmosphere
N	Chromel Ni(90) Cr(10) Alumel Ni(94) Mn(3) Al(2) Si(1)	Yellow Red	−200–1200 (1300)	±1.5°C/ ±0.004 $\lvert t \rvert$	40–55	Better in oxidizing atmosphere
R	Pt(87) Rh(13) Pt(100)	Black Red	0–1400 (1600)	±1°C/ ±0.0025 $\lvert t \rvert$	5–12	Most stable in all atmospheres
S	Pt(90) Rh(10) Pt(100)	Black Red	0–1400 (1600)	±1°C ±0.0025 $\lvert t \rvert$	5–12	Most stable in all atmospheres and a little better linearity
T	Cu(100) Constantan	Blue Red	−200–350 (500)	±0.5°C/ ±0.004 $\lvert t \rvert$	15–60	Oxidation occurs beyond stipulated range

In the Table 3.9 R, S, and B types are shown to have almost similar entries in the appropriate columns. However, the difference lies in their linearity to a very small extent and slight variation in thermoemfs stated as range in the table. Figure 3.25 shows the relative differences in *E–T* plots. The *E–T* characteristics of the types, in general, are shown in Fig. 3.26.

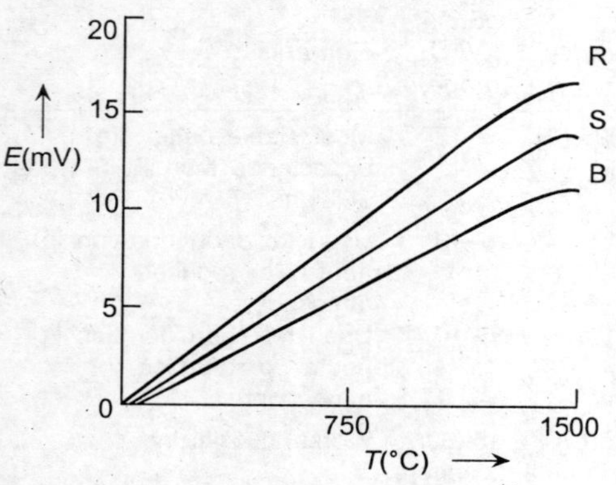

Fig. 3.25 emf–temperature characteristics of R, S, B type thermocouples.

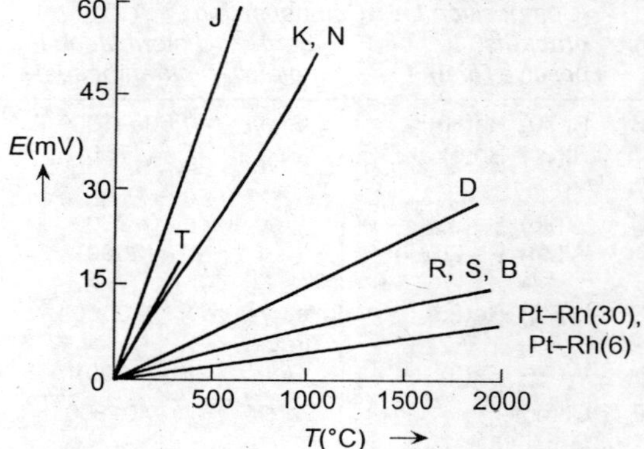

Fig. 3.26 emf–temperature curves for different types of thermocouples.

Besides the couples listed in Table 3.9, there are other thermocouples employed for short term services at high temperatures. W–Re (97–3), W–Re (75–25) is one such couple already listed in the table. It is important that materials with high melting point be employed for the purpose with the other basic requirement satisfied. Ir and Ir–Rh (60–40) is another type which can be intermittently used up to 2000°C. Long term uses cause both Ir and Rh to get oxidized in free atmosphere. Also, being brittle in nature, it breaks with prolonged use as recrystallization occurs during the process. However, it can be used under all atmospheric conditions and has a relatively better linear *E–T* characteristics upto about 2000°C.

W–Re, and W–Re thermocouples, listed already, are also used at high temperatures, mainly intermittently, but only in neutral and/or reducing atmosphere. At high temperatures, they also tend to recrystallize and turn brittle. Preparation of the junction is important to avoid this. It is welded in a protective atmosphere without being subjected to stresses. Another set of couples for high temperature applications are Mo–Re (95–5), Mo–Re (59–41) and Mo (100), Mo–Re (59–41). They are used in special cases such as temperature measurements in nuclear reactors. The *E–T* curves for such high temperature thermocouples are shown in Fig. 3.27.

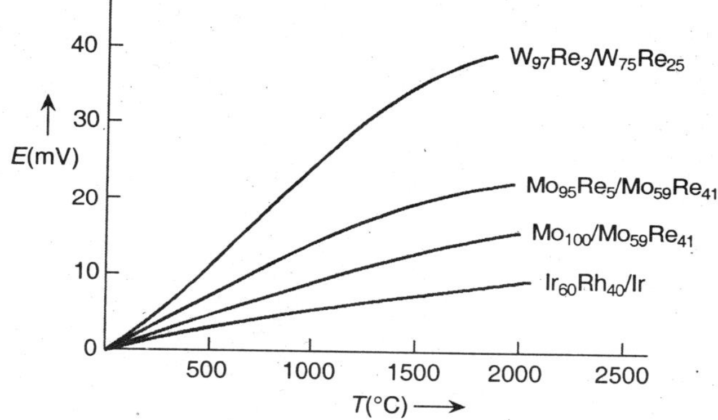

Fig. 3.27 emf–temperature curves for high temperature thermocouples.

Gold and silver, as such, have not been used as thermocouple members, although gold and gold alloys are now being increasingly used for low temperature applications. Au, Pt thermocouples are being used for calibrating other thermocouples upto about 900–1000°C. Also gold–constantan thermocouples are used for the same purpose over a slightly lower range. They are nonmagnetic and used in magnetic fields and pure noble elements (such as, Au–Pt) can be made extremely homogeneous.

Au–Fe, Ni–Cr thermocouple is being used at very low temperatures as is Au–Fe, Au–Ag thermocouple. Here, iron is found only in traces ($\leq 0.03\%$) while Ni–Cr is basically constantan. These thermocouples are ductile and care should be taken to see that no strain/stress is produced in them. Co–Au (2.11), Au–Ag (0.37) thermocouple is also employed at low temperatures. It has low electrical resistance but is easily deformable.

Nonmetallic thermocouples have been proposed to be used in atmospheres containing carbon, since metals form carbide and metals such as W and Mo become more brittle and break under such a condition. For high temperatures, upto about 2200°C, in carbon-containing conditions, B_4C, C thermocouple is used. This thermocouple is to be specially prepared, particularly the junction. It has almost a linear *E–T* response curve with a large thermoelectric power, of about 0.25 mV/°C, but is slightly dithering above 400°C.

3.10.2 *E–T* Relations

As has been shown that thermoemf is a combination of Peltier and Thomson emfs but it is not easy to obtain these values for different thermocouples at all temperatures. It is, therefore, found that thermocouple emf output can be expressed as a series function of temperature t, of the form

$$E = \sum_{j=0}^{n} a_j t^j \tag{3.61}$$

where a_j's are different for different couples. Even for the same couple they differ in different ranges. These values of a_j's are obtainable from standard institutions like NPL (India), NBS (USA) and so on.

As shown in Figs. 3.25, 3.26, and 3.27, the natures of the E–T curves do not conform to any curves of known equations and hence, by curve-fitting techniques, equations of the series form has been proposed for all with the list of values of a_j's where the value of n can be as large as 14. For example, in case of type T thermocouple in the –270–0°C range, for adequate fit, n in Eq. (3.61) is to be extended upto a value of 14.

3.10.3 Thermocouple Construction

Depending on the use under different conditions, construction of a couple complete with protection varies. Usually, the couple is kept separated by small insulator beads (single-hole, twin-hole, 4-hole types are common) to enable flexibility with the junction kept free and on the other side, the free ends of the members are passed through an insulator disc onto terminal lugs which also are shaped as per standardized recommendations. The entire thermocouple with such insulator sleevings is now enclosed in a porous ceramic tube and finally enclosed in a metal sheath for protection against the likes of mechanical shocks. The use of ceramic tube is optional for low temperatures or for non-noble metals. For salt-bath and corrosive atmospheres, the sheath materials must be properly chosen and replaced after stipulated periods.

Another way of construction is by embedding the thermocouple member in an insulating powder such as MgO, Al_2O_3 and so on, in an enclosing sheath made of prescribed materials. The insulation powder is tightly packed to allow no movement of the couple. Different shapes can be given to such thermocouples. A typical case of insulated junction type thermocouple is shown in Fig. 3.28(a). In case of Fig. 3.28(b), the junction is in contact with the sheath and the sheath, in this case, is insulated from the mounting fixtures. While thermocouple of Fig. 3.28(a) has slow response, that of Fig. 3.28(b) has a much faster response. Such thermocouples are sometimes known as mineral insulated (MI) or sheathed thermocouples. Sheaths used for different commonly used thermocouples are listed in Table 3.10.

Table 3.10 Thermocouple sheaths

Thermocouple type	Sheath material
J	Stainless steel (Ni(8), Cr(18), Fe)
K, N	Inconel (Ni (trace), Cr (15), Fe)
R, S	Stainless steel (as already discussed) Inconel (as discussed), Pt–Rh
G	Molybdenum–tungsten steel

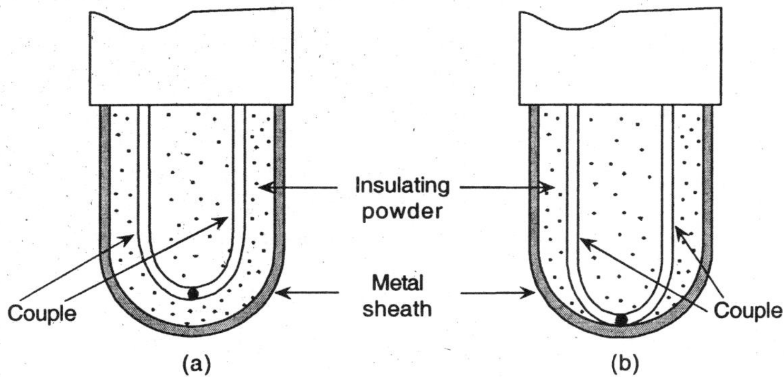

Fig. 3.28 The MI thermocouples: (a) the usual design, and (b) design with the junction in contact with the sheath.

While the basic sensing mechanism remains the same, constructional variation, particularly the end part with protective cap and the like differs for different applications and is not proposed to be taken up here.

3.10.4 Reference Temperature

Reference temperature is often referred to as the 'cold junction' temperature in temperature sensing above 0°C. Tables of thermoemf are prepared with reference junction temperature as 0°C and if it actually deviates from the specified value, appropriate correction is made. Since during measurement with thermocouples, the reference junction is at the instrument whose temperature is often not only different from 0°C, but likely to vary with ambient conditions. This variation can be accounted for by maintaining the reference junction at a fixed temperature. A metal block is heated and maintained at a temperature by thermostatic control and the insulated reference junction is attached to it. The indication is accordingly calibrated.

The loss in thermoemf due to high reference junction temperature is compensated, in some cases, by connecting a temperature sensitive element, a resistor, or a thermistor, or a transistor in a well designed circuit. A recent innovation for correction of the reference temperature in process control units with computers is to use separate temperature sensors such as PTR 100 and the computer evaluates the correct temperature t_c from the measured value t_M and the reference temperature t_R using

$$t_c = t_M + kt_R \tag{3.62}$$

where k is the ratio of the thermoelectric powers at t_M and t_R.

A couple is considered to be connected to the meter directly without separate connecting leads. But often, this is not possible because the meter is not possible to be mounted too close and the thermocouple wires when made very long for direct connection may have certain disadvantages. Thermocouple wires may have high resistances which can be avoided by making use of thick wires leading to increased cost and difficulty in handling. For noble metal varieties, the cost increases. Normally, therefore, separate lead-cables called compensating cables are used. These cables are made of materials which have the same thermal-emf characteristics as that of the associated thermocouples over the ambient variation range of 100 or 200°C and usually consist of base metals for low cost and low resistance. They are employed in association with the noble metal couples of the type B, S, R and for types C, G, and N, K. For types J, T, and E, cables of the same metals as the thermocouples are used. This extension is needed because the connection between one member and the lead must be at the same temperature as the connection between the other member and the lead to avoid measuremental errors.

Thermopiles are an aggregation of the measuring junctions of a number of thermocouples—the junctions facing the temperature to be measured and all the cold junctions maintained at a fixed reference value. Each thermocouple in this conglomeration may be made of finer junctions for better speed of response. The sensitivity is boosted up by the thermocouples being connected in series.

3.10.5 Thermosensors Using Semiconductor Devices

It must be mentioned here that the thermoelectric effects occur in semiconductors as well. In fact, Seebeck effect in silicon and its possible exploitation for thermal sensing has been investigated in some details. In fact, development and performance of integrated silicon thermopiles are now well-recorded in literature.

Mathematically stated, the Seebeck effect is given by

$$\Delta E = \alpha_s \Delta T \tag{3.63}$$

where

α_s = Seebeck coefficient,

ΔE is the open circuit emf, and

ΔT is the difference of temperature between the two junctions.

While α_s has been analytically shown to depend on many factors such as Fermi energy, conduction band edge energy, conduction band density of states, electron or hole density in doped materials, Boltzmann constant and so on, in practice it has been approximated as a function of electrical resistivity, specifically for a prescribed range of temperature. The relation is

$$\alpha_s = \frac{\lambda k}{T} \ln(\rho/\rho_0) \tag{3.64}$$

where λ is a constant of an approximate value 2.6 and ρ_0 is about $5 \times 10^{-6} \Omega$m. α_s has been found to have highest value at room temperature with least concentration difference between the donor and acceptor atoms, that is, with the smallest density of mobile charge carriers.

Integrated devices have been fabricated to function as thermal sensors. In fact, thermopiles consisting of series connected Si and Al couples, with silicon strips fabricated by any of the diffusion, epitaxy or ion-implantation processes, have been produced. It is schematically shown in Fig. 3.29. This is extended to polysilicon (B-doped)–gold thermocouples. While in the former

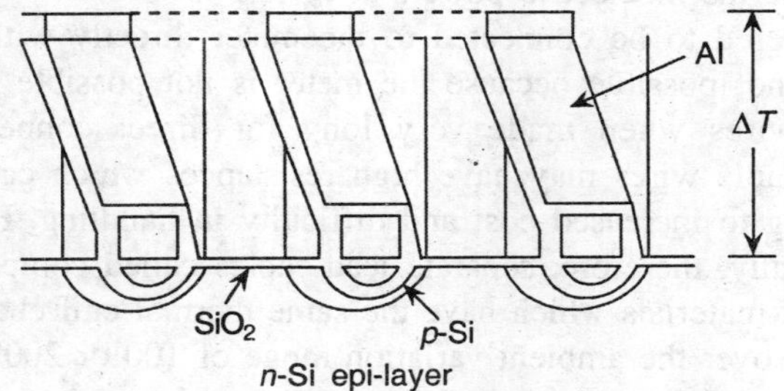

Fig. 3.29 Si–Al thermopile structure.

case, α_s has been found to vary from 250–1200 $\mu V/K$ depending on IC fabrication and materials, in the latter it has been found to be around -240 $\mu V/K$ showing that for semiconductors, a higher figure of merit may be obtained compared to metals. Figures 3.30(a) and (b) show two type of Si–Al thermopiles that can be commercially used.

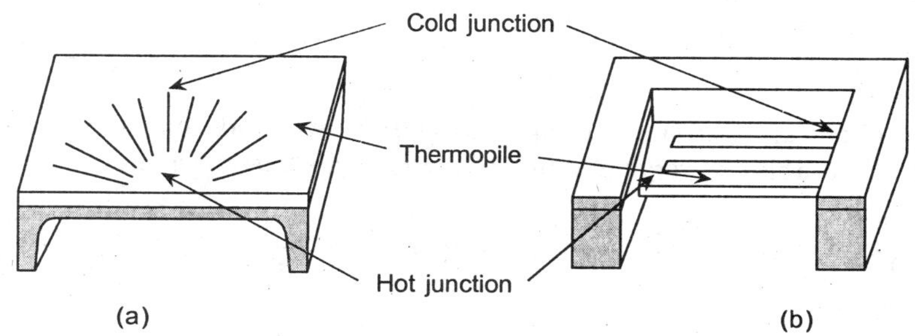

Fig. 3.30 Commercially exploitable versions of Si–Al thermopiles.

3.11 JUNCTION SEMICONDUCTOR TYPES

Junction semiconductor (Si, GaAs, Ge) diodes and transistors have their base–emitter voltage V_{BE} related to temperature T. Although best performance is below 50 K, it can normally be used over a range of 1 K–200°C and in commercial forms, specification covers a range of -50–150°C for good linearity. Over 200°C, the junction is destroyed.

The forward bias characteristics of a junction diode (Si, Ge) are such that below a certain voltage, V_F (the forward conduction voltage), practically no current flows through it. This voltage gives a measure of the minimum energy required by current carriers to cross the p-n junction, that is, junction space charge. For silicon, this has a value of 0.7 V at 20°C. The voltage has a temperature coefficient and for Si-devices, this coefficient has a value -2 mV/°C. Above V_F, current increases exponentially with V/T. In fact, the relation between V_{BE} and T is given by

$$V_{BE} = \frac{kT}{q} \ln\left(\frac{I}{I_s}\right) \tag{3.65}$$

where

q = electron charge,

k = Boltzmann constant, and

I_s is the saturation current which is proportional to the emitter area and depends on doping profile.

The term k/q is 86.17 $\mu V/K$. Unfortunately, I_s is also dependent on temperature, tending to make V_{BE}–T relationship a nonlinear one. The relation that is practically followed is, in fact, given by

$$V_{BE} = V_{g0} + \frac{kT}{q} \ln\left(\frac{I}{c_1 T^{m_1}}\right) \tag{3.66}$$

where

V_{g0} is the extrapolated band gap voltage at 0 K,

c_1 is a constant, and

m_1 is a constant normally related to the doping level.

Instead of a diode, a transistor can equally be used for the purpose, perhaps, with better flexibility. The collector–base voltage in such a case is considered zero (shorted) as this voltage, if any, is likely to affect the base width modulation which in turn affects the base–emitter voltage. In such a situation, I is replaced by I_c in Eq. (3.66). But I_c is also known to be affected by temperature and accordingly, we can write

$$I_c \propto T^{m_2} \tag{3.67}$$

Combining Eqs. (3.66) and (3.67) and using Eq. (3.66) in two cases, one at reference temperature T_r and another at an arbitrary varying value T, a little manipulation gives

$$V_{BE} = \left\{ V_{g0} + (m_1 - m_2)\frac{kT_r}{q} \right\} - (\chi T) + \left\{ (m_1 - m_2)\frac{k}{q}\left[T - T_r - T \ln\left(\frac{T}{T_r}\right) \right] \right\} \tag{3.68}$$

where

$$\chi = \frac{\left[V_{g0} + \left(\frac{kT_r}{q}\right)(m_1 - m_2) - V_{BE}(T_r) \right]}{T_r}, \tag{3.69}$$

$$F_1 = V_{g0} + (m_1 - m_2)\frac{kT_r}{q},$$

$$F_2 = \chi T, \text{ and}$$

$$F_3 = \left\{ (m_1 - m_2)\frac{k}{q}\left[T - T_r - T \ln\left(\frac{T}{T_r}\right) \right] \right\}$$

In Eq. (3.68), three terms of specific interest are F_1, F_2, and F_3 where F_1 is a constant, F_2 shows that there is a part that governs linear decrease of V_{BE} with T, and F_3 is a higher order term that brings nonlinearity in relation. Figure 3.31 shows the curve and its linear approximation with the parts governed by the terms identified. Typical magnitudinal values of the relevant terms are approximately given as

  (i) $F_1 = 1.27$ V for constant I_c with V_{g0} taken as 1.17 V and $T_r = 323$ K

  (ii) $V_{BE}\big|_{T = T_r} = 0.547$ V and $\chi = 2.24$ mV/K.

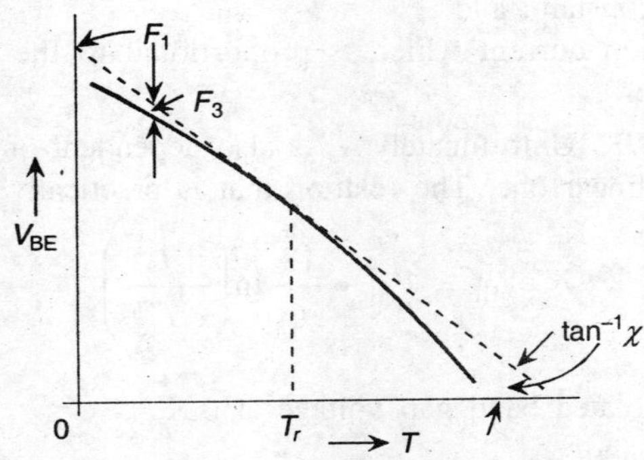

Fig. 3.31 V_{BE}–T curve of a semiconductor diode temperature sensor with the linear approximation curve.

The nonlinearity appearing due to the third term is temperature-dependent as also dependent on $(m_1 - m_2)$. For small $\Delta T = T - T_r << T_r$, the first three terms of a Taylor series expansion may be considered.

So far in the discussion, a few factors that always play their role have not been taken into account. These are:

(a) Early effect which is reduced by assuming $V_{BC} = 0$. But shorting the base to collector also would mean $I_c = I_{bias} - I_B$ and for low current operation this leads to large inaccuracy;

(b) self-heating, high level injection, as also leakage currents and noise produced nonlinearities which can be reduced by proper choice of bias current which in practice lies in the range 10–100 μA.

A single transistor sensor is a low cost sensor but for accurate linear indication, it requires complicated supporting electronic system like a constant current source for I_c and high stability reference source. For constant current I_c, $m_2 = 0$ so that Eq. (3.68) is considerably simplified. The reference voltage has to be chosen carefully for high resolution in A/D conversion, the resolution of the converter is, however, equally important. The single transistor sensor with these current and voltage biasing (providing reference) is shown in Fig. 3.32. This circuit also indicates that the technique can be adopted in differential temperature sensing by replacing V_{ref} with a second sensor of output V_{BE2} (say) so that

$$V_o = V_{BE1} - V_{BE2} = k_d (T_1 - T_2) \tag{3.70}$$

where k_d is a constant.

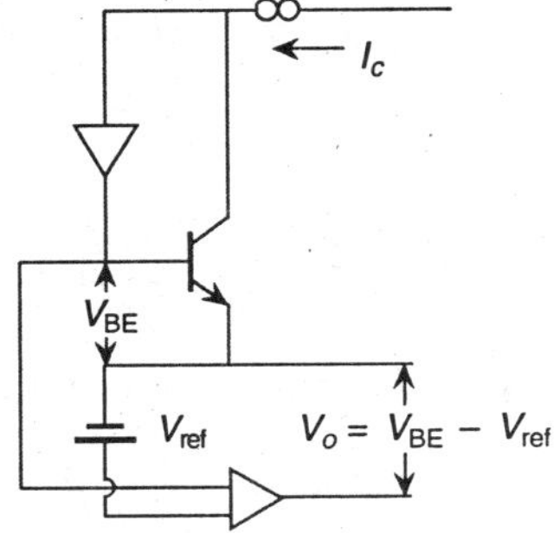

Fig. 3.32 Single transistor thermal sensor with reference biasing.

3.11.1 The PTAT Sensor

The single transistor transducer has now been extended into an IC temperature transducer which also includes biasing, amplifying, and linearizing circuits. Such an IC sensor is known as PTAT sensor, that is it provides output voltage of current 'Proportional To the Absolute Temperature'. In this, the signal obtained is the difference of V_{BE}'s of two transducers which are operated so that the ratio of their emitter current densities is constant but they are at the same temperature T. Thus,

$$\Delta V_{BE} = \frac{kT}{q} \ln\left(\frac{I_{c1} I_{s2}}{I_{c2} I_{s1}}\right) \tag{3.71a}$$

But it is known that I_s is proportional to the emitter area so that Eq. (3.71a) is given by

$$\Delta V_{BE} = \frac{kT}{q} \ln(r_1 r_2) \tag{3.71b}$$

where $r_1 = I_{s2}/I_{s1}$, is the emitter area ratio independent of temperature and $r_2 = I_{c1}/I_{c2}$ is maintained constant. A typical current source type PTAT scheme is shown in Fig. 3.33.

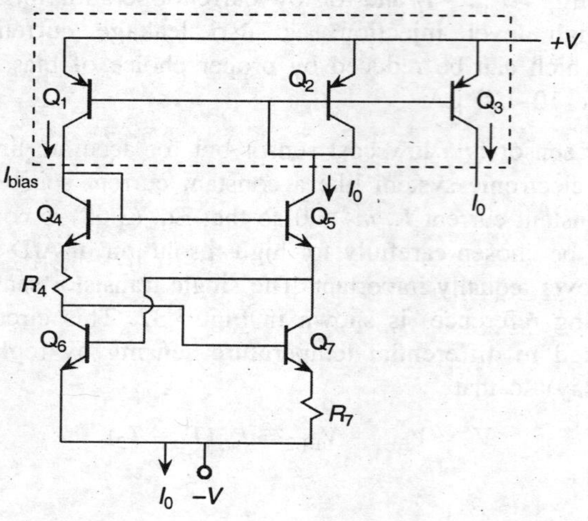

Fig. 3.33 Simplified IC-version of the PTAT.

The bases of transistors Q_6 and Q_7 are cross-connected and they have unequal emitter areas of ratio

$$r_1 = \left(\frac{\text{emitter area of } Q_7}{\text{emitter area of } Q_6}\right) = \frac{I_{s7}}{I_{s6}} \tag{3.72}$$

Also, for transistors Q_4 and Q_5 feeding the collectors of Q_6 and Q_7, the ratio r_2 is given as

$$r_2 = \left(\frac{\text{emitter area of } Q_4}{\text{emitter area of } Q_5}\right) = \frac{I_{s4}}{I_{s5}} = \frac{I_{c6}}{I_{c7}} \tag{3.73}$$

The presence of the resistors R_4 and R_7 shows that with no base currents

$$V_{BE4} + I_{E4}R_4 + V_{BE7} + I_{E7}R_7 = V_{BE5} + V_{BE6} \tag{3.74}$$

Hence, we easily obtain,

$$I_{E4}R_4 + I_{E7}R_7 = V_{BE5} - V_{BE4} + V_{BE6} - V_{BE7}$$

$$= \left(\frac{kT}{q}\right) \ln(r_1 r_2) \tag{3.75a}$$

If $R_4 = R_7 = R$ (say), then

$$\left(\frac{kT}{q}\right) \ln(r_1 r_2) = (I_{E4} + I_{E7})R = I_o R \tag{3.75b}$$

I_o being the output current. The circuit within the dotted enclosure represents the 'current mirror' biasing scheme.

The PTAT has normally a large 'offset' signal at common temperatures which is being proposed to be dispensed with 're-scaling' using an internal reference. Usually a PTAT at a temperature of $T_x K$ requires a resolution of $(T_x/\Delta T_x) : 1$ for detecting a temperature of T_x. In degree Celsius scale, this requirement reduces to $(T_x - 273)/\Delta T_x : 1$ which means that for lower range, if the transducer has a zero closer to the range of interest, it is much better. This can be obtained by providing an internal reference, the basic scheme of which is shown in Fig. 3.34. The current source type PTAT with an internal resistor R_i produces a current $I_o = V_P/R_i$ which, across R, produces a drop $V_P R/R_i$ so that the output voltage

$$V_o = V_{BE3} - V_P R / R_i \tag{3.76}$$

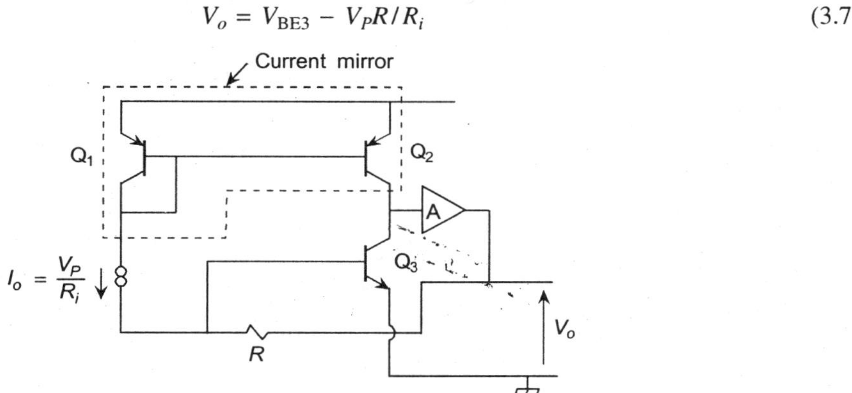

Fig. 3.34 Modified semiconductor sensor with internal reference.

where the bias current for the transistor Q_3 is provided by the current mirror and amplifier A, having a very high gain which keeps the collector current of Q_3 to be the current of the current mirror. As given in Eq. (3.66)

$$V_{BE} = V_{go} - k_1 T \tag{3.77}$$

so that Eq. (3.76) can be written (for a reference 'zero' temperature T_0 which is adjustable by trimming R) as

$$V_o = -V_{go} \frac{T - T_0}{T_0} \tag{3.78}$$

This is explained in Fig. 3.35. In fact, the output curve rotates about V_{go} because change of R_i changes the slope of $V_P R/R_i$.

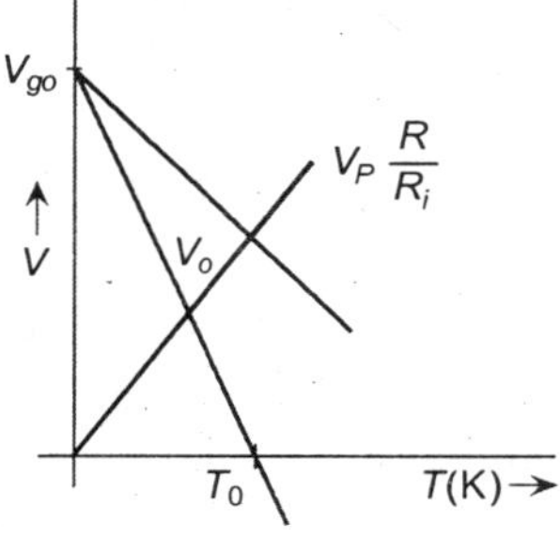

Fig. 3.35 Method showing adjustment procedure for reference zero.

Figure 3.36 represents the schematic diagram of a commercial type IC temperature sensor (AD 590). In these, transistors Q_1 and Q_2 produce the voltage which is converted into current by resistances R_2 and R_1. The collector current of transistor Q_3 tracks the collector current of transistors Q_2, and Q_4 supplies the bias current and the substrate leakage current as shown so that the total current between the supply terminals remains proportional to temperature. The current mirror is shown as CM as discussed earlier. R_1 and R_2 are laser-trimmed for an output of 1 μA/°C at 25°C.

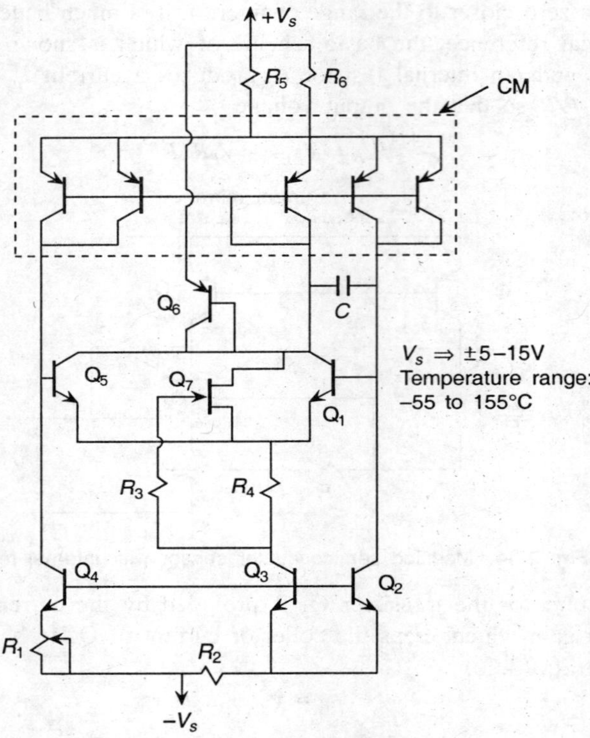

Fig. 3.36   AD 590 sensor.

It will not be out of context here to mention that the current source type characteristic of the sensor allows some of these to be connected in series for selection of minimum temperature and in parallel for average temperature measurement. As mentioned in an earlier section, it can also be used for reference junction compensation in thermocouples. AD 580 is a precision reference voltage source and is often used as such with this type of sensor.

3.12   THERMAL RADIATION SENSORS

While most of the previously described thermal sensors receive 'quantum' of heat by conduction, the radiation types receive this by radiation and hence, this type requires no physical contact with radiating system. Thermal radiation sensors are guided by basic laws of black body radiation such as Planck's law and Stefan–Boltzmann Law.

Planck's law of black body radiation follows Wien's spectral radiance law which states that "the radiation flux emitted by a black body per unit solid angle per unit area in a direction normal to it in the wavelength range λ to $\lambda + d\lambda$ is given as $L_\lambda d\lambda$, where L_λ is called the spectral radiance and is given by

$$L_\lambda = \frac{C_1}{\lambda^5 \exp\left(\dfrac{C_2}{\lambda T}\right)} \tag{3.79}$$

where C_1 and C_2 are constants defined in terms of the fundamental constants as $C_1 = 2hc^2 = 3.742 \times 10^{-16}\,\text{Wm}^2$, and $C_2 = hc/k = 1.4388 \times 10^{-2}\,\text{mK}$."
Planck deduced the relation given by Eq. (3.79) as

$$L_\lambda = \frac{C_1 \mu^2}{\lambda^5 (e^{C_2/\lambda T} - 1)} \tag{3.80}$$

(Planck's law of black body radiation)

where μ is the refractive index of the medium in which the radiation is emitted and normally, taken as unity for air. The prime quantity in radiation-based temperature transduction is the radiant energy Q which is propagated as waves or photons. The radiant flux $\phi = dQ/dt$ gives its power. The flux emitted per unit solid angle by a source is called intensity $i = d\phi/d\omega$, expressed in watts per steradian, however, flux emitted by a surface is called radiant exitance $M = d\phi/dA$. Radiance L, the flux per unit area per unit solid angle, is given by $L = d^2\phi/(dA\cos\phi\,d\omega)$ shown schematically in Fig. 3.37. The flux received by a surface of unit area is given, however, by irradiance E.

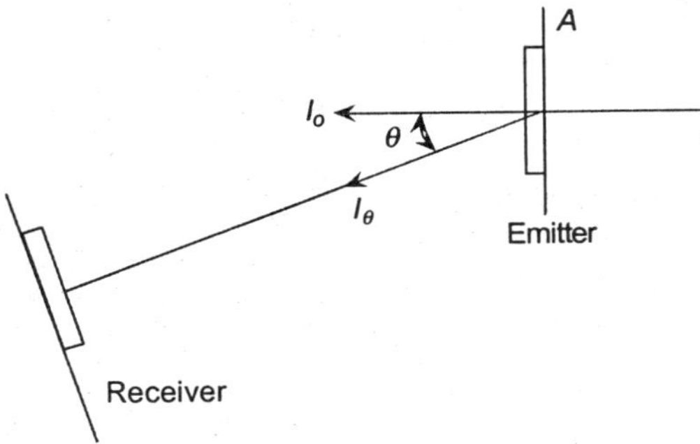

Fig. 3.37   Radiation emission and reception at angles other than perpendicular to the source.

The total radiance L of a black body is obtained by integrating L_λ over the range of λ; thus,

$$L = \int_0^\infty L_\lambda\, d\lambda = \frac{\mu^2 \sigma T^4}{\pi} \tag{3.81}$$

which, in terms of M, the radiance exitance is

$$M = \pi L = \sigma T^4 \qquad \text{for } \mu = 1 \tag{3.82}$$

(Stefan–Boltzmann law)

σ is the Stefan-Boltzmann constant and has a value $5.67 \times 10^{-8}\,\text{Wm}^{-2}\text{K}^{-4}$. σ is actually given by

$$\sigma = \frac{2\pi^5 k^4}{15h^3 c^2} \tag{3.83}$$

where k is the Boltzmann constant.

If L_λ versus λ is plotted for different values of T, as shown in Fig. 3.38, the radiance is seen to have a peak value for a specific temperature T_m at a wavelength λ_m and it is also seen from the curves that

$$\lambda_i T_i = \lambda_h T_h = \lambda_m T_m = 2898 \ \mu m \ K \tag{3.84}$$

(Wien's displacement law)

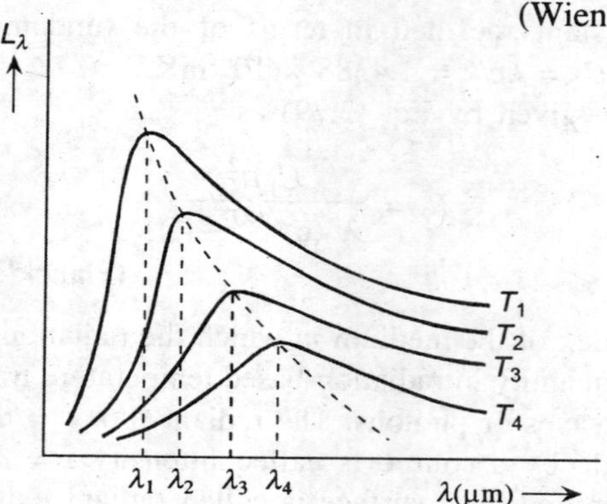

Fig. 3.38 L_λ–λ curves for varying temperatures.

From Fig. 3.37, it is also seen that if I_o is the radiant intensity in the direction normal to the source, and I_θ is the intensity at an angle θ as shown, then

$$I_\theta = I_o \cos \theta \quad \text{(Lambert's cosine law)} \tag{3.85}$$

In case of black bodies, some more properties are true. These include—it absorbs all radiant energy incident on it and it emits the maximum amount of flux per unit area at any temperature and its emission is independent of direction, that is, it is an ideally diffuse radiator. Thus, the absorptivity α and emissivity ε, or the spectral absorptivity α_λ and spectral emissivity ε_λ for a black body are unity.

Thus,
$$\alpha = \varepsilon = \alpha_\lambda = \varepsilon_\lambda = 1 \quad \text{for a black body.} \tag{3.86}$$

Physically realizable radiators do not have these conditions fully satisfied. If a body has an absorptivity α, reflectivity ρ and transmittivity τ, then from the law of conservation of energy

$$\alpha + \rho + \tau = 1 \tag{3.87}$$

For an opaque material, as are practically encountered, $\tau = 0$, so that

$$\alpha + \rho = 1 \tag{3.88a}$$

For a specified wavelength λ, this can be written as

$$\alpha_\lambda + \rho_\lambda = 1 \tag{3.88b}$$

For real bodies, the emissivity is less than 1 and its value changes with the body in question. As a consequence, the black body laws are required to be modified accordingly. In fact, Planck's law given by Eq. (3.80) is now rewritten as

$$L_{a\lambda} = \frac{\varepsilon_{a\lambda} C_1}{\lambda^5 (\exp(C_2/\lambda T) - 1)} \tag{3.89}$$

where $\varepsilon_{a\lambda}$ is the actual spectral emissivity being a function of the body surface condition as also of temperature and the direction of emission. Its spectral emissivity $\varepsilon_{a\lambda}$ is often loosely approximated by

$$\varepsilon_{a\lambda} = k_2 \sqrt{\frac{\rho_r}{\lambda}} \quad \text{(Drude's law)} \tag{3.90}$$

where ρ_r is the resistivity of the material.

Strictly speaking, $\varepsilon_{a\lambda}$ depends on some other factors as has already been mentioned. Also, state of polarization of the radiation is important. As Drude's law states—'with increasing wavelength, $\varepsilon_{a\lambda}$ decreases and this is true for polished metals but it is opposite with ceramics, glasses, and paints. In some semi-transparent materials such as plastics, the behaviour is a little irregular with peaks and troughs in $\varepsilon_{a\lambda}$ with changing λ. In others such as oxidized metals or carbides, $\varepsilon_{a\lambda}$ is fairly constant with λ over a narrow band. Figure 3.39 shows comparative curves of emissivities of different materials discussed above. Actual or real body spectral emissivity ε_λ with λ for general situation is shown in Fig. 3.40 by plotting $L_\lambda - \lambda$ curves for ε_λ, $\varepsilon_{a\lambda}$ and $\varepsilon_{g\lambda}$ where $\varepsilon_{g\lambda}$ is the emissivity of a grey body and is assumed constant as the average value of the $\varepsilon_{a\lambda}$'s.

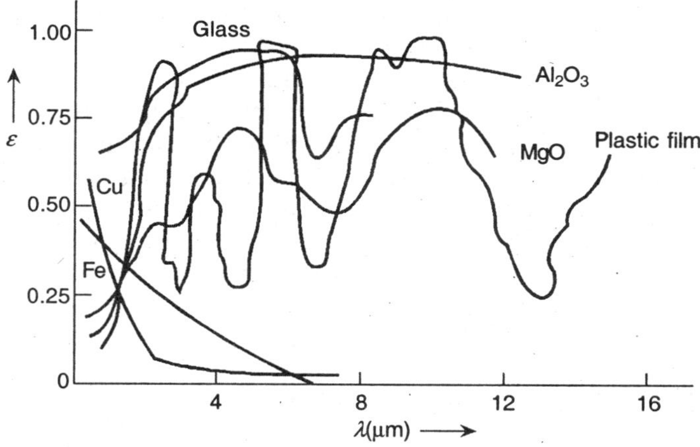

Fig. 3.39 Emissivity versus wavelength plots for various materials.

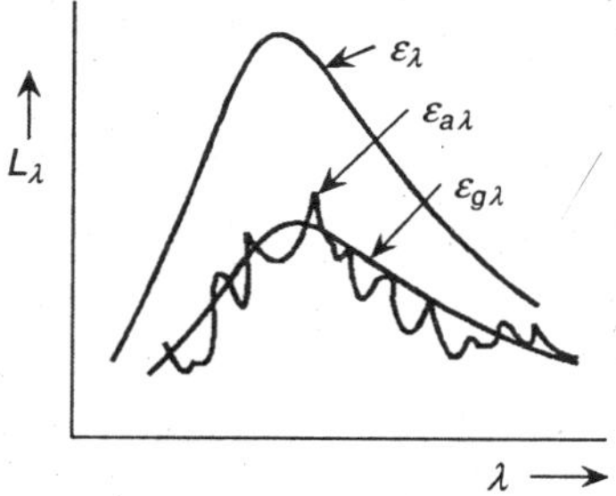

Fig. 3.40 Different types of emissivities.

When the surface roughness increases or the degree of oxidation on the surface increases, emissivity increases. This is true for conducting materials. For dielectric materials, the change in surface condition does not very much affect the emissivity. Angle of emission greatly affects emissivity—in case of metals all through the angular range, whereas for dielectric materials upto about 60–65° no changes are observed but beyond this angle, sharp reduction occurs. With temperature, emissivity, particularly the total emissivity, increases to a slight extent.

Radiation thermometers developed on the laws discussed just now can be classified broadly under:

   (i)   Total radiation type,
   (ii) Multiwaveband type,
  (iii) Single waveband type or spectral radiation type, and
  (iv) Ratio type.

Of these the first, third, and the fourth types are quite commonly used in industries.

All radiation thermometers have three basic components, namely (i) the optical system, (ii) the detector, and (iii) the signal processing unit.

The majority of the commercially adopted thermometers of this type use lens or mirror optics for the optical system. After the advent of fibre optics, it is also being used in certain special systems. In the conventional system, a diaphragm is located at a suitable distance between the detector and the target to produce an aperture determining the angular field of view. This technique is gradually being discontinued because it limits the target size, irradiance at the detector, and sensitivity at lower temperatures.

As such, radiation has to be collected from a reasonable size of the target for representative irradiance at the detector over the entire temperature range. This is usually done by 'lens and mirror' type transfer optics, two typical schemes of which are shown in Figs. 3.41(a) and (b) with a single lens and a single mirror respectively. In Fig. 3.41(a), the operation is straightaway—the lens acts as the aperture of the system which determines the cone-size of radiation accepted from an axial point of the target while the detector acts as the field stop determining the field of view. But the lens has a limited transmittance τ, less than 1 in value, and hence the irradiance at the detector for a radiance L of the source is given by

$$E = \pi\tau L \sin^2\theta \qquad (3.91)$$

where θ is the 'field cone half angle' as shown in the figure.

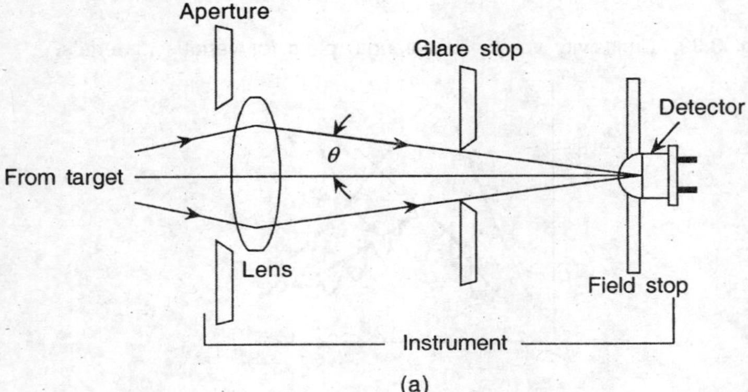

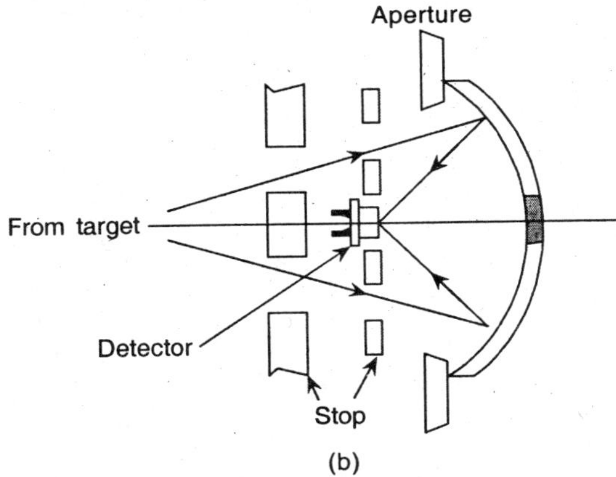

Fig. 3.41 Radiation pyrometer: (a) with lens optics, and (b) with mirror optics.

The optical system must be well-designed to stop stray radiation from outside the target that may enter the measurement system by reflection or scattering. 'Glare stop' is used at suitable position which suppresses the reflected and scattered radiation but passes the rays from the aperture stop. With the lens system, the choice of the lens material plays an important role as it has varied transmittance for different wavelengths of radiation. Transmittance versus wavelength curves for some of the commonly used materials are shown in Fig. 3.42. The properties on which they are selected other than transmittance are high refractive index with low temperature coefficient, low scattering and dispersion, good homogeneity, uniform density, good strength and hardness yet easy workability, high melting point, non-corrosive nature and easily coatable when necessary. Mirror optics does not require all these considerations to be specified; besides, the additional advantage it provides is that it can be used in a number of wavebands because it is free from chromatic aberration. However, to retain the reflecting characteristic of the mirror, it must be protected from the smoke, dust, moisture and the like, and a front window is to be provided with a 'lens' material causing similar problems as with lens optics. Also, it is very difficult to provide glare protection in mirror systems.

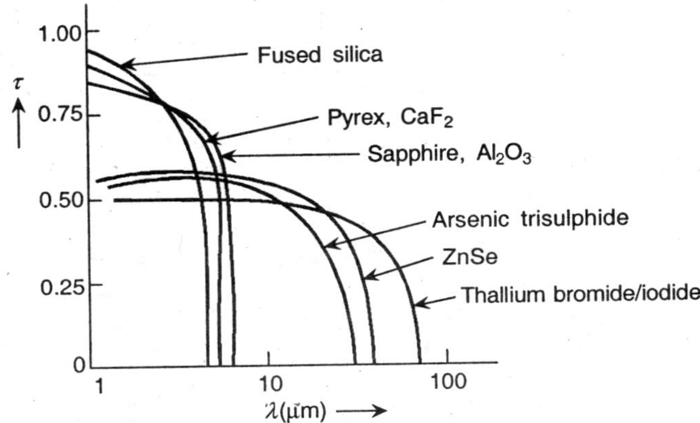

Fig. 3.42 Transmittance characteristics of different lens materials.

In total radiation pyrometry, another important consideration is the absorption of radiation in the intervening space between the target and the optical system. This absorption is selective. The most important constituents in the atmosphere which are responsible for such absorption are CO_2 and water vapour. A typical relative absorption with wavelength band is shown in Fig. 3.43.

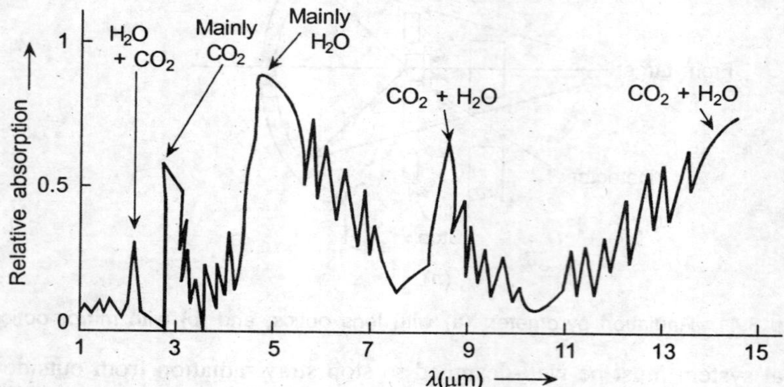

Fig. 3.43 Relative absorption versus wavelength of different media gases.

Lens and mirror optics, in some situations, can be replaced by fibre optics. A major advantage with fibre optic cables is their flexibility which allows them to be used in situations where (i) direct sighting of the target is not possible, (ii) absorbing and corrosive atmosphere is prevalent, (iii) electromagnetic interference or reactions due to nuclear radiation requires the measuring system to be away from the place, (iv) small area optical head is essential, and many more. In all such cases, the fibre optic cable is used to transmit the radiation from the optical head to the detector located away from the head. It can, however, be used as aperture optics as well without using a lens or a mirror. Two cases are illustrated in Figs. 3.44 (a) and (b). The use of the scheme of Fig. 3.44(a) is limited by the optical head which can be used upto a certain temperature only. Si or glass fibre are used for transmission—the radiation range is between 0.5–2 μm. Length of cable is usually between 50 cm–20 m and temperature upto 1000°C can be measured easily.

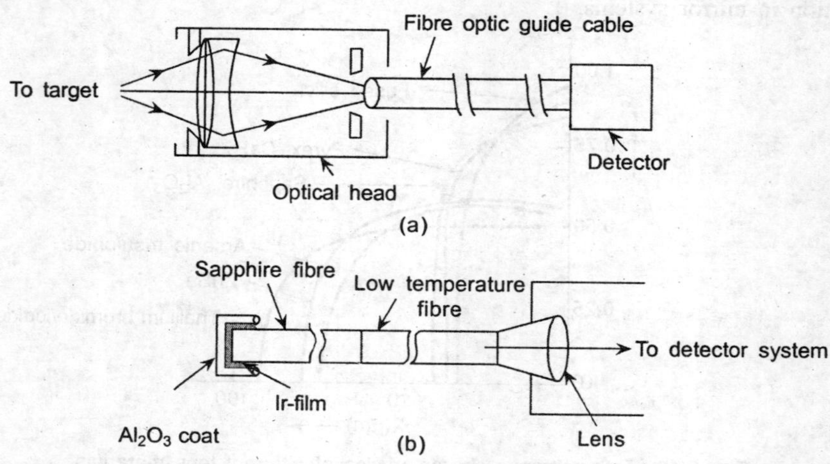

Fig. 3.44 (a) Pyrometer with optical fibre transmission, (b) fibre-optics thermometer.

In Fig. 3.44(b), the high temperature fibre cable (sapphire) is terminated at one end in the form of a cavity by sputtering with a thin coating of metal like iridium or platinum which is then covered with a film of Al_2O_3. This is particularly useful for target materials of low emissivity. The cavity is thus, a black body cavity and this is immersed in the thermal system where temperature is to be measured. The radiation is then transmitted first through the high temperature optical fibre and then through a low temperature fibre onto the detector system.

3.12.1 Detectors

All total radiation thermometers use a few 'standard' detectors. Many of these detectors are, however, wavelength/frequency selective and are not, therefore, suitable for general purpose. Such detectors are quite conveniently adapted in single waveband thermometry and also in ratio thermometry. The factors that are taken into account while selecting a detector for a specific purpose are: (a) responsivity, (b) spectral range coverage, (c) noise equivalent power (NEP), (d) speed of response, (e) linearity, (f) stability, (g) operating temperature of the detector, and (h) operating mode.

A new factor is defined as detectivity which takes into consideration the NEP and the detector area. *Responsivity* is very simply defined as output voltage/current per watt of incident radiation and should, obviously, be high. NEP is the radiation in watts that makes signal-to-noise ratio of the detector unity and this should be kept as low as possible. *Detectivity*, on the other hand, is defined as

$$D_\lambda = \left(\frac{1}{NEP}\right)\sqrt{\left(\frac{A\,\Delta\omega}{2\pi}\right)}\ \mathrm{cm\,Hz}^{1/2}\mathrm{W}^{-1} \tag{3.92}$$

where

A is the detector area and

$\Delta\omega$ is its frequency bandwidth or that of the entire measuring system.

Spectral responsivity is the responsivity with respect to wavelength of incident radiation. Some very commonly used detectors are wavelength selective. Detectivity is, in fact, a measure of spectral responsivity normalized with respect to noise equivalent of power. Figure 3.45 shows the detectivity of a number of radiation detectors; their characteristics are given in Table 3.11 to follow. In Fig. 3.45, InSb and InAs are actually photovoltaic or barrier photocells while PbS, HgCdTe are the examples of photoconductors.

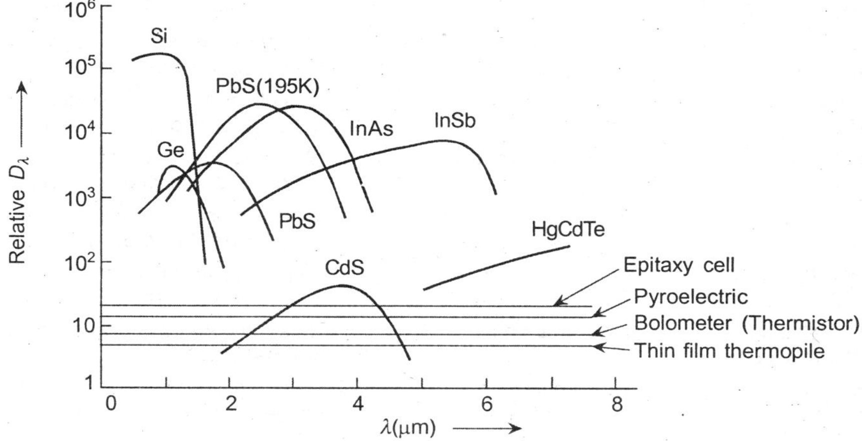

Fig. 3.45 Relative detectivity versus wavelength for different types of detectors.

Stability is of two kinds, namely (i) short term and (ii) long term. Short term stability is essentially due to ambient condition variation and a sensor with low temperature coefficient would be better in this respect. Silicon has a low temperature coefficient and is, therefore, better. However, it is often suggested to compensate for the ambient temperature variation either by

- controlling the ambient condition,
- correcting for the variation automatically or manually, and
- comparing with a standard reference source like GaAs lamp. Long term stability is checked by frequent calibration and is found in most cases to be within 0.2°C per year. In such cases also, silicon is a better choice.

Speed of response is necessary for dynamic measurement but for thermal processes, the speed need not be very high. 1 μs is a good value even though a few detectors have a response time as low as a few nanoseconds. Response time for thermistor-bolometers and thermopiles is about a few milliseconds. Some photon detectors like PbS have low response speed, being as large a time as 100–300 μs, but is still better than the thermopiles. Good linearity contributes to less effort in calibration and extrapolation or interpolation. Silicon has a better linearity compared to Ge and InSb type photon detectors. Operating temperature of the detector is usually the room temperature and Si happens to be useful in this respect as well. But other photon detectors have better resolution at lower temperatures, that is, decreasing temperatures increasing their resolution.

Most of the detectors can be used in continuous (dc) or intermittent mode but pyroelectric ones are dependent on change of temperature and can be used only in intermittent mode of operation. This is achieved by chopping the radiation.

Table 3.11 Comparison of detectors

S.No.	Detector (type)	Used in thermometer	General remarks
1	Photomultiplier (photoemissive)	Primary standard	Selective in waveband, continuous mode, costly.
2	Si (photoconductive (PC), as well as Photovoltaic (PV))	Primary or secondary standard; Single waveband and ratio type	Selective, IR-band, range above 400°C, linearity high, detectivity and stability high, dc and ac mode, low cost.
3	Ge (Both PC and PV)	Single waveband and ratio type	Selective, good linearity range 200°C and above, dc and ac mode, low cost.
4	PbS (PC)	Single waveband and ratio type	Selective, usually ac mode range 100°C and above, but limited by cost.
5	InAs (PV)	Single waveband	Selective, ac mode operation, range 0°C and above, costly.
6	InSb (PV)	Single waveband	Selective, ac mode operation, range 0°C and above, costly.
7	PbSe (PC)	Single waveband and ratio type	Like PbS but with lower detectivity, range starts from 50°C.
8	HgCdTe (PC)	Single waveband	Selective, ac operation, range starts from −50°C, costly.
9	Pyroelectric (thermal detector)	Total and single waveband	Total, ac operation, range starts from −50°C, medium cost.
10	Bolometer thermistor (Thermal)	Total and single waveband	Total, ac operation at room temperature, range from 0°C, low cost.
11	Thermopile; metal film type (Thermal)	Total and single waveband	Total, ac operation at room temperature, range from 0°C, low cost.

The signal processing part is used to amplify the low level signal from the detector, that is, the detector output and then process it, as required, for indication, recording or control or for any other purpose. The schemes of total radiation thermometers have been shown in Figs. 3.41(a) and (b).

Of the remaining three types, the single band or the monochromatic thermometers are most widely used. Starting from the same basic equation but integrating now between two wavelengths λ_1 and λ_2 where $\lambda_1 \sim \lambda_2 \leq 10$ nm only, the single band radiance can be obtained. With the band made narrower, effectively a single wavelength thermometer may be designed which is often called the effective wavelength. In this case, the detector irradiance is given by

$$E_\lambda = g_\lambda L_\lambda \tag{3.93}$$

where g_λ is a factor denoting the geometrical effect and responsivity. For low value of λT, Planck's law can be written as

$$L_\lambda = \frac{C_1}{\lambda^5} \exp\left(\frac{-C_2}{\lambda T}\right) \tag{3.94}$$

from which one can write

$$\frac{\Delta L_\lambda}{L_\lambda} = \frac{C_2}{\lambda T}\left(-\frac{\Delta T}{T}\right) \tag{3.95}$$

and the normalized sensitivity, defined as $(\Delta L_\lambda/L_\lambda)/(\Delta T/T)$, decreases as λ increases which, in turn, means accuracy decreases at longer wavelengths. Hence, with single wavelength thermometer, measurement should be made at the shortest possible wavelength. For a short band, effective wavelength is the 'effective' mid-wavelength and the selection of the wavelength or band is done by an appropriate filter such as a narrow band interference filter having a bandwidth, as stipulated, of 10 nm.

Single waveband thermometers are often designed to be used as primary and secondary types. The primary type uses an effective wavelength of 660 nm or more recently a wavelength between 900 and 1000 nm. Two detectors, of which one is a photomultiplier, are used for 660 nm and the other a Si-photodiode is used for the 900–1000 nm range. A typical primary scheme is shown in Fig. 3.46.

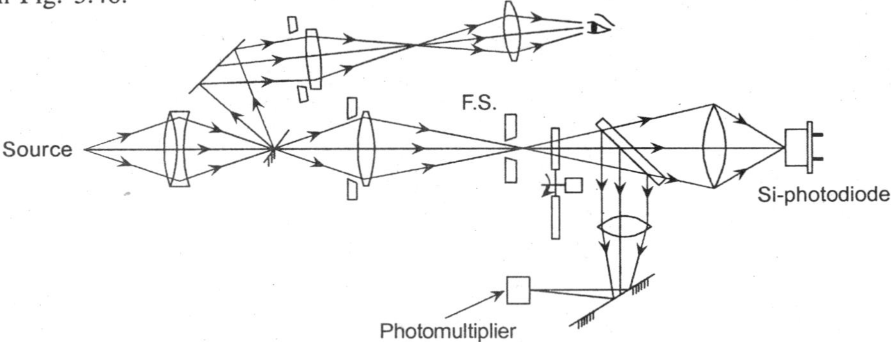

Fig. 3.46   Primary scheme of a single waveband thermometer.

Secondary standards are similarly developed but are simpler and compact. Single waveband thermometers are also called infrared thermometers where the wavelength range is 0.5–14 µm in steps. These have accuracy from ± 0.5–4% of scale, response time 0.1–2 s, and a resolution from 0.1–1°C.

In the single waveband variety, a typical class, often called the *optical pyrometer* or the *disappearing filament type pyrometer,* is very popular and operates in a range from 700°C onwards. In this type, eye becomes the detector element. A typical diagrammatic representation of this type is shown in Fig. 3.47. Radiation from the source is focussed on to a filament which is separately heated by an electric supply of voltage V_s. The field is viewed by the eyepiece, through a band-selecting filter and when the brightness of the field of vision is identical to that of the filament at the specified wavelength of the filter, the filament disappears. The brightness of the filament is controllable by the adjustment of a potentiometer and the ammeter in series, A, can be calibrated directly in °C. It is portable but not usable as a control component.

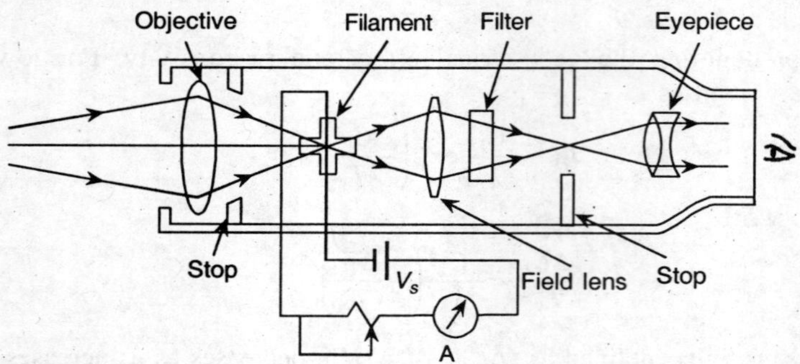

Fig. 3.47 An optical pyrometer.

When radiance from a target is measured, the true temperature of the target T_t is not indicated because of target emissivity ε_λ being less than 1 in most cases and absorption by the optics and the intervening medium. If the transmittance is τ_λ and $\tau_\lambda \varepsilon_\lambda = \chi_\lambda$, we get

$$T_t = \frac{1}{\dfrac{1}{T_\lambda} + \dfrac{\lambda}{C_2} \ln \chi_\lambda} \tag{3.96}$$

where T_λ is the measured radiance temperature. Plots of $(T_t - T_\lambda)$ versus λ for various values of χ_λ are shown in Fig. 3.48.

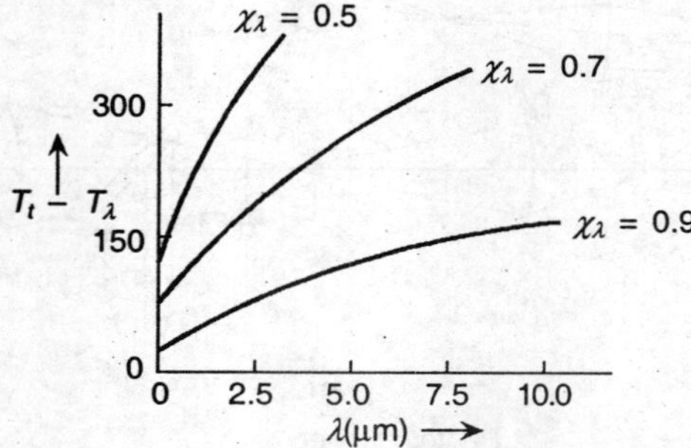

Fig. 3.48 Optical pyrometer error versus wavelength for different χ_λ.

Multiwaveband thermometer is not of much practical importance although it is a means for solving emissivity problem, but, that too through ratiometric technique. In ratio thermometers, temperature is obtained by measuring the ratio of the detector signals at two different wavebands of the radiation of effective wavelengths λ_1 and λ_2 (say). Thus,

$$r(T) = g_t \left(\frac{L_{\lambda 1}(T)}{L_{\lambda 2}(T)} \right) \tag{3.97}$$

where $r(T)$ is the ratio obtained and g_t is constant dependent on the spectral response. From Eq. (3.97), one easily derives

$$\frac{\Delta r}{r} = \frac{-C_2}{T} \left(\frac{1}{\lambda_1} - \frac{1}{\lambda_2} \right) \frac{\Delta T}{T} \tag{3.98}$$

Comparing this with Eq. (3.95), it appears that the sensitivity increases because the effective wavelength is now $\lambda = (\lambda_1 \lambda_2)/(\lambda_2 - \lambda_1)$ which is larger than λ_1 or λ_2 and sensitivity is related as shown in Eq. (3.96). At different wavelengths, responsivity changes because of spectral emissivity, transmittance, and so on inducing certain amount of uncertainty. Thus, with $\tau_{\lambda 1,2} = 1$,

$$r(T) = g_t \left(\frac{\varepsilon_{\lambda 1}}{\varepsilon_{\lambda 2}} \right) \left(\frac{L_{\lambda 1}(T)}{L_{\lambda 2}(T)} \right) \tag{3.99}$$

Only on the assumption of grey body condition does the system accuracy improve. From Eq. (3.99),

$$T_t = \frac{1}{\dfrac{1}{T_m} + \dfrac{\lambda_1 \lambda_2}{C_2'(\lambda_2 - \lambda_1)} \ln\left(\dfrac{\varepsilon_{\lambda 1}}{\varepsilon_{\lambda 2}} \right)} \tag{3.100}$$

where T_t is the true temperature and T_m is the measured temperature. Typical scheme of a ratio thermometer is shown in Fig. 3.49. It has an accuracy of $\pm 1\%$ of scale and a resolution of $1°C$.

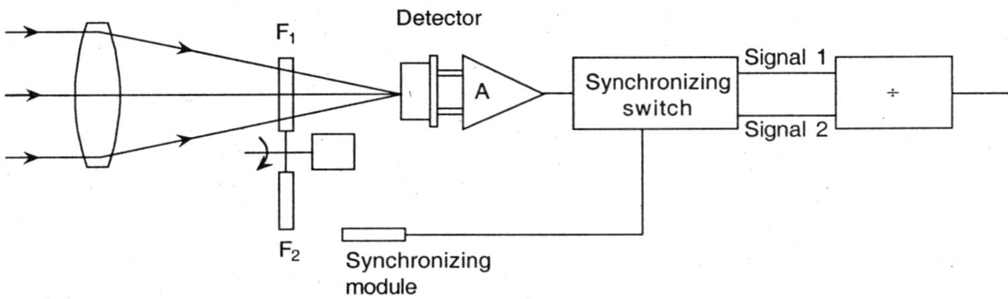

Fig. 3.49   Scheme of a ratio thermometer.

Other special thermometers have also been made and some are commercialized as well but these are different from those already described in view of the attempt to compensate for emissivity of radiation surface, the sensing mechanism remaining the same. The methods used for the purpose are as follows:

(a)   auxiliary reflector methods where emissivity is enhanced by placing a good reflector close to it,

(b)   hot source methods where the emissivity of the surface is enhanced by measuring signals corresponding to two temperatures of a black body irradiating the surface,

(c)   polaradiometer methods which involve the determination of polarizations of radiation emitted by the target and that from a black body reflected by the target. If the temperatures of the black body and the target are the same, the resultant radiation is unpolarized in complying with the Kirchoff's law. Polarization is measured by a rotating polarizer, and

(d) reflectance methods which also establish a value for emissivity by reflectance measurement.

3.12.2 The Pyroelectric Thermal Sensors

The pyroelectric thermal sensor is comparatively a new entrant in the area of thermal/temperature detection. It comprises a type of ferroelectric material. Ferroelectric materials are non-centrosymmetric and their ferroelectricity is attributed to the spontaneous electric polarization on a polar axis. The direction of this polarization can, however, be changed by the application of electric field. Also, there occurs a remanent polarization because of permanent electric dipole in the primitive unit cell of the crystal.

If the permanent dipoles in the material exhibit electric polarization with temperature, the characteristic property is called pyroelectricity. Materials with polar point symmetry are likely to exhibit this property. There are 32 possible point group symmetries; materials which possess 10 out of these 32, exhibit pyroelectricity.

Materials of this category are mainly ceramics. The dipoles, normally, are in random orientation in the material and net electric output is zero, and at ambient temperature these orientations are also fixed. If the temperature is now raised above a certain value, often called the Curie temperature or 'critical' temperature, which is again a characteristic of the material, the molecules with the dipoles are free to rotate.

When a slice of pyroelectric ceramic is placed between a pair of electrodes and the electrodes given an electric field, with its temperature raised above Curie point, the molecules in the material orient themselves in the direction parallel to the applied field with opposite polarity of the dipoles. This state, however, persists even when the field is removed (see Fig. 3.50). The amount of polarization thus affected is proportional to the applied field. Thus, if P is the polarization and E, is the applied field, then

$$P = \sigma E \qquad (3.101)$$

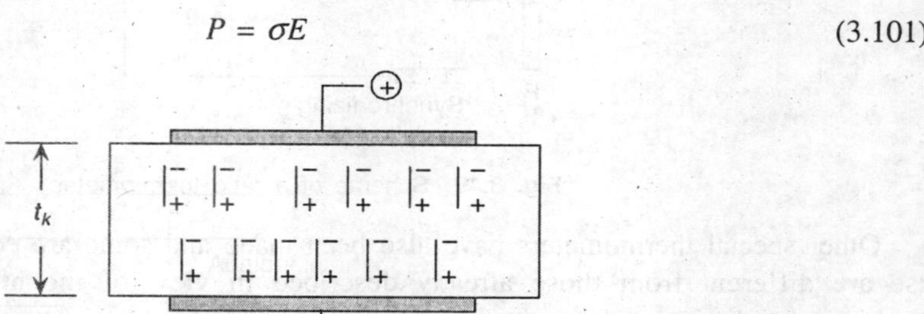

Fig. 3.50 Sketch of a pyroelectric thermal detector.

where σ is a constant, which is a function of the material. The dipoles, however, oscillate about their parallel orientation and if now the temperature is raised the oscillation angle increases. If the dipole length is l with charges $\pm q$, it has an electric moment m such that

$$m = ql \qquad (3.102)$$

If the dipole oscillates with an average angle θ, its effective length is given by

$$l_e = l\cos\theta \tag{3.103}$$

Thus, the electric moment of polarization reduces as it becomes

$$m = ql\cos\theta \tag{3.104}$$

If electrode area is A and slice thickness t_k, then the magnetic moment of the entire slice is the sum of moments of all the dipoles so that

$$M = P_v A t_k \tag{3.105}$$

where P_v is the dipole moment per unit volume. Because of this polarization moment, charge Q at the ceramic surface as collected by the electrodes is expressed as

$$Q = P_v A \tag{3.106}$$

since $M = Qt_k$ = dipole moment.

With rising temperature, P_v is lessened as also Q. Now if temperature change ΔT occurs, charge changes by ΔQ. In fact, for small change in temperature, one can write

$$\frac{dQ}{dT} = \left(\frac{dP_v}{dT}\right)A = \phi A \tag{3.107}$$

where dP_v/dT is known as the pyroelectric coefficient ϕ. Because of the dielectric nature of the ceramic material, the capacitance developed due to the ceramic slice put between a pair of electrodes of area A has a value C so that the voltage change measured between the electrode pair, because of change in temperature, is given by

$$\frac{dV}{dT} = \frac{1}{C}\frac{dQ}{dT} = \frac{\phi A}{C} \tag{3.108}$$

ϕ is also a nonlinear function of temperature that becomes zero at the Curie temperature. Figure 3.51 shows the transducer scheme.

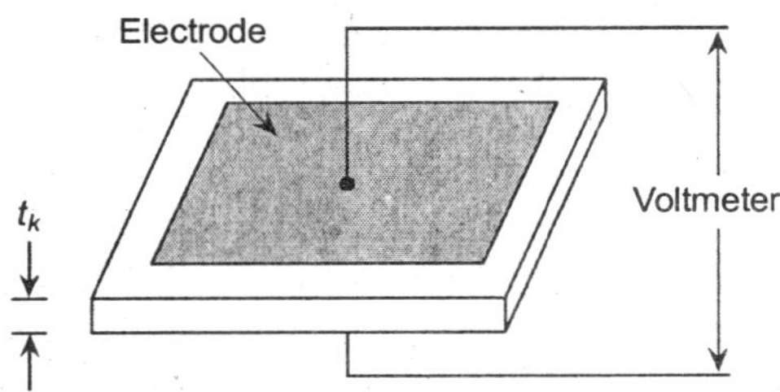

Fig. 3.51 Pyroelectric detector for voltage output.

Equation (3.108) shows ac nature of the pyroelectric device, which is more evident when the equation is recast as

$$\frac{dV}{dt} = \frac{\phi A}{C}\frac{dT}{dt} \tag{3.109}$$

Expressing this so as to measure current

$$i = \phi A \frac{dT}{dt} \tag{3.110}$$

This means that for measuring a constant radiance, it is necessary to chop the radiation to generate the variation required. A schematic view of the arrangement is shown in Fig. 3.52.

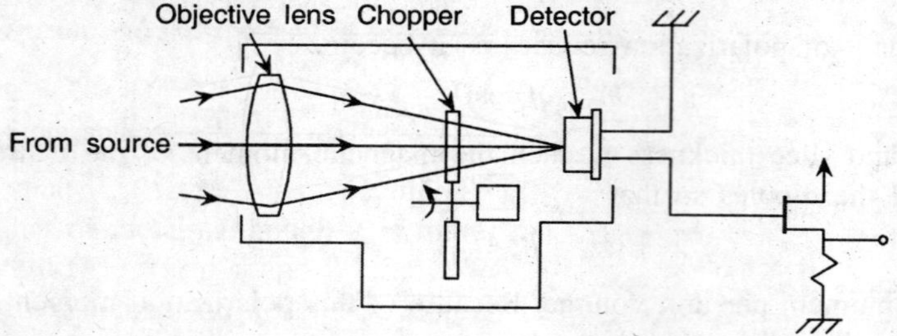

Fig. 3.52 Arrangement for a pyroelectric pyrometer.

The material would have its polarization retained within Curie temperature and it would not work as temperature sensor beyond that temperature. Some important and commonly used materials with constants are listed in Table 3.12.

Table 3.12 Characteristics of pyroelectric materials

Material	Density gcm^{-3}	Sp. heat C_v Jcm^{-3} K^{-1}	Dielectric constant (ε)	$\phi(\times 10^{-8})$ (Coulomb $cm^{-3}K^{-1}$)	Voltage responsivity $\phi/\varepsilon C(\times 10^{-10})$	ac resistivity (ohmcm)
Triglycerine Fluoberytate	1.66	1.68	15	2.1	8.3	—
Triglycerine sulphate (TGS)	1.66	2.5	35	4.0	4.6	1.7×10^{10}
$BaTiO_3$	6.0	3.0	200	7.0	0.57	6×10^6
Plumbum Zinc Titanate (PZT)	7.5	3.0	1400	2.3	1.35	3.2×10^8
Li_2SO_4, H_2O	2.06	2.3	12.3	1.17	4.1	9.8×10^{10}

The material is classified in terms of its Figure of merit F, defined as

$$F = \frac{\phi}{C_v \rho \varepsilon} \tag{3.111}$$

Figure of merit, however, is defined in different ways for different applications. For example, for application in video camera it is defined as

$$F_v = \frac{\phi}{\rho C_v \sqrt{\varepsilon} \tan \sqrt{\delta}} \tag{3.112}$$

where δ is the dielectric loss of the material. Such materials are widely used for photoradiation detection in IR region.

Other materials often used these days as temperature sensors are lithium tantalate ($LiTaO_3$) and polyvinyl fluoride plastic films. The pyroelectric sensors have good spectral response covering 0.001–1000 μm, that is, soft x-rays through infrared rays and have fast response time, sometimes of the order of nanoseconds.

3.13 QUARTZ CRYSTAL THERMOELECTRIC SENSORS

Single crystal SiO_2, otherwise known as quartz, is available in various modifications. The commonly available modification is α-quartz, consisting of three SiO_2 molecules in its elementary cell and is defined in terms of symmetry as class 32 or D_3. Depending on the fact that around its axis of symmetry called the optic axis, the plane of polarization of light rotates anticlockwise or clockwise and accordingly the quartz is named left or right. Usually, in certain form of coordinates, Z-axis is the optic axis, X-axis is the direction of electric axis and Y-axis is the mechanical axis.

Quartz has elastic, piezoelectric and resonating properties. Of these elastic as well as resonating properties are very important for using it as a thermal sensor. Quartz has low acoustic absorption and good elastic, mechanical, and chemical properties. The general conditions of propagation of elastic waves in an anisotropic, elastic body are due to Christoffel and a set of homogeneous equations were obtained as

$$\rho v^2 \varepsilon_j = \sum_{i=1}^{3} \psi_{ji} \varepsilon_i \tag{3.113a}$$

where

ε = particle elongation,
ρ = density of the material,
v = velocity of propagation, and
ψ = elastic stiffness proposed by Christoffel defined as

$$\psi_{ji} = \sum_{p=1}^{3} \sum_{q=1}^{3} C_{jpiq} \alpha_p \alpha_q \tag{3.113b}$$

where α's are direction cosines of propagation with respect to the crystal axis. For solution of Eq. (3.113a), it is necessary that

$$\begin{vmatrix} \psi_{11} - \rho v^2 & \psi_{12} & \psi_{13} \\ \psi_{21} & \psi_{22} - \rho v^2 & \psi_{23} \\ \psi_{31} & \psi_{32} & \psi_{33} - \rho v^2 \end{vmatrix} = 0 \tag{3.114}$$

The resonance frequency is given by

$$f_{nr} = \frac{n}{2h} \sqrt{\frac{C_r'}{\rho}} \left(= \frac{n v_r}{2h} \right) \tag{3.115}$$

where

suffix r indicates one of the three roots obtained from Eq. (3.114),
n is the order,
elastic stiffness C_r' depends on crystal cut, and
h is the thickness of quartz crystal plate.

Equation (3.115) indicates that the frequency of plate depends on C_r and h, and also on density and thickness of the metal electrodes. All these factors depend on temperature as does f. A truncated polynomial representing the f–T relation is

$$\frac{f(T) - f(T_0)}{f(T_0)} = \sum_{j=1}^{3} \alpha_{fj}(T - T_0)^j \tag{3.116}$$

where

T_0 is the reference temperature,

α_{fj} is the jth order temperature coefficient of frequency, and

f is the resonance frequency.

The coefficient α_{fj} can be obtained, as usual, as

$$\alpha_{fj} = \frac{1}{f_0}\frac{1}{j!}\frac{\partial^j f}{\partial T^j} \tag{3.117}$$

These coefficients are dependent also on the cuts of the crystal. A particular cut that has a maximum value α_{f1} is called the *HT-cut* which is a Y-cut of the type $(Y \times l)\ \theta = -4°$. Typical values are $\alpha_{f1} \approx 9 \times 10^{-5}\,\mathrm{K}^{-1}$ with $\alpha_{f1}/\alpha_{f2} = 1.5 \times 10^3$ K and $\alpha_{f1}/\alpha_{f3} = 3 \times 10^6$ K^2. Nonzero values of α_{f2} and α_{f3} produce nonlinear f–T relation. However, a coefficient cut is followed in practice for temperature resonating quartz sensor and this is, in short, known as *LC-cut*. For such a cut, $\alpha_{f1} \approx 4 \times 10^{-5}\,\mathrm{K}^{-1}$ and $\alpha_{f2} = \alpha_{f3} \approx 0$. This cut is represented as $(Y \times wl)\ \phi/\theta = -9.2°$. The HT- and LC-cut schemes are shown in Figs. 3.53 (a) and (b) respectively. LC-cut is obviously a double-rotated one, l is rotated by ϕ and w is rotated by θ.

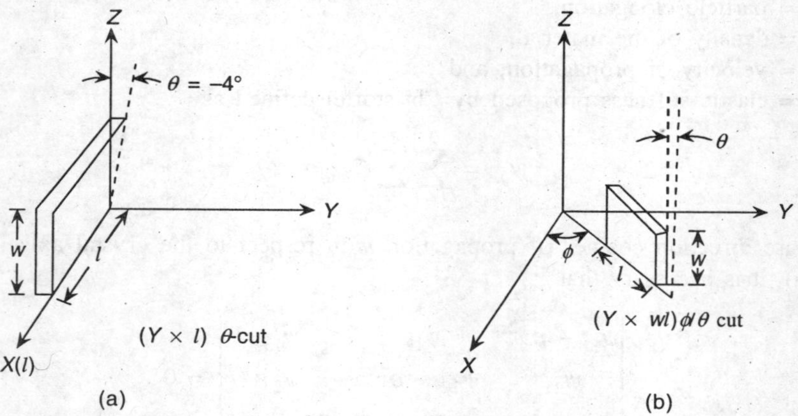

Fig. 3.53 Schemes of (a) the HT-cut crystal, and (b) the LC-cut crystal.

The crystals so cut are used as resonators with resonance frequency varying with temperature. Taking the first order variation only for LC-cut types, for example, the change in the $\{f(T) - f(T_0)\}/(T - T_0)$ slope is given by $f(T_0)\,\alpha_{f1} = f(T_0) \times 4 \times 10^{-5}$ Hz/K. For a resonant frequency of 20 MHz at $T_0 = 273$ K, the slope becomes 800 Hz/K.

The resonator is often in the form of a tuning fork. A resonator is coupled to an oscillator producing a base frequency $f(T_0)$. Its output is compared with a standard/reference oscillator and the beat or sum counted. The scheme is shown in Fig. 3.54. Another method is to divide the oscillator frequency down to Hz level so that it can be transferred by ordinary cable. High frequency requires coaxial cable for transmission. Figure 3.55 is a depiction of the same. The frequency is measured by counting the frequency 'pulses' (after a pulse former). The relevant equation is obtained from Eq. (3.116) as

$$t_{\text{int}} = \frac{N}{f(T_0)\left[1 + \displaystyle\sum_{j=1}^{3} \alpha_{fj}(T - T_0)^j\right]} \tag{3.118}$$

where N is the number of divisions or dividers and t_{int} is the time interval.

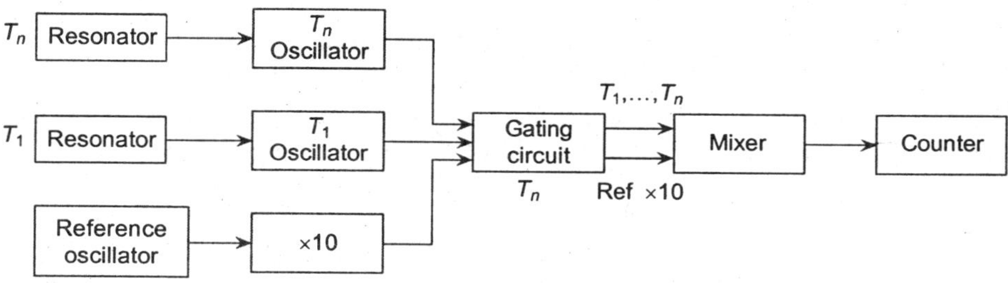

Fig. 3.54 Basic scheme of an n-channel crystal resonator type temperature meter.

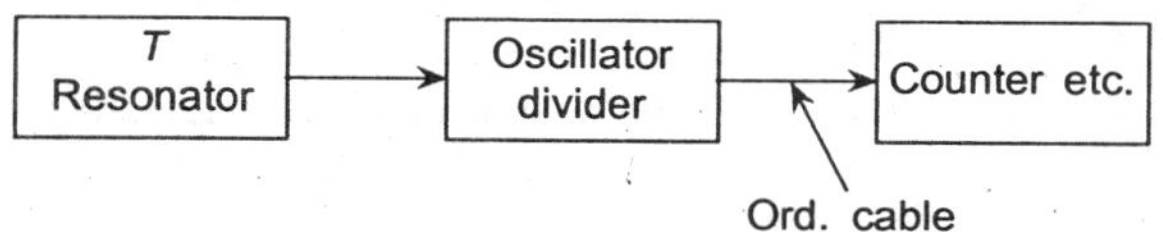

Fig. 3.55 High frequency transmission using oscillator, divider, and ordinary cable.

3.14 NQR THERMOMETRY

The basic principle of nuclear quadrupole resonance (NQR) thermometer is that the NQR frequency of certain nuclei varies with temperature. Only very specific nuclei are known to observe this. Common example is ^{35}Cl in KClO_3. The nucleus possesses an electric quadrupole moment which interacts with the electric field gradients generated by the surrounding ions in the compound lattice and the valence electrons of the nucleus. These interactions can cause atomic energy level to split into two components which can be excited by radio-frequencies and detected by the standard NQR technique.

The frequency corresponds to the separation of energy of the two components and is given by the relation

$$\nu_0 = \frac{eQ\dot{E}_a}{2h} \tag{3.119}$$

where

e = electron charge,

Q = electrical quadrupole moment,

$\dot{E}_a$ = electric field gradient tensor component along the principal axis, and

h = Planck's constant.

This is the frequency near 0 K and is seen to be independent of temperature; but with rising temperature, for example in KClO_3, ClO_3 ion has torsional vibrations and there occurs fluctuation in the orientation of the electric field gradient tensor resulting in variation of $\dot{E}_a$ and resonant frequency v. The frequency v is given by

$$v = v_0\left[1 - \sum_i \frac{3M_{ri}h}{4\pi\omega_i}\left\{\frac{1}{2} + \frac{1}{\exp(h\omega_i/(2\pi kT)) - 1}\right\}\right] \qquad (3.120)$$

upto about 50–60 K. In Eq. (3.120),

M_{ri} = reciprocal of the moment of inertia of the ith lattice node and
ω_i = angular frequency of the ith node of vibration of the lattice.

At higher temperatures, the lattice expands and anharmonic oscillations predominate. Equation (3.120) is then suggested to be modified for the range 60–300 K, as

$$v = v_0\left[1 - \frac{3k}{2M_{ri}\omega_i}\left\{\sum_i \frac{h\omega_i/(2\pi k)}{\exp(h\omega_i/(2\pi kT)) - 1}\right\}\right] - q(T) \qquad (3.121)$$

where $q(T)$ is a quartic polynomial in T. Table 3.13 shows some figures with sensitivity v at various temperatures for ^{35}Cl in $KClO_3$.

Table 3.13 Frequency sensitivity to temperature

Temperature (K)	Sensitivity (v/T) (kHz/K)
15	0.1
80	2.5
400	7.0

3.15 SPECTROSCOPIC THERMOMETRY

For special cases of temperature measurement in heated gases, plasma, flames, and stellar objects, spectroscopic techniques have long been in use. The principles may be different depending on the media and also in the way they can be probed. In some cases, atomic, ionic, or molecular spectral lines or bands are produced which have intensities that vary with temperature. Also, emission and absorption characteristics may change with temperature but measurement can be made provided a local thermodynamic equilibrium exists within the volume of the medium concerned. The range of such instruments is from 500 to 10^5 K.

Intensities of atomic, ionic or molecular lines are measured only relatively and the relevant relation is

$$\frac{I_1}{I_0} = \frac{A_1}{A_0}\left[\exp\left(-\frac{E_1 - E_0}{kT}\right)\right] \qquad (3.122)$$

where

I_1 represents the intensity of the line to be measured,
I_0 is the intensity of the reference line and A's are corresponding transition probabilities for the line as it is due to level changes, and
E_1 represents the energy of the level before the change.

Thus, knowledge of E's and A's is essential for computing T by relative measurement I_1/I_0. It can be shown that the measured temperature is given by

$$T = \frac{E_1/k}{\left\{ C - \ln\left(\dfrac{I_1}{A_1 v_1^4} \right) \right\}} \tag{3.123}$$

where C is a constant dependent on energy, intensity, and frequency of the reference line and v_1 is the frequency of the measured line.

One method that is used quite effectively is known as the method of line reversal and is similar to the disappearing filament type optical pyrometer described earlier. The spectral radiance from a standard source is spectroscopically viewed through the measured medium. This standard source is a calibrated variable temperature source. The spectral line arising in the medium would appear brighter or darker depending on whether the source temperature is lower or higher than the medium temperature. The correct temperature is obtained when the reversal occurs. Standard lines are selected from alkali metal resonance lines such as Na-D-line.

Comparison of spectral radiances of two beams arising from the same source—one moving along a single path through the medium, the other crossing the medium twice obtained by using a mirror (see Fig. 3.56), can measure temperature of the medium which is calculated through successive application of Planck's law. In another, variation of emission from the medium is compared with the absorption of a laser beam by the medium for the temperature to be computed. The emission must be at laser wavelength.

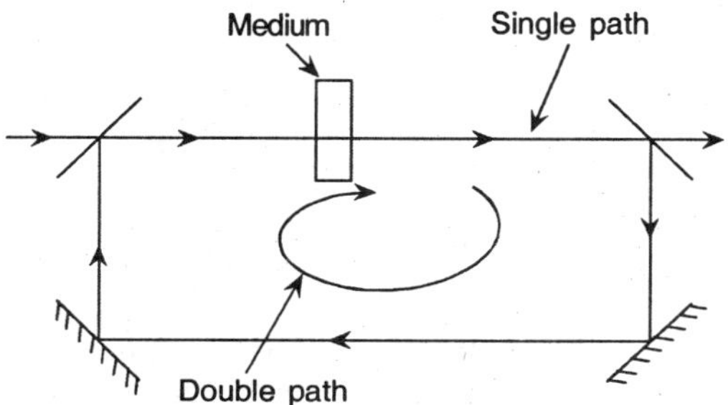

Fig. 3.56 Scheme of a double path spectrometric thermometer.

A very interesting measuring technique is Doppler broadening of lines. This is caused by the increased thermal motion of the radiating particles due to increasing temperature of the gaseous medium. The width of the line $\Delta\lambda_w$ is related to temperature as

$$T = \frac{M \Delta\lambda_w^2}{2.05\lambda^2} \times 10^{12} \tag{3.124}$$

where M is the molar mass and the numerical constant squared, $10^{12}/2.05$, is made up of gas constant and velocity of light. The technique is usable only in special cases.

Laser induced scattering is also being used to measure temperature and the 'coherent anti-Stokes Raman spectroscopy' (CARS) is specifically suitable for the purpose although it is very expensive as it requires coherence and three laser beams, two of identical frequencies, v_1, and the other of tunable lower frequency ν_2. The three beams intersect at the medium and when $\nu_1 \sim \nu_2$ is arround the Raman active resonance frequency, a coherent anti-Stokes signal of frequency $2\nu_1 - \nu_2$ is produced. By using wide band tunable frequency beam, anti-Stokes spectrum is generated and by measuring line widths, temperature can be computed comparing with standard file data.

It needs to be stressed that the transducers for spectroscopic detection are often the same as used in the total radiation or spectral radiation thermometry and optics is also similar.

3.16 NOISE THERMOMETRY

Noise temperature sensors are basically metal conductors in which random statistical thermal agitation of the electrons in the conduction band is measured following a thermodynamic relationship proposed by Nyquist. It states that with temperature above $0°K$, a voltage fluctuating statistically around zero is obtained across a passive element/network and its mean square value $\Delta \overline{V}^2$ is given by

$$\Delta \overline{V}^2 = 4kT \, \mathrm{Re}\{Z(f)\}\Delta f \tag{3.125}$$

where

$\mathrm{Re}\{Z(f)\}$ is the real part of the complex impedance $Z(f)$ of the element/network,
f is the frequency,
T is absolute temperature, and
k is the Boltzmann constant. The power spectrum or spectral power density of thermal noise is given by

$$P_s = \frac{\Delta \overline{V}^2}{\Delta f} = 4kT \, \mathrm{Re}\{Z(f)\} \tag{3.126}$$

The thermal noise or Johnson noise, is independent of the chemical composition, physical state of the substance, and the nature of the charge carriers. Hence, solid or liquid metals, semiconductors, electrolytes, and carbon films are all suitable for making sensors.

Practicability suggests that $\mathrm{Re}\{Z(f)\}$ becomes independent of frequency so that P_s is also independent of frequency. This is approximated by fixing an upper and a lower frequency limit f_u and f_d respectively so that $\Delta f = f_u - f_d$. It is to be noted, in such a situation, that replacing $\mathrm{Re}\{Z(f)\}$ by R and with all other terms in Eq. (3.125), it is the resistance R which is affected by environmental changes including material properties and a simple technique of measuring R would provide the scope for determination of environmental conditions. Because, these external influences can be taken care of by measurement, the noise thermometer (NT) shows no changes or drifts and ensures accuracy and stability.

Unfortunately, a major disadvantage with noise thermometer is its very low output and constraints in signal processing part. For a reasonable value of R and a frequency band of 0.1 MHz, the output is less than a microvolt. Also, the measured signal is stochastic in nature and its average/mean value is obtainable being a measure of the temperature. This averaging needs a delay.

Choice of resistance material is not arbitrary but its behaviour must be related, that is, matched with the transmission cable behaviour as also of electronic 'circuitry'. It has been found that over a wider temperature range and a considerable spatial distance between the sensor and the electronic circuitry, materials with low temperature coefficient of resistance show better performance. Ni–Cr (80–20) resistance alloy is one such material with a specific resistance variation of not more than 3% over a temperature change 5–1600 K. It provides good stability as well, because of a stable oxide layer over the surface in use. For higher temperature range, alloys with metals of high melting points are used such as W, Re, Pt, Mo and so on. These have high α-values and need be properly matched.

Possibility of designing a combined thermometer (CT), using a thermocouple (TC) and a noise thermometer (NT) has also been exploited quite extensively. This type of a sensor is known as *TC–NT sensor*. It consists of a pair of sheathed thermocouples with two hot junctions and a noise resistor between the hot junctions as shown schematically in Fig. 3.57.

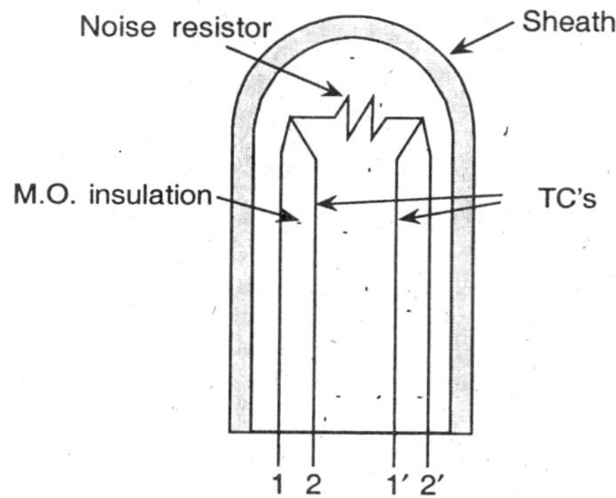

Fig. 3.57 Scheme of a TC–NT sensor.

Each individual couple measures the temperature in usual fashion. The noise resistor has, on the other hand, 4 lead wires (the thermocouple wires) and while measuring noise temperature, they can be arranged to eliminate the resistance of these wires by cross correlation. The thermocouple wires can have considerable resistances. Since noise voltage is fluctuating in nature, it can be separated out from its dc thermocouple output by using a suitable capacitor.

The resistance noise of the thermocouple wires and the noise of the amplifiers used produce uncorrelated signals and the cross-correlation of the signals so derived is eliminated. A typical scheme of measurement is shown in Fig. 3.58.

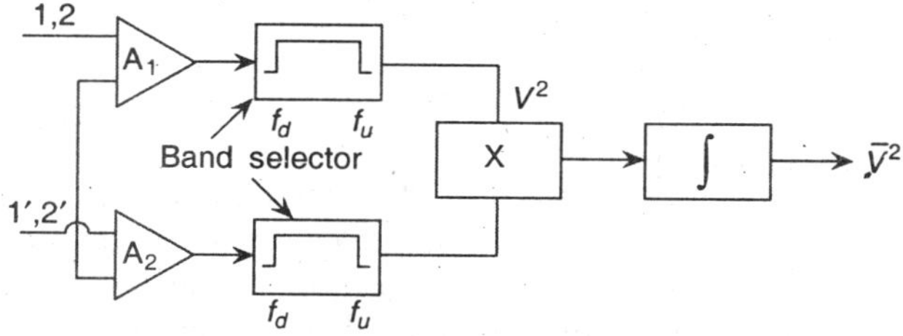

Fig. 3.58 Measurement scheme of the TC–NT thermometer.

For TC–NT, the resistor materials are required to be compatible with thermocouple materials. Table 3.14 shows the list of NT materials as compatible with corresponding thermocouples. Also shown in the list are the insulation materials (metal oxides).

Table 3.14 NT materials for given TC's

TC	NT materials	Insulation materials
Ni–Cr	Ni–Cr alloy	$\left\{\begin{array}{l} Al_2O_3,\ MgO \\ BeO,\ HfO_2 \\ BN \end{array}\right.$
Pt–Rh	Pt, or Rh, or Pt–Rh alloy	
W–Re	W, Re, or W–Re alloy	

As the noise voltage is stochastic in nature, measurement is susceptible to statistical errors. For the measurement bandwidth $\Delta f = f_u - f_d$, and measurement time t_M, the relative error is calculated as

$$E_R = \frac{\Delta T}{T} \sim (\Delta f t_M)^{-1/2} \tag{3.127}$$

Thus, for increasing precision, t_M is to be increased although the increase in precision is in proportion to square root of t_M. It should, however, be kept in mind that the equivalent noise resistance contributed by the measuring amplifier plays a crucial role and attempts have been made (and are still on) to keep this optimally low by special arrangement making, in the process, the system an elaborate laboratory method only. A practical noise thermometer system with a reference resistor R_R and a measuring resistor R_M is schematically shown in Fig. 3.59. Cable length is restricted on error consideration and for an error in the range 10^{-3} for a bandwidth range of 50 kHz. The upper limit of cable length is seen to be 50 m where an averaging time of 10 min is considered.

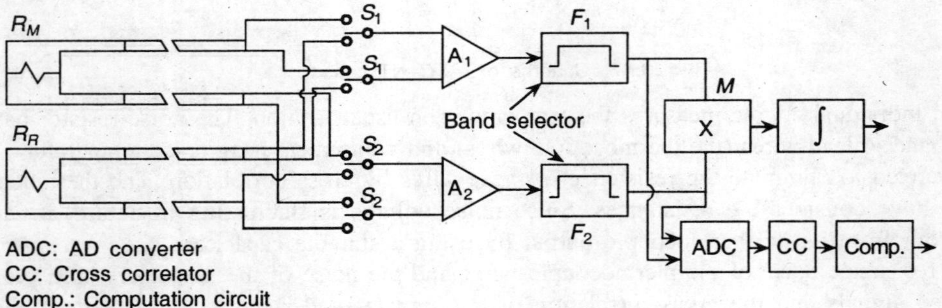

Fig. 3.59 Practical scheme of a noise thermometer.

3.17 HEAT FLUX SENSORS

There are certain special problems where instead of temperature, total heat flow or heat flux needs to be measured as in the case when heat is transmitted through a wall, specific heat, heat of melting or solidification, heat of hydration, heat of reaction and so on. Heat flux (or heat flow) measurement often involves measurement of temperature and computation is done to give the heat flow/flux.

One of the principles of heat flux measurement is dependent on the measurement of temperature difference across a thin layer added to the slab of a 'homogeneous' material through which the heat is transmitted. Figure 3.60 shows a scheme of heat flux transmission. If q is the heat flux, ΔT is the temperature difference, and λ is the thermal conductivity of the material of the thin layer of thickness w, then

$$q = \left(\frac{\lambda}{w}\right)\Delta T \tag{3.128}$$

Knowing λ, w and measuring ΔT, q can be easily calculated.

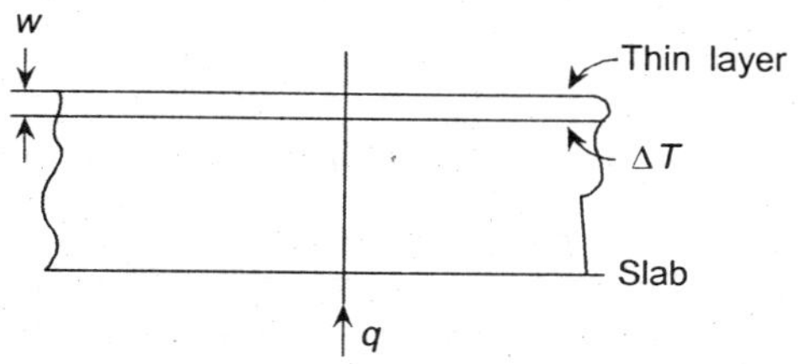

Fig. 3.60 Heat flux transmission.

One of the types of heat flux sensors consists of serially connected thermopiles over or embedded in a thin layer of rubber or plastic. If n is the number of thermocouples forming the thermopile, ΔT is the temperature difference across the layer, and E is the thermoelectric power of each thermocouple, then the thermopile output voltage V is given as

$$V = nE\Delta T \tag{3.129}$$

so that

$$q = \left(\frac{\lambda V}{wnE}\right) = K_1 V \tag{3.130}$$

More recent embedded/encapsulated type or the older 'half-plated wound wire' type sensors are shown in Figs. 3.61(a) and (b) respectively. Another type consists of a plastic or rubber disc with thermopile embedded in it as shown in Fig. 3.62.

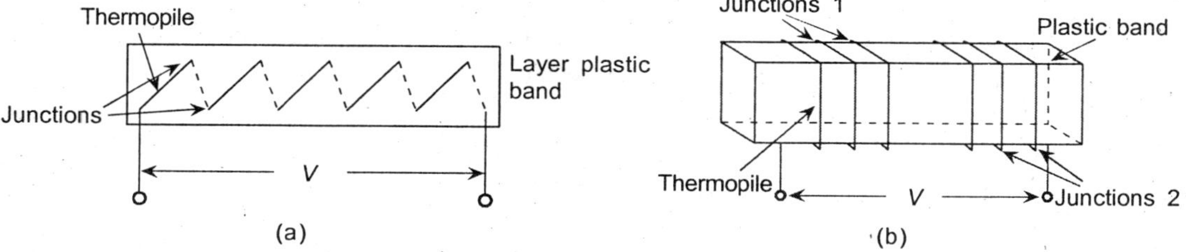

Fig. 3.61(a) Embedded type heat flux sensor, (b) half-plated wound-wire type heat flux sensor.

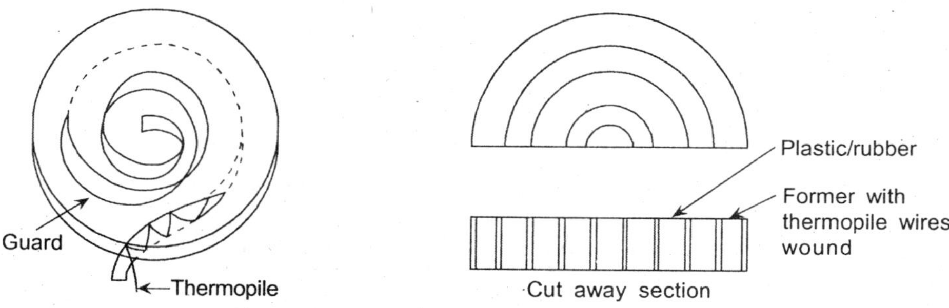

Fig. 3.62 Plastic rubber disc with embedded thermopile.

Thermopiles are, however, made using more recent techniques rather than winding dissimilar metal wires. There are substrate materials or core materials such as epoxy glass laminate, silicon, polyurethanes, and ceramics. Copper-constantan, copper-nickel or copper-silver thermocouples/thermopiles are produced on these by

(i)   plating and photoetching,

(ii) etching metallization layers or metal foils like strain gauges,

(iii) producing thick film and so forth. Such heat flux sensors have temperature range of 150–800°C and a sensitivity range of 5×10^{-9}–10^{-3} Vm²/W.

For measurement of heat flow rate across a surface, often a metal slug is embedded in it. If mass of this slug is m, a is the surface area, C is the specific heat of the slug, and temperature rise is ΔT measured by a thermocouple as shown in Fig. 3.63, then heat transfer rate q is given by the relation

$$aqdt = mCdT \tag{3.131}$$

$$(\text{cm}^2)(\text{W/cm}^2)(\text{s}) = (\text{kg})\ (\text{Ws/kg°C})(\text{°C})$$

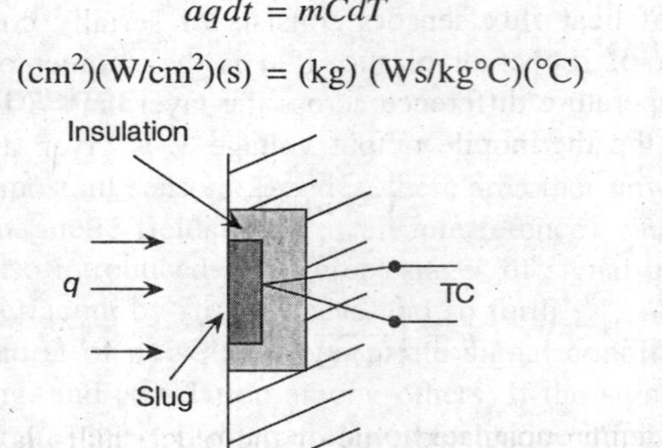

Fig. 3.63 Heat sensor using 'slug' technique.

However, heat losses are considered, an additional term for heat transfer rate due to heat loss given by $K_2(t_{sl} - t_{ca})$ is taken, where K_2 is the loss coefficient, t_{sl} is the slug temperature and t_{ca} is the casing temperature. Hence,

$$q = \left(\frac{mC}{a}\right)\left(\frac{dT}{dt}\right) + K_2(t_{sl} - t_{ca}) \tag{3.132}$$

Gordon gauge is another heat flux sensor and is not much different from the types described previously. Figure 3.64 shows a scheme of such a sensor. A thin constantan disc fitted to a heat

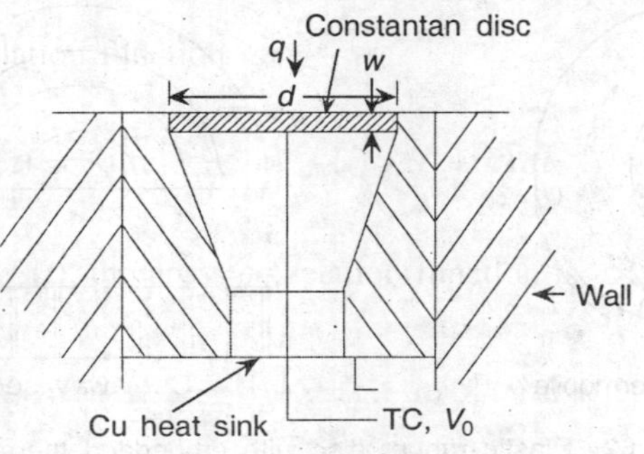

Fig. 3.64 Scheme of a Gordon gauge.

sink of copper is mounted as shown and a thermocouple is formed with this disc and the copper heat sink. Here, also the thermocouple voltage V is directly proportional to heat flux at steady state. However, the response of the sensor, as such, is of first order type, and is given by

$$\frac{V_0(s)}{q(s)} = \frac{Ed^2/(16w\lambda)}{\dfrac{\rho C d^2}{16\lambda}s + 1} \tag{3.133}$$

where disc thickness, diameter, mass density, specific heat, and thermal conductivity are w, d, ρ, C and λ respectively and E is the thermoelectric power of the thermocouple. Such a sensor can operate upto a temperature of 300°C and with a sensitivity as already specified.

REVIEW QUESTIONS

1. How did temperature scale evolve? What are the important primary temperature sensors? Are they all useful in industrial practice?

2. What are the commonly used liquids for vapour pressure thermometers? What are their ranges? Is the scale of such a thermometer linear?

3. A helix type bimetal thermometer uses equally thick strips of two metals of coefficients of expansion 1.7×10^{-6}/°C and 2×10^{-4}/°C respectively. The helix has four turns of mean diameter 1 cm. If the strip thickness is 1 mm, by what angle would the tip of the bimetal rotate for 1°C increase in temperature. Assume the other end of the bimetal element is fixed, and the Young's moduli of the strip materials are nearly equal.
   [*Hint:* One material has very low expansion coefficient than the other, so that, neglecting the lower one, increase in length would be
   $$\Delta l = 3.14 \times 4 \times 1 \times 2 \times 10^{-4} = 0.0025 \text{ cm } (= \pi n D \alpha)$$
   Hence, angle $= \Delta l/r = 0.0025/0.5 = 0.005$ rad $= 17$ min.]

4. How does an acoustic temperature work? What are the different types of acoustic sensors? Describe a pulse echo transit time technique of measuring temperature with requisite diagrams.

5. Describe two techniques of low temperature thermometry elaborating their principles of operation.
   [*Hint:* Discuss the nuclear type for $T < 0.1$ K, and, magnetic type for 0.01 K $\leq T \leq 80$ K.]

6. (a) How does conductivity change in a metal or in a semiconductor with change in temperature? Discuss with appropriate analysis.

   (b) Why platinum is preferred to other materials for making resistance thermometers?
   If at 25°C, a platinum wire has a resistance of 100 Ω, what length would be required for a wire of diameter 0.005 cm? The resistivity is 10.60 $\mu\Omega$cm. What would be its resistance at 500°C?
   [*Hint:* $R_{25} = 100 = 10.60 \times 10^{-6} \times (l/(3.14 \times 6.25 \times 10^{-6}))$, giving $l = 185$ cm.
   $R_{500} = R_{25}(1 + 0.00397 \times 475) = 288.575 \ \Omega$]

7. (a) What changes in the construction of the resistance thermometer are suggested when the wire diameter is lowered (say) to 0.001 cm? Sketch three convenient designs for such a case.

   (b) What material is preferred and how the construction is changed for low temperatures (0.5–30 K)? Describe with diagrams and response.

   (c) A platinum resistance thermometer of a resistance value of 100 Ω at 25°C is used to measure 520°C for which a current sent through the wire is seen to be 1 mA, but the temperature shown is 520.5°C. At some other current, it shows 520.2°C, what should this other value of current be?
   [*Hint:* $520 = 520.5 - (520.5 - 520.2)\{1^2/1^2 - I_x^2)\}$ or, $I_x = 0.632$ mA.]

8. (a) What is Seebeck effect? How does it develop and how has it been commercially exploited?

   (b) What different material-pairs are used for making commercial thermoemf generators? How are they designated?

   (c) What are MI thermocouples? What special advantage do these thermocouples have and what are their disadvantages?

9. How junction semiconductor diodes are used as temperature sensors? What are the factors on which the output of the sensors depend? How is the current output linearized in such a transducer? Obtain an expression for current output of a simplified commercial transducer of this type to show that it is proportional to the absolute temperature. What is a PTAT?

10. What are the different radiation laws on which radiation thermometers are based? How is emissivity important in ascertaining correct temperature of a target surface? What is the relation between emissivity of a surface and the radiation wavelength? Is there a fixed relationship? Discuss with diagrams.

11. What are the three important aspects of a radiation thermometer? Discuss their involvement in the measurement of temperature.

12. (a) What are the important detectors in a total radiation pyrometer? How are they characterized?

    (b) What are detectivity and noise equivalent power? How is detectivity related to wavelength?

13. (a) What are single waveband thermometers? Draw the optical scheme of such an instrument and discuss its operation.

    (b) An optical pyrometer is focussed to measure a target temperature of 2000°C but shows only 1950°C. Assuming no absorption losses in the intervening media, obtain the target emissivity. Assume $\lambda = 6.5$ μm and $C_2 = 1.434 \times 10^{-2}$ mK.
    [*Hint:* $\ln \varepsilon_\lambda = (1/T_t - 1/T_\lambda) C_2/\lambda = (1/2273 - 1/2223) \times 1.434 \times 10^{-2}/6.5 \times 10^{-6}$
    $= -0.0218$; hence, $\varepsilon_\lambda = 0.978$]

    (c) Describe the principle of ratio thermometer.

14. How does a pyroelectric ceramic work as a thermal sensor? A pyroelectric ceramic of coefficient 7×10^{-8} coul/cm^3/K, $l \times w \times t = 10$ cm $\times$ 4 cm $\times$ 2 cm, is placed between a pair of electrodes of size 10 cm $\times$ 4 cm. If it is heated at a rate of 0.5°C/s, what would be the change of voltage per unit time across the electrodes? Assume $\varepsilon = 200$.
[*Hint:* $dV/dt = 7 \times 10^{-8} \times 40 \times 2 \times 0.5/(200 \times 40) = 0.35$ nV/s.]

15. How are quartz crystal resonators used as temperature sensors? How is resonant frequency related to temperature?
What do you mean by HT–cut and LC–cut crystals? Draw the schematic diagram of a crystal resonator thermometer used in practice.

16. Describe the principle of spectroscopic thermometry. How are intensities of molecular lines measured and related to temperature? What is coherent anti-stokes Raman spectroscopy?

17. Describe with diagram a TC–NT sensor and show how is it used in temperature measurement?
What are the materials used for noise resistors and what are the corresponding TC materials?

18. Describe two types of heat flux sensors and briefly state how do they operate. Where are such sensors used in practice?

Chapter 4

Magnetic Sensors

4.1 INTRODUCTION

Magnetization serves a strong impact in changing the properties of certain materials. Magnetization changes or produces effects which are mechanical or electrical in nature and which are measurable. Also, optical energy may produce changes in magnetization characteristics of the materials. Not all such changes are easily transducible, perhaps, not at this stage of the state-of-the-art. Nevertheless quite a few are being conveniently used in developing sensors. Some of these are discussed here.

1. *Magnetic field sensors*—developed following 'ΔY effect' which, in effect, is observed as the change in Young's modulus with magnetization. The sensors are often termed as Acoustic Delay Line Components (ADLC).
2. *Magneto-elastic sensors*—based on the fact that in a longitudinal field, torsion given in a ferromagnetic rod changes its magnetization. This is known as 'Matteucci effect'.
3. *Magnetic elastic sensors*—produced using 'Villari effect' in which a tensile or compressive stress changes magnetization or affects magnetization in some way.
4. *Torque/force sensors*—'Wiedemann effect' is used to develop the torque/force sensors. In such sensors, torsion is produced in a ferromagnetic rod carrying a current when subjected to a longitudinal field.
5. *Magnetoresistive sensors*—becoming increasingly popular, are developed on the basis of 'Thomson effect' which is basically a change in resistance of specified materials with magnetic field impressed.
6. *Hall effect sensors* or *magnetogalvanic sensors*—perhaps, the most common and widely used type magnetic sensors. These operate on the fact that a crystal carrying a current when subjected to a magnetic field perpendicular to the direction of the current, produces a transverse voltage.
7. *Distance* or *proximity sensors*—developed based on 'skin effect' in which eddy current forces the current flowing through the interior of a material to move to its surface level.
8. *Wiegand* and *pulse wire sensors*—a specific type of material when subjected to pulse voltages under stress shows switching effect which occurs due to Barkhausen jump. This

is utilized to produce such sensors. The effect is called 'Sixtus-Tonks effect' after the experimenter who demonstrated the effect.

9. *Superconducting Quantum Interference Devices (SQUIDs)*—used for varying application areas, are based on the superconducting state specifically, 'flux quantization and Josephson effect'. These types of sensors have a resolution of the order of a few femtoTesla (fT).

10. *Magnetostriction*—a phenomenon, known over a century and a half now, has also been used in combination with piezoelectric elements for field measurements. This effect is also known as 'Joule effect' in which magnetization changes the shape of a ferromagnetic material body.

4.2 SENSORS AND THE PRINCIPLES BEHIND

In effect, the ΔY effect is an outcome of magnetostriction. Change in dimension due to magnetostriction in a material is actually caused by rotation of the magnetization. A demagnetized ferromagnetic material, when undergoes a mechanical stress, develops two types of stresses in it, namely (i) the plain mechanical elastic strain ε_s and (ii) the magnetoelastic strain ε_m which is the result of reorientation of magnetic domains by the applied stress S_a; thus, giving the Young's modulus of the demagnetized material as

$$Y_{dm} = \frac{S_a}{\varepsilon_s + \varepsilon_m} \tag{4.1a}$$

However, for a saturated sample no magnetoelastic strain is produced because no further reorientation is possible, and hence, the Young's modulus becomes

$$Y_{sm} = \frac{S_a}{\varepsilon_s}$$

where Y_{sm} is the Young's modulus of the saturated material.
So that,

$$\frac{\Delta Y}{Y_{dm}} = \frac{Y_{sm} - Y_{dm}}{Y_{dm}} = \frac{\varepsilon_m}{\varepsilon_s} \tag{4.1b}$$

Magnetoelastic strain is obviously a function of applied stress and the amount of anisotropy in the material.

The ΔY effect occurs in Ni–Fe based crystalline alloys and in some amorphous alloys such as Fe(40) Ni(38) Mo(4) B(18). Such an alloy can be used for making magnetic field sensors. One typical construction is shown in Fig. 4.1.

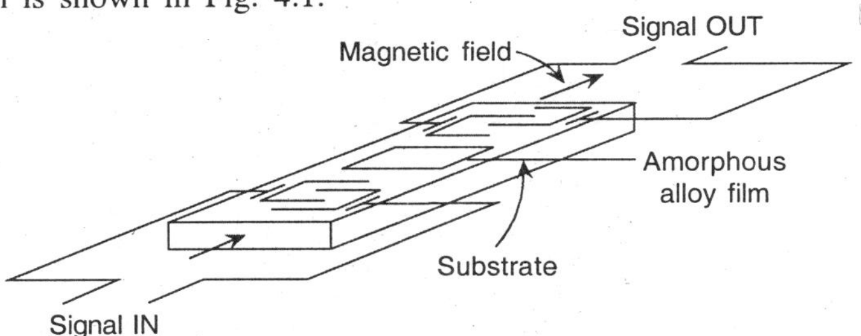

Fig. 4.1 Sketch of a magnetic field sensor using ΔY-effect.

A delay line using tunable surface acoustic wave components as transducers has been designed where the basic substrate is a piezoelectric material and an amorphous film spread as depicted in the figure. The sound velocity is given by the relation

$$v = \sqrt{\frac{Y}{\rho}}$$

where ρ is the density of the material.
The change in velocity Δv, in the film, is given by

$$\frac{\Delta v}{v} = \sqrt{\frac{\Delta Y}{Y}}$$

For the material mentioned, this figure varies depending on the annealing condition (temperature) of the material. A characteristic set of curves for the annealed alloy is presented in Fig. 4.2. The critical temperature is t_c which usually is close to the annealing temperature. Curve I in the figure corresponds to the magnetized state while curve II, the nonmagnetized state. In the alloy mentioned, a change in v of about 10% can be obtained at room temperature.

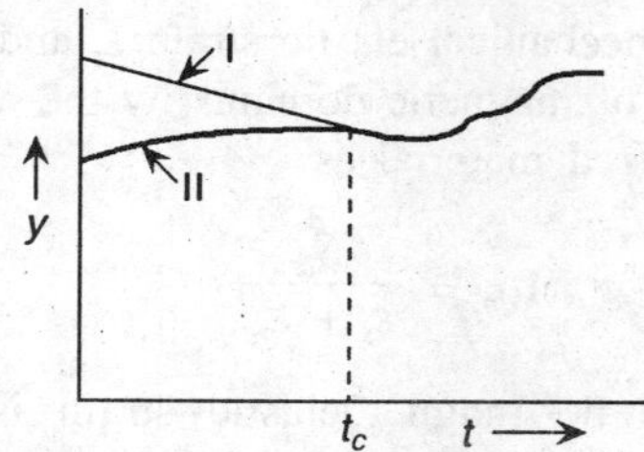

Fig. 4.2 Characteristic curves for annealed alloy: curve I corresponds to magnetized state, and, curve II to the nonmagnetized state.

Matteucci effect is closely related to a non-general form of Wiedemann effect. Villari effect is, to an extent, the reciprocal of the ΔY effect and is close to Wiedemann and Matteucci effects. Sensors developed on the basis of these effects are of similar kind in operation and construction. Wiedemann effect has subsequently been demonstrated to be a consequence of magnetostriction in a material and is tensorially related to this property. If longitudinal and transverse magnetostriction constants in a material are λ_l and λ_t, and the longitudinal and circular magnetic fields are H_l and H_r respectively, then the twist angle θ in a rod of length l and diameter d subjected to these fields, as shown in Fig. 4.3, is given by the relation

$$\theta = (\lambda_l - \lambda_t)\frac{4l}{d}\frac{H_l H_r}{H_l^2 + H_r^2} \tag{4.2}$$

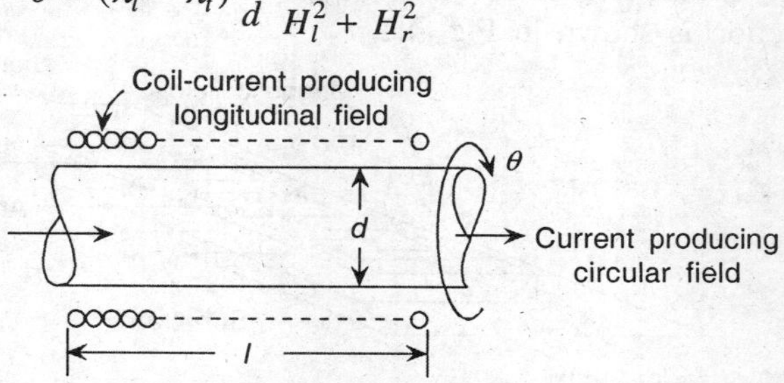

Fig. 4.3 Twist in a material, subjected to magnetic fields.

Wiedemann effect is used to make torque/force sensors, as has already been mentioned. The design principles are explained in Figs. 4.4(a) and (b). With a current I passing in direction as shown in figures and a torque produced in the rod of Fig. 4.4(a), an output voltage V_{ot} is obtained that gives a measure of the torque. Similarly in Fig. 4.4(b), V_{of} is the output voltage for the force in the balanced condition.

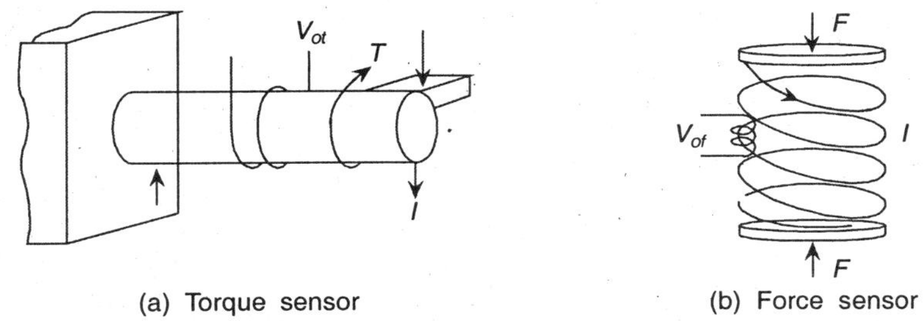

(a) Torque sensor   (b) Force sensor

Fig. 4.4   (a) A torque/force sensor using Wiedemann effect. (b) A typical force sensor using magnetostrictive effect.

The Wiedemann effect has two inverse effects:

(a) when a ferromagnetic rod which is circularly magnetized, is twisted, a longitudinal magnetic field is produced in it and

(b) when such a rod with longitudinal magnetization is twisted, a circular magnetic field is produced in it which essentially is the Matteucci effect.

Basically, for magnetoelastic sensors, there occur magnetoelastic interactions and conversion of elastic energy E_s into magnetic energy E_m takes place or vice versa. Various other types of energies are involved in a magnetic material apart from these two forms. In fact, E_m is proportional to the product of field strength and polarization and is often called the *magnetic field energy*; E_s is related to magnetostriction and is, therefore, called the *elastic stress energy*. In crystalline materials, depending on the material, there appears a *crystalline energy* E_{cr}. During magnetic annealing in materials, both crystalline and amorphous, an energy E_{ua} arises which is called the *uniaxial anisotropy energy* and depending on the demagnetization factor N, another type, the *shape anisotropy energy* E_N is developed.

The magneto-mechanical coupling factor K_{33} is defined as the ratio of the elastic stress energy to the total stored energy such that

$$K_{33} = \frac{E_s}{E_s + E_m + E_{cr} + E_{ua} + E_N} \tag{4.3}$$

For large values of K_{33}, E_m, E_{cr}, E_{ua}, and E_N must be small. In fact, soft magnetic materials such as Co–Fe and Ni–Fe alloys have very small E_{cr} and E_{ua} and a proper selection of the shape may render a small E_N so that effectively

$$K_{33} = \frac{E_s}{E_s + E_m} \tag{4.4}$$

If the saturation polarization is J_s, H is the field, that is, the coercive force, S is the stress applied to a strip, λ_s is the saturation magnetostriction, μ_0 is a magnetic constant, and α is the angle between the magnetization and specimen axis, then

$$E_m = -J_s H \cos \alpha \qquad (4.5a)$$

and

$$E_s = \frac{3}{2} \lambda_s S \sin^2 \alpha \qquad (4.5b)$$

If now the total energy $(E_s + E_m)$ is minimized, Eq. (4.4) can be written as

$$S = \frac{1}{\mu_r} \frac{J_s^2}{3 \mu_0 \lambda_s} \qquad (4.6)$$

where μ_r is the relative permeability.
From Eq. (4.6), we obtain

$$\mu_r = \frac{J_s^2}{3 \mu_0 \lambda_s Y \varepsilon} \qquad (4.7)$$

with strain ε and Young's modulus Y.

Figure 4.5 shows the theoretical (solid lines) and practical curves for soft alloy strip material with tensile load. In the ideal situation, μ_r decreases, becoming inversely proportional to S. Materials are available for which K_{33} can be as large as 0.95, for example, in amorphous Fe(81){B, Si, C}(19).

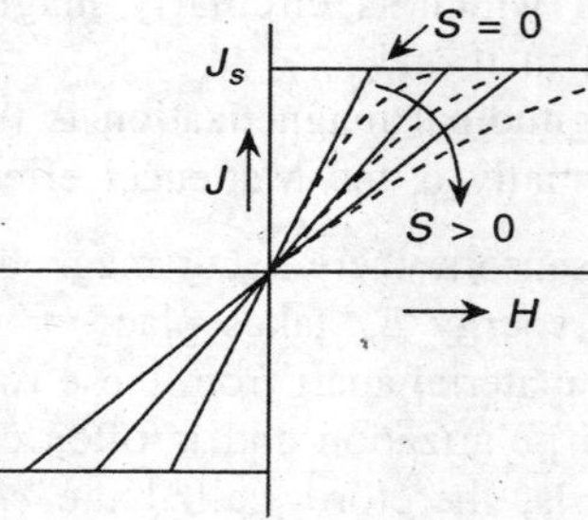

Fig. 4.5 J_s–H curves for soft alloy strip materials (soild lines denote theoretical values and dotted lines indicate practical curves).

Based on Villari effect, three basic types of magnetoelastic sensors may be designed, namel·

(a) the type in which mechanical loading is unidirectional so as to produce compression or tension and this changes the inductance or permeability with the specimen having predefined magnetic flux path, as in choke or coil type design,

(b) one in which mechanical loading changes the flux in two directions or in a plane as in circular rings or laminated cores, and

(c) the third in which loading changes the flux spatially, that is 3-dimensionally in torque transducers for shafts.

In the first type, strips or pot core sensors are used as shown in Figs. 4.6(a) and (b) when inductance variation is actually measured.

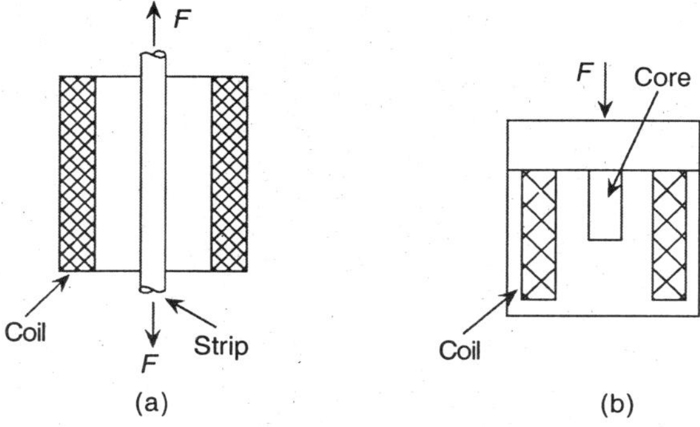

Fig. 4.6 Inductance variation sensors (a) strip and (b) pot core sensors.

The circular ring in the second type is deformed into an elliptical form as shown in Fig. 4.7 and a change in inductance of the ring or a change in voltage in the secondary winding ΔV gives the value of the load.

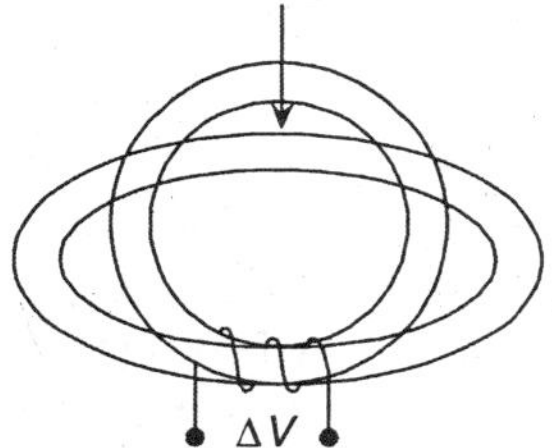

Fig. 4.7 Ring type sensor.

In case of laminated core load cells, isotropic magnetic materials are used which become anisotropic under stress due to varying deformation in longitudinal and transverse directions relative to the load axis and a change in voltage can be derived in the ring type design.

By far, the most important are the torque sensors. If the shaft material does not have the requisite magnetic properties such as magnetostriction, an additional magnetic coating on the shaft surface produces the desired mechanical stress on this surface that is to be measured.

In the solid or hollow cylindrical shafts, stress develops in two principal orthogonal directions, one compressive and the other tensile, each at an angle of $\pm 45°$ with the shaft axis in a screw-like fashion around the shaft as shown in Fig. 4.8. For a hollow shaft of inner and outer diameters D_i and D_o, the angle of torsion ϕ, the length of the shaft l, torque produced is given by

$$T = \frac{C\pi\phi}{32l}(D_o^4 - D_i^4) \tag{4.8}$$

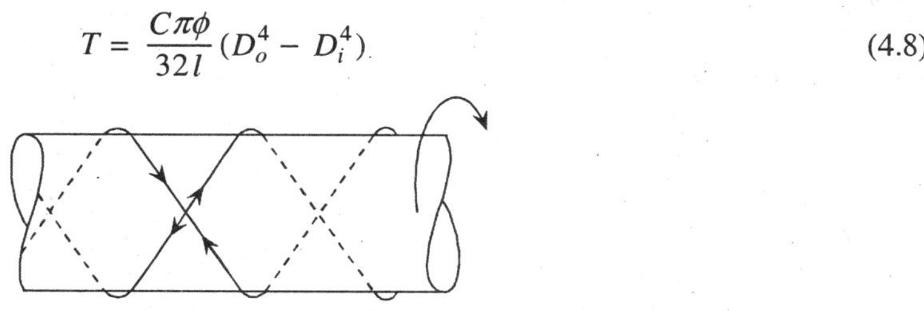

Fig. 4.8 Stress directions in hollow cylindrical shaft.

The maximum stress on the surface of the shaft is

$$S_m = \frac{16 D_o T}{\pi (D_o^4 - D_i^4)} \tag{4.9}$$

and maximum strain ε_m is

$$\varepsilon_m = \frac{S_m}{Y}(1 + v) = \frac{16 D_o (1 + v)}{\pi (D_o^4 - D_i^4) Y} T \tag{4.10}$$

where v = Poisson ratio.

The magnetoelastic interactions because of the complex stress–strain paths can be picked up by complex arrangement of the pick-up coils as it is not possible to arrange detectors so that change in permeability μ, along the principal stress axis is directly picked up. But it must be remembered that there is symmetry in stress–strain paths and coils can be arranged so that field and signal detection at ±45° to the principal stress axis can be made. Two types of designs are known for the purpose—(a) Yoke coil type and (b) the cylindrical coil type—which are mounted coaxially with respect to the shaft.

4.2.1 The Yoke Coil Sensors

The yoke coil type sensors can further be subdivided into (i) The cross transductor shown in Figs. 4.9(a) and (b) the four branch type torque sensor shown later in Fig. 4.10(a). In the former, the U-shaped magnetic pole pieces are mounted in a crossed fashion facing the shaft surface to provide narrow air gaps. One U-branch is used as primary with two coils, P_1 and P_2 excited by an ac supply, the poles are arranged to be along the shaft axis and the other branch is mounted at right angles to it with two secondary sensing coils S_1 and S_2 and poles at right angles to shaft axis. Without torsion, the surface magnetic flux pattern so obtained is symmetrical as shown in Fig. 4.9(b) and with torsion, if magnetostriction is positive (λ_s positive), flux density increases in the direction of tensile stress and decreases in the direction of compressive stress and the flux pattern gets distorted as shown in Fig. 4.9(c). The secondary pole pieces (S_{1p} and S_{2p}) then face unbalanced magnetic potential and a flux difference, resulting in an induced voltage in the coils S_1 and S_2, occurs which obviously is a function of the torque.

The equivalent magnetic circuit of the system is shown in Fig. 4.9(d). Inductive impedances are represented as X, subscripts t and c refer to tensile and compressive stress conditions and l is used for leakage fluxes in the primary pole pieces. In the balanced condition, X_t's and X_c's are equal hence, balancing the bridge. Under torque when X_t's $\neq X_c$'s the balance of the bridge is disturbed. Ohmic resistances allow for eddy losses.

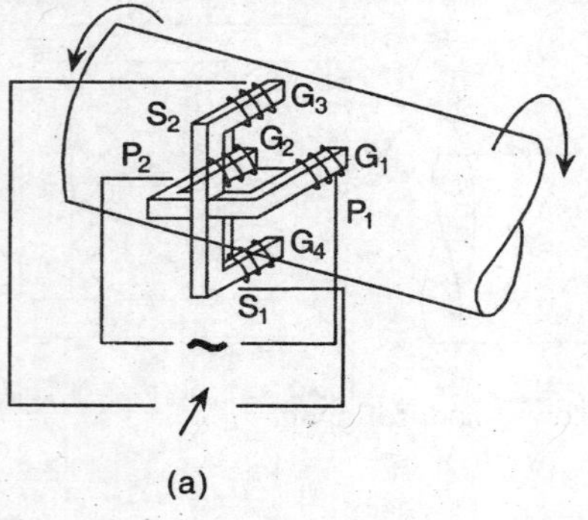

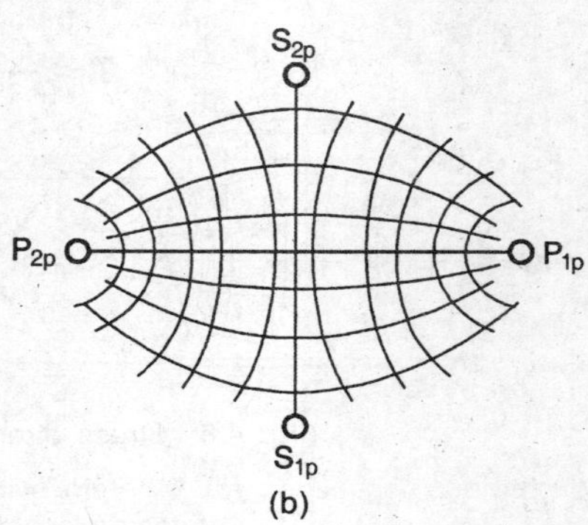

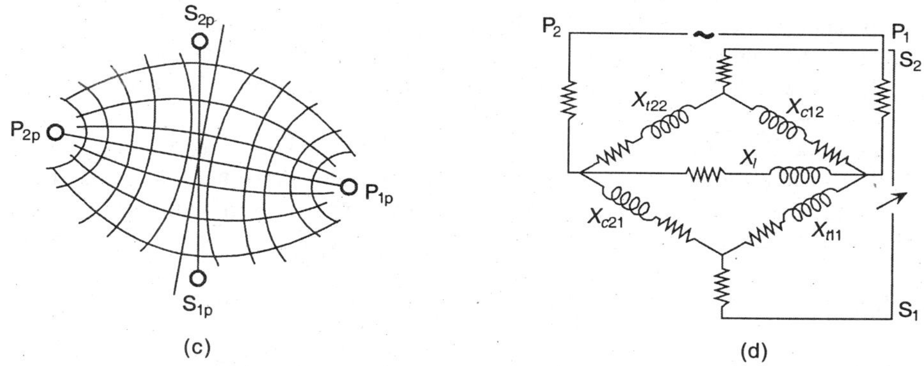

Fig. 4.9 (a) The cross transductor yoke coil type sensor, (b) The surface magnetic flux pattern without torsion, (c) The pattern with torsion, (d) Equivalent 'magnetic' circuit.

The four branch design is shown schematically in Fig. 4.10(a). At centre is the excitation pole provided by the exciter coil E_x. Four corner poles, two on the tensile stress lines and two on the compressive stress lines, are arranged with the sensing coils 1, 3 and 2, 4. Air gap between the coil cores is less than 1 mm. The permeability dependent reluctances shown by X_t's and X_c's and eddy losses by resistances are represented in the magnetic circuit diagram of Fig. 4.10(b). The four sensing arms are now arranged so that outputs from coils 2 and 4 are deducted from outputs of coils 1 and 3.

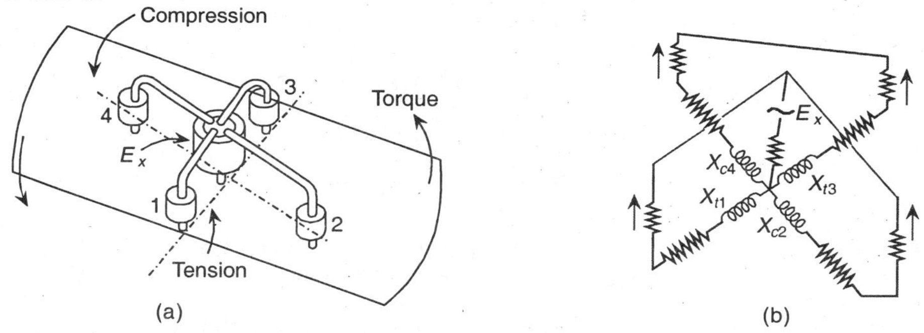

Fig. 4.10 (a) The four branch type sensor, (b) The equivalent magnetic circuit.

Depending on the material, the sensitivity of the transducer system changes. For example, for the four branch type design, sensitivity varies from 1 mV/Nm to 2.5 mV/Nm when material is C15 steel or Cr–MoV steel. In general, torque sensitivity decreases with increasing hardness for a ferromagnetic material. This varying sensitivity problem can be solved by coating the shafts or providing a sleeve on the shafts with a ferromagnetic material of satisfactory and uniform magnetoelastic property.

Other parameters governing the sensitivity are the excitation frequency—which is chosen on the basis of (a) transient requirements, (b) shaft speed, (c) shaft material, (d) oscillator power and, of course, (e) the output signal, besides, the air gap between the pole faces and shaft surfaces.

Frequency can be chosen on the basis of the counts specified in the three different ranges:

- Power range: 50–60 Hz,
- Audio range: 400–20,000 Hz, and
- Radio range: 100–200 kHz.

The gap changes due to radial shaft displacement. Also, magnetic inhomogeneities at the shaft surface tend to change the torque signal. This is overcome to a large extent by ring type torque transducers which can again be designed either on the principle of (i) cross type, or, (ii) four branch type sensors.

The cross type design consists of three pole rings with coils arranged on it for excitation and pick-ups. The middle ring holds the excitation coils and the outer ones hold the sensing coils. Figure 4.11(a) shows the arrangement of the rings on the shaft with 8 poles per ring as shown in Fig. 4.11(b) whereas Fig. 4.11(c) depicts the pole faces on the evolved ring and also the winding schemes. P poles are displaced by a 1/2 polepitch relative to S poles and are equidistant so that the functioning as in the cross type of Fig. 4.9(a) can be maintained.

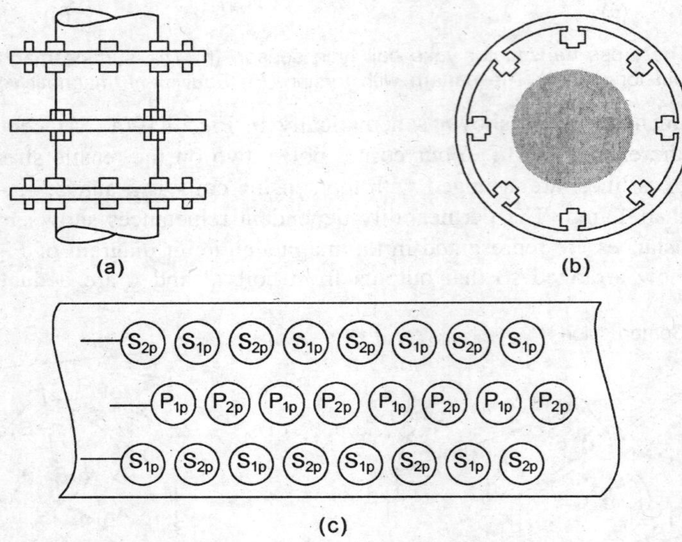

Fig. 4.11   Ring type torque transducer: (a) arrangement of the rings on the shaft, (b) the cross sectional view showing 8 poles per ring, and (c) pole faces on the evolved ring.

The four branch type design may, similarly, consist of a number of coils enclosed in common ring held to surround the shaft. A scheme with four branch type design is shown in Fig. 4.12(a) with the excitation poles in the centre portion and the sensing poles on the outside portions of the ring (as shown in Fig. 4.12(b)).

All magneto-elastic sensors are required to be calibrated with requisite standards.

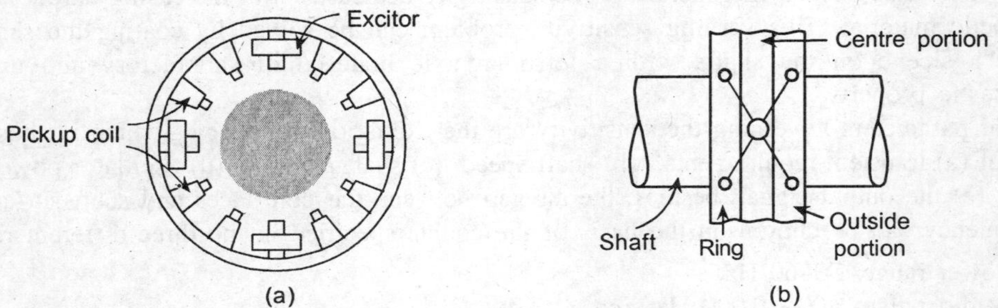

Fig. 4.12   (a) Scheme of four-branch type ring design, (b) sensing poles shown on the outside portion of the rings.

4.2.2 The Coaxial Type Sensors

The coaxial type design largely overcomes the effects of material inhomogeneity and gap length variation with less complexity than the ring type design. Over the last two decades, since its inception in the late seventies, this type has undergone many modifications and improvements.

The torque sensor, in its current improved form, is based on surface magnetic anisotropy of the shaft, intentionally introduced in such a way that the direction of such anisotropy is in the direction of the main torsional stresses to be produced in the shaft under loading, that is, effectively at ±45° to the principal stress axis of the shaft. This imposed anisotropy on the shaft surface is being excited by a coil around the shaft and magnetic flux is produced in this 'preferred' direction. Two identical secondary coils connected differentially would now be induced by the linkage from the primary through this ±45° anisotropy-induced magnetization. Under no torsion, equal induced voltages produce zero output from the secondary. In the presence of torsion, however, tensile and compressive stresses occur and flux conduction is either aided for compression or opposed for tension or vice versa depending on the negative or positive magnetostriction and the arrangement of the anisotropy direction in the two parts as illustrated in Fig. 4.13. Consequently, there is increase in flux in one of the secondary coils compared to the other giving a non-zero output voltage V_o for a given V_i. The magnitude of this voltage is dependent on the torque while its sign depends on the direction of the torque for a given set up.

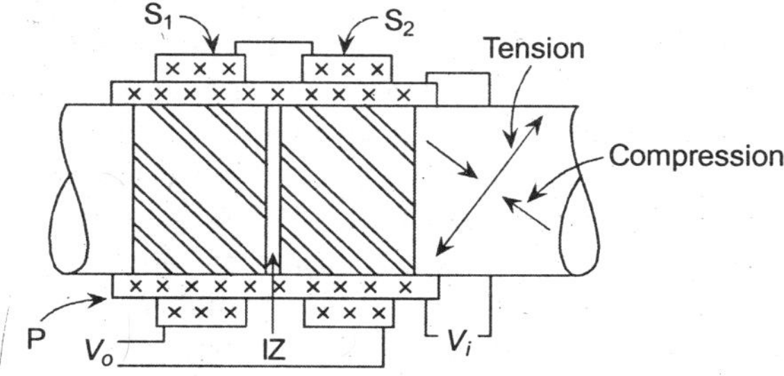

Fig. 4.13 Coaxial torque sensor.

The surface anisotropy, as mentioned, is produced by

  (i)   mounting magnetic strips on the shaft at ±45°,
 (ii)   mounting magnetic foil on the shaft in a pre-stressed condition of the shaft, or
(iii)   pre-stressing a magnetic foil and then mounting it on the shaft by a special ring arrangement, and so on.

This method has been applied for shafts with diameters varying from 1 cm to 10 cm and a torque range of 10–5000 Nm producing a strain in the range 0.005–0.057 with symmetry and linearity within ±2%.

The magnetic strip material may have a composition of amorphous soft materials such as Co(75) Si(15) B(10), Co(68) Ni(10) B(14) Si(8), and so forth.

As already mentioned, torque measurement is influenced by many factors—two major ones being variation of air gap and shaft surface inhomogeneity (as has already been discussed). The other factors that affect torque measurement are:

(i)   shaft speed, which tends to decrease output signal because of enhanced skin effect but this, in turn, depends on shaft surface material, excitation frequency, and amplitude. Measuring circuit can be designed to compensate for the variation, the effect is insignificant for high frequency excitation and ferromagnetic amorphous surface coating,

(ii)   axial displacement for coaxial design, which is often reduced by separating the active surface zones by an inactive zone (IZ), in Fig. 4.13, splitting the primary coil into two, and placing them closer to the secondary coils,

(iii)   temperature: mainly comes as a temperature coefficient of magnetization and expansion and reduced largely by choice of material. Circuit compensation is easier in this case.

4.2.3   Force and Displacement Sensors

Magneto-elastic sensors are easily adapted to force-sensing in the load cell form. Or in the complex structure, a component can be considered as the sensing element.

Magnetic load cell may be a stress sensitive solid cylinder made of a Mo–Permalloy, Ni–Fe alloy, or Al–Fe alloy. A winding on the cylinder is done on concentric grooves made in it. With lengthwise compression in the cylinder, the cross-sectionwise stress distribution is considered to be uniform which is ensured with large l/d ratio. The decrease in inductance under compression is measured by an ac bridge.

Of more recent designs is the transformer type magneto-elastic force transducer named as the *pressductor*. It essentially consists of a stack of transformer core sheets provided with four holes symmetrically made on the diagonals of the stack. Each diagonal pair of holes is provided with windings so that the two windings cross each other at right angles. One winding acts as the primary with a supply given to it. With the stack unloaded, there is no magnetic linkage between the two windings while on loading the stack, a compressive stress is developed in the core sheets. As a result, the sheet permeability becomes anisotropic and the flux paths are distorted. As a consequence, the secondary winding is magnetically linked with the primary and a voltage is induced in it which, obviously is proportional to the load or force. The situation is explained in Figs. 4.14(a) and (b).

Fig. 4.14   Schemes of pressductor (a) without force, and (b) with force applied.

The shape of the core sheets is important for proper distribution of the force/stress so that magneto-elastic sensing is appropriate. Various shapes are proposed for different applications. Round and rectangular designs are most common. For shearing force measurement, a design shown in Figs. 4.15(a) and (b) has also been used. Hole locations are also changed here. The frequency usually is 50 Hz, but with special material, it can lie in the range 1–10 kHz.

Pressductors are available in single unit upto 25 kN to 5 MN. Multiple unit designs exist for higher loads upto about 15 MN. Good quality elasto-sensing load cells are available with accuracy of ±0.1%, linearity ±0.1%, and, precision 0.1%. Operating temperature range is 20–70°C and the cells can take 3 times the nominal load without damping the performance characteristics.

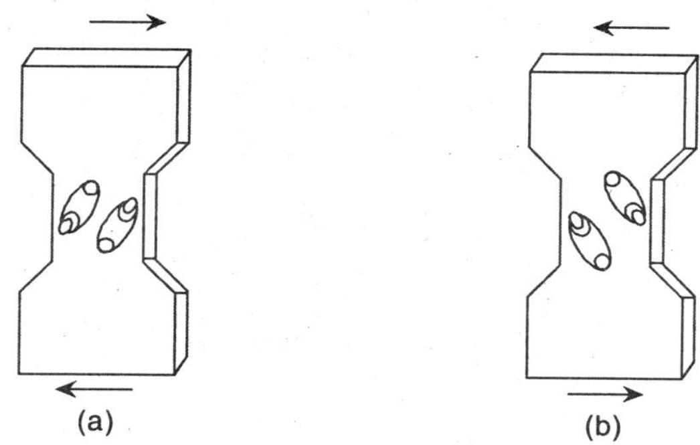

Fig. 4.15 Pressductors used for shearing force sensing (a) and (b) force in two directions opposite to each other.

It must be mentioned at this stage that the cross yoke type or the four branch type sensors can be used for force measurement or stress analysis with the sensing element making no contact with test specimen. A typical scheme is shown in Fig. 4.16. Displacement and position sensors are quite common in process industries these days which use differential transformer principles as in LVDT, proximity sensors, and other inductive sensors, discussed briefly in Chapter 2.

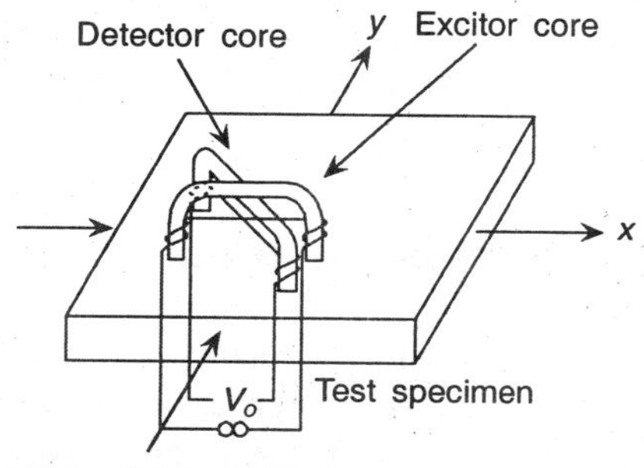

Fig. 4.16 Four branch type displacement sensors.

A ring type design, somewhat similar to that of Fig. 4.7, utilizes, however, the magnetostrictive effect for measurement of displacement or position in relation to force working the ring within elastic limit. The ring is made of amorphous metal. Figure 4.17 schematically represents this design. A circular ring made of single/multilayer ribbons of diameter 5–10 mm with, usually, positive magnetostriction is used for the purpose. The ring is used as a transformer and as a spring. When it is deformed into an ellipse, its hysteresis loop is also flattened and hence, for an input V_i the output V_o reduces considerably. When the material works within the elastic limit and the load is released, the shape is regained by the spring and the secondary voltage returns to the original value. Extensometers can be developed on this principle. The material Fe(78) Mo(2) B(20) is often used for the purpose, a four-layer core ring of diameter

10 mm is common with a displacement upto 5 mm. Linearity for this sensor is about 2% and temperature stability as good as 0.02%/°C, that can be used upto 90°C.

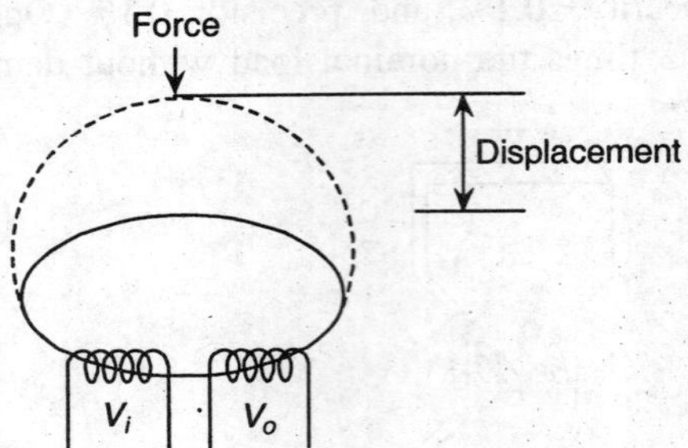

Fig. 4.17 Ring type displacement sensor.

4.3 MAGNETORESISTIVE SENSORS

Magnetoresistive effects are observed in metals specially in ferromagnetic types, and in such cases, it is known as *anisotropic magnetoresistive effect*. In semiconductors, this effect is sometimes called the *geometrical magnetoresistive effect* particularly for short samples. In the former case (ferromagnetic materials), the effect is due to the fact that under magnetic field, there occur different shifts of energy levels for electrons having negative and positive spins resulting in a change in density of states at the Fermi level. In the latter, the magnetic field restricts the travel length of a carrier between the collisions along the direction of the field because the current density vector is rotated with respect to the field and hence, the current lines become longer and the resistivity increases.

4.3.1 Anisotropic Magnetoresistive Sensing

This type of magnetoresistive effect can be analyzed taking into account the complex ferromagnetic behaviour. However, for sensing purposes, knowledge of the relations between the direction of magnetization and resistivity is sufficient as also between the magnetization-direction and external fields.

If the angle between the direction of internal magnetization **M** and that of current in the sample **I** is ϕ, then, the resistivity is given by

$$\rho(\phi) = \rho_\alpha + (\rho_\beta - \rho_\alpha) \cos^2\phi \tag{4.11}$$

where

ρ_α is the value of ρ for $\phi = 90°$ and

ρ_β is the value of ρ for $\phi = 0°$.

The quantity $(\rho_\beta - \rho_\alpha)/\rho_\alpha$ specifies the magnetoresistive effect or its coefficient which, in general, is positive and quite large.

Obviously, $\rho(\phi)$ is not scalar and should produce an electric field $\mathbf{E}_a$ perpendicular to the external field $\mathbf{E}_b$ to generate a current density $\mathbf{J}_b$. $\mathbf{E}_a$ is in the J–M plane which is the plane of the ferromagnetic material, but perpendicular to **J** so that

$$\mathbf{E}_a = J_b\, \Delta\rho\, \sin\phi\, \cos\phi \tag{4.12}$$

where

$$\Delta\rho = \rho_\beta - \rho_\alpha$$

If now, a bar of length l, width w and thickness t is considered with current I flowing along the length, from Eq. (4.11), the resistance of the bar is

$$R(\phi) = \frac{\rho_\alpha l}{wt} + \left(\frac{\Delta\rho l}{wt}\right)\cos^2\phi = R + \Delta R\cos^2\phi \tag{4.13a}$$

and the voltage drop V_b is

$$V_b = \frac{\rho_\alpha Il}{wt} + \left(\frac{\Delta\rho Il}{wt}\right)\cos^2\phi \tag{4.13b}$$

From Eq. (4.12), V_a is calculated as

$$V_a = \frac{\Delta\rho I}{t}\sin\phi\,\cos\phi \tag{4.14}$$

The effect that produces $\mathbf{E}_a$ is called the planar Hall effect. V_a is planar Hall voltage which is dependent on the sign of ϕ while the magnetoresistive voltage V_b does not depend on the sign of ϕ. If the swing ratio $l/w \gg 1$, the ratio of the instantaneous parts of V_b and V_a or the swing ratio l/w is much larger. This makes $V_{bi} \gg V_{ai}$ so that planar Hall voltage is important in small length samples only.

Ferromagnetic materials have high internal magnetization because of exchange coupling at the quantum mechanical level. This makes the electron magnetic moments parallel particularly in small areas known as domains which are separated by walls since different domains have parallelism in different directions. In thin films with thickness of film smaller than domain length, magnetization lies in the film plane only unlike that in the bulk material. This is specially true for soft ferromagnetic materials.

Under the influence of an external field $\mathbf{H}$, magnetization $\mathbf{M}$ rotates which can be calculated by evaluating the energy density ε in terms of the angle ψ between $\mathbf{M}$ and the axis of lowest energy called the *easy axis*.

For all practical purposes, the film can be considered to possess an elliptical shape so that Fig. 4.18 gives the geometry of the assumed structure with the major axis of the ellipse as the

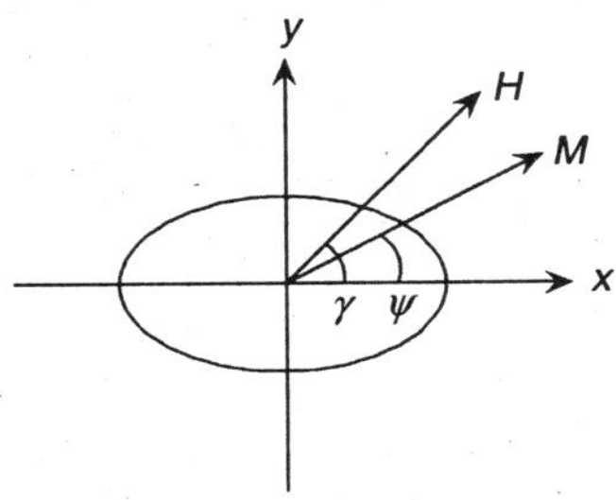

Fig. 4.18 Assumed geometry of the thin film structure.

easy axis. Energy ε has three components:

(i) the field energy

$$\varepsilon_H = -HM\cos(\gamma - \psi) \tag{4.15}$$

(ii) the anisotropy energy

$$\mathcal{E}_A = K_m \sin^2 \psi \tag{4.16}$$

where K_m is a material constant.

(iii) the demagnetization energy (because free poles are created at the edges which are enumerated by demagnetizing field, H_d)

$$H_{dx} = -N_x M \cos \psi \tag{4.17}$$

$$H_{dy} = -N_y M \sin \psi \tag{4.18}$$

where $N_x + N_y + N_z = 1$, with N_j's as the demagnetizing factors in the j direction. The demagnetization energy is

$$\mathcal{E}_d = \frac{1}{2} N_x M^2 \cos^2 \psi + \frac{1}{2} N_y M^2 \sin^2 \psi$$

$$= \frac{1}{2} N_x M^2 + \frac{1}{2} N_d M^2 \sin^2 \psi \tag{4.19}$$

where $N_d = N_x - N_y$

Thus, the total energy density ε is

$$\varepsilon = \mathcal{E}_H + \mathcal{E}_A + \mathcal{E}_d \tag{4.20}$$

The anisotropy material constant K_m is given by anisotropic field H_a as

$$H_a = \frac{2K_m}{M} \tag{4.21}$$

and H_a, combined with demagnetization field $H_d = -N_d M$ gives the characteristic field

$$H_a + H_d = H_0 = H_a - N_d M \tag{4.22}$$

so that combining Eqs. (4.20), (4.21), and (4.22)

$$\varepsilon = K_m \sin^2 \psi + \frac{1}{2} N_d M^2 \sin^2 \psi + \frac{1}{2} N_x M^2 - HM \cos(\gamma - \psi)$$

$$= \frac{1}{2} H_a M \sin^2 \psi + \frac{1}{2} N_d M (M \sin^2 \psi) + \frac{1}{2} N_x M^2 - HM \cos(\gamma - \psi)$$

$$= \frac{M}{2} \sin^2 \psi (H_a + H_d) + \frac{1}{2} N_x M^2 - HM \cos(\gamma - \psi)$$

$$= \frac{1}{2} H_0 M \sin^2 \psi + \frac{1}{2} N_x M^2 - HM \cos(\gamma - \psi) \tag{4.23}$$

Differentiating ε with respect to ψ and setting it equal to zero, we get

$$\sin \psi = \frac{H \sin \gamma}{H_0 + H \cos \gamma / \cos \psi} = \frac{H_y}{H_0 + H_x / \cos \psi} \tag{4.24}$$

for the angle that M makes with the lowest energy.

The permissible rotation of magnetization is less than 90°. In fact, it varies as ±30°.

The basic magnetoresistive sensor is a thin film of appropriate alloy which is obtained by etching to get a length l, width w, and thickness t and electrodes are then attached at length ends. Usually, $l \gg w \gg t$. Current I is allowed to flow along the easy axis, with the length as in Fig. 4.19. For a long flat rectangle, the demagnetization factor N may be given by

$$N = \frac{t}{w} \tag{4.25}$$

for $N \ll 1$.

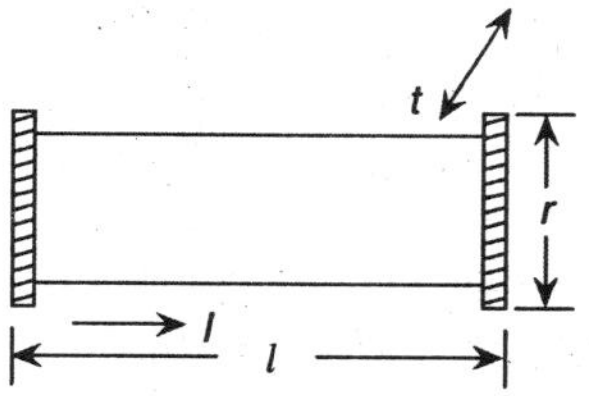

Fig. 4.19 Basic scheme of a magnetoresistive sensor.

The sensor resistance is calculated using Eqs. (4.13a) and (4.24). If $\phi = \psi$, and $H_x = 0$, then

$$R(H) = R + \Delta R(1 - \sin^2 \psi)$$

$$= R + \Delta R - \Delta R \left(\frac{H_y}{H_0}\right)^2$$

$$= R_0 - \Delta R \left(\frac{H_y}{H_0}\right)^2 \tag{4.26}$$

Similarly,

$$V_a = \Delta \rho I \left(\frac{1}{t}\right)\left(\frac{H_y}{H_0}\right)\sqrt{1 - \left(\frac{H_y}{H_0}\right)^2} \tag{4.27}$$

If $H_x \neq 0$, H_0 should be written as $H_0 + H_x/\cos \psi$.

Equations (4.26) and (4.27) are represented in Figs. 4.20(a) and (b) respectively.

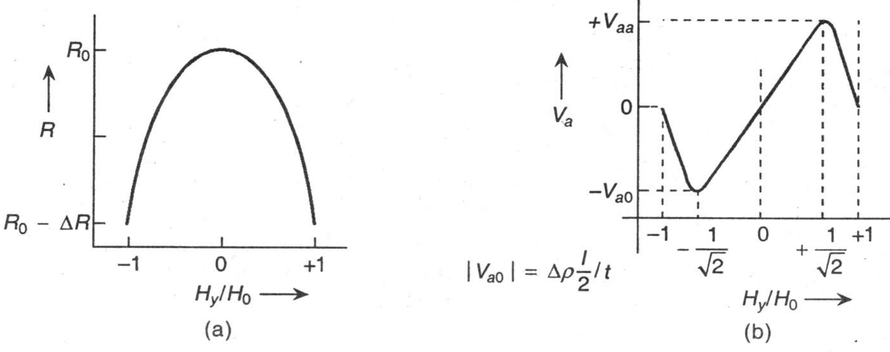

Fig. 4.20 Representation of (a) Equation (4.26), (b) Equation (4.27).

It must be mentioned that the squared characteristic given by Eq. (4.26) is due to the assumption of I flowing along the easy axis. Linearization can be effected by 'inclined' elements where the current is not along the easy axis, that is, I is rotated or M is rotated in the rest state. The barbar poles type design, where the film is spirally coated with good conductivity strips looking like shop sign, is very common for fabrication where current I is at an angle with respect to the magnetization M and magnetization itself is along the length and easy axis. This is shown in Fig. 4.21. The barbar poles will be inclined at $\delta + 90°$, with δ usually being $\pm 45°$, so that now,

$$R(H) = R \pm \Delta R \left(\frac{H_y}{H_0} \right) \sqrt{1 - \left(\frac{H_y}{H_0} \right)^2} \tag{4.28}$$

and the nature of R versus H_y/H_0 is like that shown in Fig. 4.20(b). A linear portion is obtained for $-1/\sqrt{2} < (H_y/H_0) < 1/\sqrt{2}$ or at least $-1/2 < (H_y/H_0) < 1/2$.

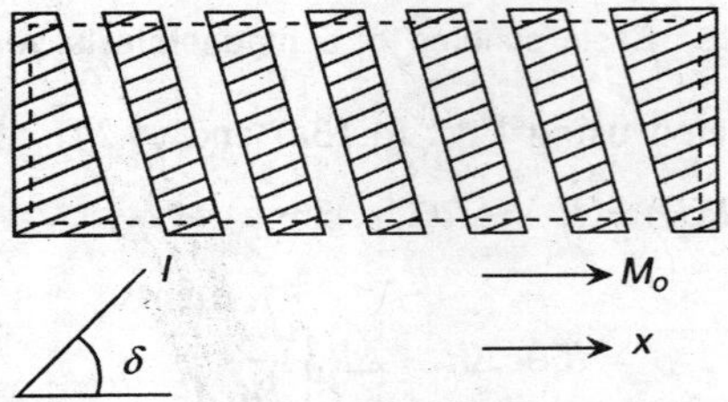

Fig. 4.21 The barbar pole design of magnetoresistive sensor.

Because the conducting strips partly short the area of the film and the current path is broader in parts (spirals), the resistance per unit length of the sensor is less than usual.

Materials for such sensors are chosen from crystalline binary and ternary alloys of Ni, Te, and Co—of which Ni(81) Fe(19) (permalloy) is most common that has relatively high $\Delta\rho/\rho$ and ρ, low H_a, and zero magnetostriction ($\lambda_s = 0$). Ni–Co (50–50) has other parameters almost the same as Ni–Fe except H_a which is 10 times larger. Also, coercivity should be small, for permalloy, in fact, it is 10 times smaller than Ni–Co alloy. Ni(60) Fe(10) Co(30) gives comparable values of $\Delta\rho/\rho$ and ρ, quite low λ_s but H_a in the alloy is nearly as high as Ni–Co alloy although coercivity is 1/4th that of Ni–Co. Amorphous Co(72) Fe(8) B(20) has been tried which has high ρ but poorer $\Delta\rho/\rho$ and except for H_a, other characteristics are better. Temperature coefficient of resistance is similar to that of the metals used, varying in the range 3×10^{-3}–$4 \times 10^{-3}/°C$.

4.3.2 Semiconductor Magnetoresistors

Generally, when a semiconductor material is exposed to magnetic field, its resistance increases. It is subsequently seen that Lorentz force acts perpendicular to the velocity **v** of a free charge carrier and the magnetic induction **B**, and the charge carrier eventually collides with crystal lattice to lose its velocity. This is attributed to the Hall angle θ_h between the electrical field $\mathbf{E}_x$ and the direction of the current. This change in the direction of the current or its rotation increases the path length of the current flow as mentioned earlier but is observed as an increase in the resistance of the material.

When the field is weak, the change in resistivity is proportional to the square of the component of the field perpendicular to the current vector, $\mathbf{B}_p^2$, thus,

$$\rho_B = \rho_0(1 + H_R B_p^2) \tag{4.29}$$

H_R being the coefficient of magnetoresistance, which is obtainable by solving the general transport equation with appropriate boundary conditions.

If the magnetic field is large, θ_h approaches $\pi/2$ and ρ_B shows a linear dependence on field B. In fact, the geometry of the semiconductor plate and magnetic field are included in the resistance ratio of the plate with and without field. Thus, for $\theta_h \leq 25°$,

$$\frac{R_B}{R_0} = \frac{\rho_B}{\rho_0}[1 + C_a(\mu_H B)^2] \tag{4.30}$$

and for $\theta_h \to \pi/2$

$$\frac{R_B}{R_0} = \left(\frac{\rho_B}{\rho_0}\right)(C_b + C_c \mu_H B) \tag{4.31}$$

where

C_a, C_b, and C_c are factors dependent on the plate geometry, specifically length l and width w and

μ_H is the Hall mobility.

It is seen that with increasing field, resistance is linearly proportional to field B. For a magnetoresistor to be of effective use, the factors C_a, C_b, and C_c should be optimized and appropriate materials chosen for optimum Hall mobility.

Effect of geometry

It has been shown that in an isotropic material shaped as a square plate subjected to a magnetic field perpendicular to the plate surface and current supplying electrodes mounted on the width side, Hall angle θ_h is significant only in the regions close to the electrodes. And in the central part, the current paths are almost parallel to the length edges where the Hall field tends to reduce the magnetoresistive effect. This is, to a certain extent, countered by reducing the l/w (Fig. 4.22) ratio which is shown in Fig. 4.23.

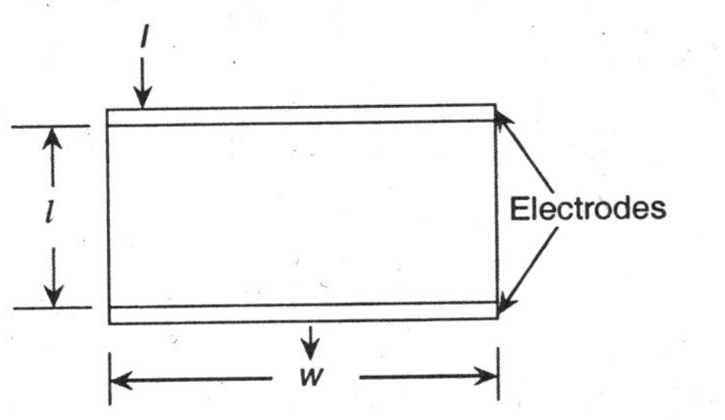

Fig. 4.22 The magnetoresistive sensor with electrodes.

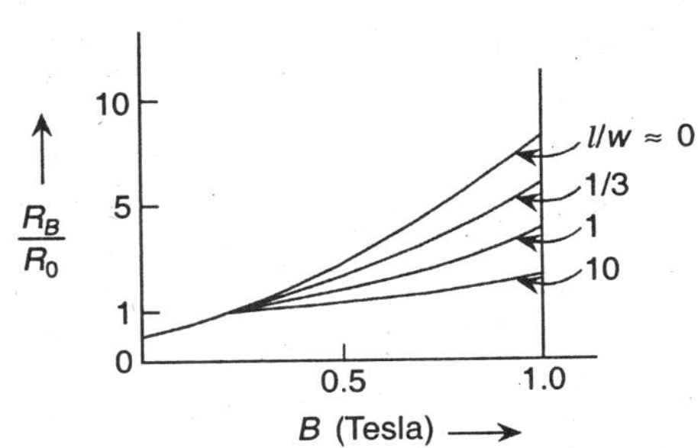

Fig. 4.23 Effect of changing l/w.

The ratio $l/w \approx 0$ is obtained in a special design known as the *Corbino disc* which is a round plate of semiconductor material with one electrode at the centre of the round plate on whose surface, the other electrode is formed. As a result, the width is practically infinite and in absence of nonconducting boundaries unlike that of Fig. 4.22, the Hall field does not build up and **E** is radial as shown in Fig. 4.24 with angle θ_h between current density **J** and **E** being the same everywhere.

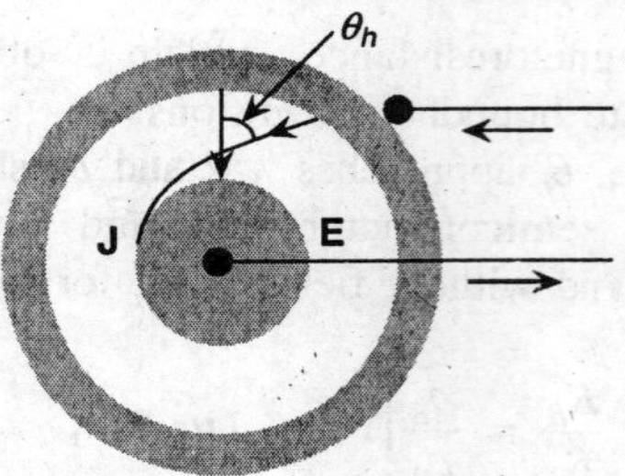

Fig. 4.24 Design of magnetoresistor with radial *E*.

The curves of Fig. 4.23 can be used for the determination of the constants C_a, C_b, and C_c which are functions of l/w so that Eqs. (4.30) and (4.31) fit the curves. The empirical results show that

$$C_a = 1 - \frac{0.54l}{w} \tag{4.32a}$$

$$C_b = 0.1 + \frac{0.25l}{w}, \quad \frac{l}{w} > 1 \tag{4.32b}$$

and

$$C_c = \frac{1}{l/w} \tag{4.32c}$$

Effect of material

As already mentioned, Hall mobility μ_H is dependent on the material. It should be larger, however, for better measurement. Of the four possible base materials GaAs, InAs, Si, and InSb, the last one has the highest mobility of about 7.7×10^4 cm^2/V^{-1}/s^{-1} which is 1.5 times larger than that of InAs, 30 times larger than that of Si, and nearly 10 times larger than GaAs. However, temperature coefficient of resistance is also higher in InSb, being -2×10^{-2}/°C. For the other materials, the values are as follows:

$$\alpha_{\text{GaAs}} = 8 \times 10^{-4}\text{/°C},$$

$$\alpha_{\text{InAs}} = 10^{-3}\text{/°C, and}$$

$$\alpha_{\text{Si}} = 5 \times 10^{-3}\text{/°C}$$

Further, α's are also functions of temperature and experimental results have shown that normalized resistance of GaAs or InAs do not substantially vary over a range of -20°C to about 150°C whereas it decreases exponentially with InSb and increases almost linearly with Si. However, appropriate doping can shift the intrinsic conductivity range to higher temperature. Tellurium doping of the order of 10^{22}–10^{23}m^{-3} for InSb is recommended.

There are various ways of obtaining magnetoresistors. A convenient resistance range is 0.1–10 kΩ which is generally obtained by connecting a large number of small l/w ratio elements in series, as shown in Fig. 4.25. The arrangement is good for producing short circuit straps such as Corbino disc.

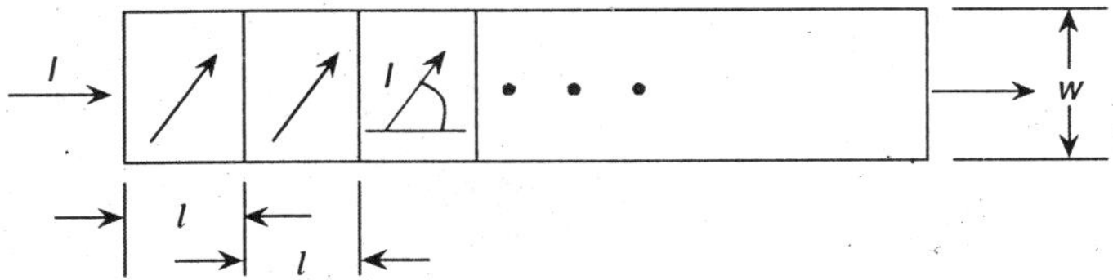

Fig. 4.25 Cascading a number of small *l/w* elements.

Another technique of attaining this is by forming NiSb needles in InSb with a mass percentage of about 1.8 of the former in the latter. NiSb is hundred times more conducting than InSb and InSb so doped would produce a good short-circuiting effect even at room temperature. The shape of a magnetoresistor is something like that of a conventional resistance strain gauge obtained by etching wafers of 25 μm thickness produced by photolithography. One such typical sensor is schematically depicted in Fig. 4.26.

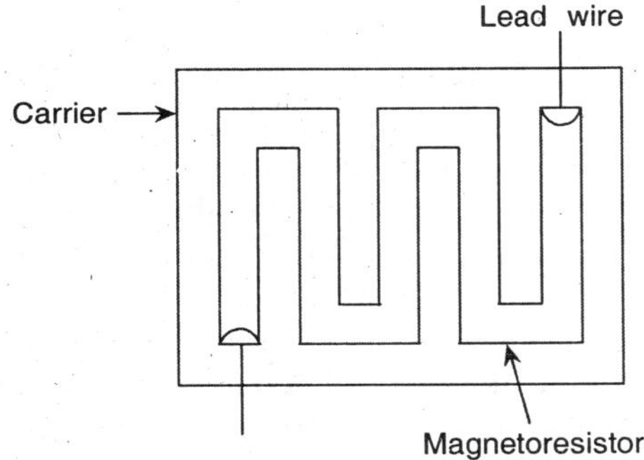

Fig. 4.26 Wafer type magnetoresistor.

Depending on the doping level, conductivity of InSb varies and accordingly the material is classified under specific types. This is particularly true for Tellurium doped semiconductors. Table 4.1 lists the sensor types and their conductivities.

Table 4.1 Sensor types and conductivities

S.No.	Material	Conductivity $(\Omega cm)^{-1}$
1.	D-type	
	Doped upto $2 \times 10^{22} m^{-3}$	200
2.	L-type	
	Doped upto $6 \times 10^{22} m^{-3}$	550
3.	N-type	
	Doped upto $2 \times 10^{23} m^{-3}$	800

The R_B/R_0 versus B plots for the three cases enlisted in the table are shown in Fig. 4.27. As already mentioned, temperature dependence of InSb, doped or undoped, is an important factor which can, however, be compensated by arranging a pair of sensors in a differential configuration.

A separation between the two would allow the field gradient to produce a difference in changes in resistances so that the two, when connected in a bridge, produce an output voltage.

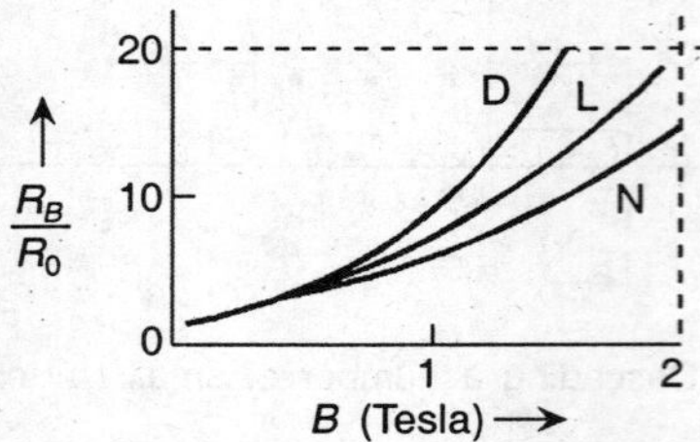

Fig. 4.27 Normalized resistance versus *B* plots for three types of magnetoresistors having different doping levels.

Often, the two sensors are initially biased magnetically and the resistances of both raised to value say, R_m; actuating now with an external assymetric field, the two magnetoresistors show different values say, R_1 and R_2.

4.3.3   Active Semiconductor Magnetic Sensors

Active semiconductor devices such as diodes and transistors are found to be sensitive to magnetic field which modifies the carrier transport condition of the devices and changes their electrical characteristics, although not significantly in normal conditions. When subjected to special optimized condition with respect to this field, these devices can offer some utility as sensors.

The realization of such sensors, keeping in view the phenomena involved in their operation (such as carrier injection, Hall effect for carrier deflection, recombination/generation of carriers (for diodes), carrier deflection injection modulation, and those that are present in a diode), is not very operation-friendly. Effectively, application of a magnetic field in a transistor, for example, changes the collector current as

$$I_c = I_s \exp\left(\frac{q\left(V + \frac{V_h}{2}\right)}{kT}\right).$$

where $V_h = h_c B I_E / t,$

with h_c = Hall coefficient,

I_E = emitter current,

t = device thickness, and

B = magnetic flux density.

Such devices are yet to be commercially exploited mainly because of small output level or small sensitivity they provide.

4.4   HALL EFFECT AND SENSORS

By far, the most important of the magnetic sensors are the Hall effect sensors. They have wide scale commercial and scientific applications. Similar to magneto resistive type, Hall effect sensors are also galvanomagnetic effect sensors. As Hall effect is observed both in metals and semiconductors, both kinds are employed though the latter material is more popular and used extensively in developing these.

4.4.1 The Hall Effect

When a current is sent through a very long strip of extrinsic homogeneous semiconductor in the x (long) direction and across the plane xy perpendicular to it, a magnetic field is applied to produce a flux density B_z, then an electric field E_y in the direction of y is produced which is called the Hall field. With electrodes across the strip in the y direction, a voltage V_H, called the Hall voltage, can be collected which approximately is given by

$$V_H \approx B_z I_x \tag{4.33}$$

Galvanomagnetic effects, in general, arise because of the action of the Lorentz force on the charge carrier transport phenomena in condensed medium. The Lorentz force is expressed as

$$\mathbf{F} = e\mathbf{E} + e[\mathbf{v} \times \mathbf{B}] \tag{4.34}$$

where

    e is the charge of the carrier,
    $\mathbf{E}$ is the electrical field,
    $\mathbf{v}$ is carrier velocity, and
    $\mathbf{B}$ is the magnetic induction.

If $\mathbf{J}$ is the total current density, then the carrier transport equation is

$$\mathbf{J} = \mathbf{J}_0 + \mu_H[\mathbf{J}_0 \times \mathbf{B}] \tag{4.35}$$

where μ_H is known as the Hall mobility and $\mathbf{J}_0$ is the current density due to the electric field $\mathbf{E}$ and the carrier concentration gradient ∇n. A magnetic field also affects the electrical field potential and carrier concentration and hence, it is not justified to write $\mathbf{J} = \mathbf{J}_0$ for $\mathbf{B} = 0$, as is apparent from Eq. (4.35). If σ is the conductivity, and $\mathbf{D}$, the diffusion coefficient, $\mathbf{J}_0$, in general, can be written as

$$\mathbf{J}_0 = \sigma\mathbf{E} - e\mathbf{D}\nabla n \tag{4.36}$$

which takes account of the drift (first term) and diffusion (second term) and the transverse transport caused by the magnetic field is taken care of by the second term in Eq. (4.35).

Therefore, the transport coefficients μ_H, σ, and $\mathbf{D}$ are dependent on the electric and magnetic fields and are determined by the carrier scattering process. The galvanomagnetic effects such as magnetoresistive and Hall effects can be derived from the solutions of Eq. (4.35) with appropriate boundary conditions.

It must be mentioned that Hall mobility μ_H is the product of the drift mobility of the carrier μ and the Hall scattering factor r, which is given by the appropriate ratio of relaxation time averages of the carriers over their energy distribution. Thus,

$$r = \frac{\langle \tau^2 \rangle}{\langle \tau \rangle^2} \tag{4.37a}$$

and

$$\mu_H = r\mu \tag{4.37b}$$

$r = 1$ for degenerate semiconductors or metals,
$r = 1.93$ for scattering with ionized impurities, while
$r = 1.18$ for acoustic phonons

If a long strip of extrinsic and homogeneous semiconductor material with xy-plane is considered as the strip plane, where length l is in the x-direction, width w in the y-direction, and **B** in the z-direction, so that $\mathbf{B} = (0, 0, B_z)$ as shown in Fig. 4.28, and, if an external electric field in the x-direction $\mathbf{E} = (E_x, 0, 0)$ is imposed, a current I with density $\mathbf{J} = (J_x, 0, 0)$ will flow in it so that a transverse electric field E_y builds up for countering the second part of the Lorentz force given by Eq. (4.34). As $J_y = 0$, from Eq. (4.35), we may obtain

$$E_y = -\mu_H B_z E_x \tag{4.38}$$

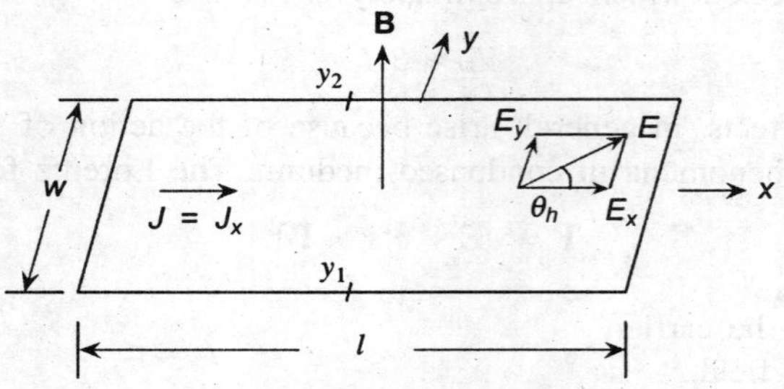

Fig. 4.28   Basic scheme of a Hall device.

E_y is the Hall field often represented as E_H and this field would produce a voltage across the width of the strip. This transverse voltage called the Hall voltage, V_H, is given by

$$V_H = \int_{y_2}^{y_1} E_H \, dy = -\mu_H B_z E_x w \tag{4.39}$$

The effect leading to the phenomenon just described is called the Hall effect.

Another parameter that sometimes acquires importance in the discussion of Hall sensors is the Hall angle and is given by (Fig. 4.28)

$$\tan \theta_h = \frac{E_y}{E_x} = -\mu_H B_z \tag{4.40}$$

The Hall effect has varying intensity in different materials. The materials for this effect are characterized by Hall coefficient which is defined as

$$h_c = -\frac{\mathbf{E}_H}{\mathbf{J} \times \mathbf{B}} \tag{4.41}$$

For Fig. 4.28, this becomes

$$h_c = -\frac{E_y}{J_x \times B_z} = \frac{\mu_H E_x}{J_x} \tag{4.42}$$

For a special case of zero carrier concentration gradient for homogeneous material ($J_x = \sigma E_x$ and conductivity σ is given by $e\mu n$), the Hall coefficient is

$$h_c = \frac{r}{en} \tag{4.43}$$

The Hall voltage can be expressed in terms of Hall coefficient h_c, using Eqs. (4.39) and (4.42) as

$$V_H = -h_c J_x B_z w \tag{4.44}$$

Hall sensor is based on the availability of this voltage with the element subjected to a magnetic field for a given current passing in it. Hence, the sensitivity of the sensor as well as its usability is determined by the magnitude of V_H and its stability. If electrons and holes are present as free carriers in a material, the coefficient h_c is reduced. If n_p and n_n are the concentrations, r_p and r_n are the Hall scattering factors, and μ_p and μ_n are the mobilities for holes and electrons respectively, then the coefficient is given by

$$h_c = \frac{1}{e} \frac{r_p n_p - r_n n_n (\mu_n/\mu_p)^2}{(n_p + n_n(\mu_n/\mu_p))^2} \tag{4.45}$$

Such situations arise in case of intrinsic semiconductors and those under high injection conditions. In the former case, the intrinsic carrier density n_i equals n_p and n_n which is easily calculable from standard equation

$$n_i = AT^{\frac{3}{2}} \exp\left(\frac{-E_g}{2kT}\right) \tag{4.46}$$

where

    A is a coefficient,

    T is absolute temperature,

    k is Boltzmann constant, and

    E_g, is the band-gap energy. Often Hall sensors are made of drift-mobility materials which are intrinsic in nature.

4.4.2 The Hall Effect Sensor

The long strip of material with negligible thickness assumed in Fig. 4.28 has to be made a practically feasible structure which is a rectangular plate of semiconductor material with four electrodes two covering the width-thickness faces (the supply electrodes (SUE)) and two covering small parts of length-thickness faces (the sensor electrodes (SEE)), as shown in Fig. 4.29. A voltage V_c across the supply electrodes produces a current I_x which flows along the length and for a magnetic induction B_z across $l \times w$ face as shown

$$V_{Hi} = \frac{h_c B_z I_x}{t} \tag{4.47a}$$

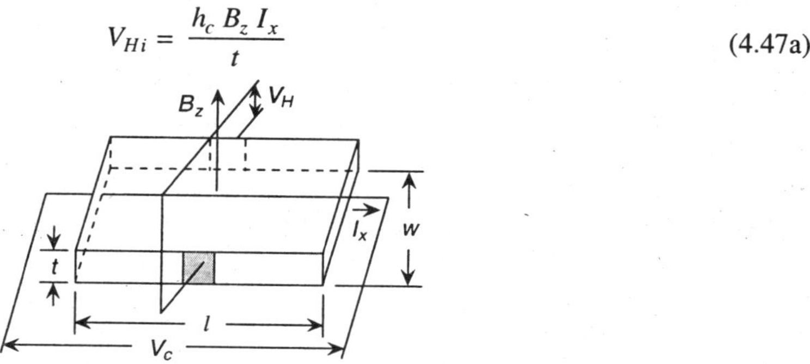

Fig. 4.29 A practical hall sensor.

since $J_x = I_x/(t \times w)$. Subscript i has been used to indicate a infinitely long material strip. For finite length, a geometrical correction factor needs to be introduced which is defined as $K_g = V_H/V_{Hi}$ so that

$$V_H = K_g \frac{h_c B_z I_x}{t} \tag{4.47b}$$

K_g varies because of the 'short-circuiting effect' since the current is finite and the area of the sensing electrodes is not negligible. If electrode contact areas are very small, K_g approaches unity and if these are large, K_g approaches zero.

In general, electrons are the main carriers which have higher mobility, so that, in Eq. (4.45), $(\mu_n/\mu_p)\, n_n \gg n_p$, and as a consequence

$$h_{cn} = -\frac{r_n}{en_n} \tag{4.48a}$$

and

$$V_H = -\frac{r_n K_g B_z I_x}{en_n t} \tag{4.48b}$$

Hall voltage magnitude is, thus, inversely proportional to plate thickness t and carrier concentration n_n both of which must be made small, that is, the Hall sensor plate must be very thin and material must be chosen to have low carrier concentration. These choices, however, make the resistance of the Hall strip high which is given as

$$R_h = \frac{\rho l}{wt} = \left(\frac{1}{e\mu n_n}\right)\frac{l}{wt} \tag{4.49}$$

resulting in large voltage drop along the strip which may not be acceptable as a sensor. This voltage drop V_c is given by

$$V_c = R_h I_x = \left(\frac{I_x}{e\mu n_n}\right)\left(\frac{l}{wt}\right) \tag{4.50}$$

Combining Eqs. (4.37b), (4.48b), and (4.50), V_H can now be written as

$$V_H = -\frac{\mu_{Hn} K_g B_z V_c w}{l} \tag{4.51}$$

Multiplying Eqs. (4.48b) and (4.51) both sides and then taking square roots and writing $V_c I_x = P_d$, where P_d is the power dissipated in the strip, we obtain

$$V_H = r_n K_g B_z \sqrt{\frac{\mu_n w P_d}{en_n t l}} \tag{4.52}$$

It is, thus, seen that the sensor material should be so chosen that carrier mobility is high which would reduce V_c and power dissipation P_d. This would also increase the Hall output.

Sensor geometry and fabrication

For a finite-sized Hall device, its shape is of little importance as far as electrical efficiency is concerned, provided the shape is simple such as circle, square and so on, and the sizes and positions of electrodes are appropriate. Priority is then given to shapes which is advantageous to technology and applications. In this context, another important aspect is the 'rotational invariance' or invariance with respect to SUE and SEE as in the cases of circle, cross, square, octagon, and the like where interchange of SU and SE electrodes does not affect operation in any way. Some of these are shown in Figs. 4.30(a), (b), (c), (d) and (e).

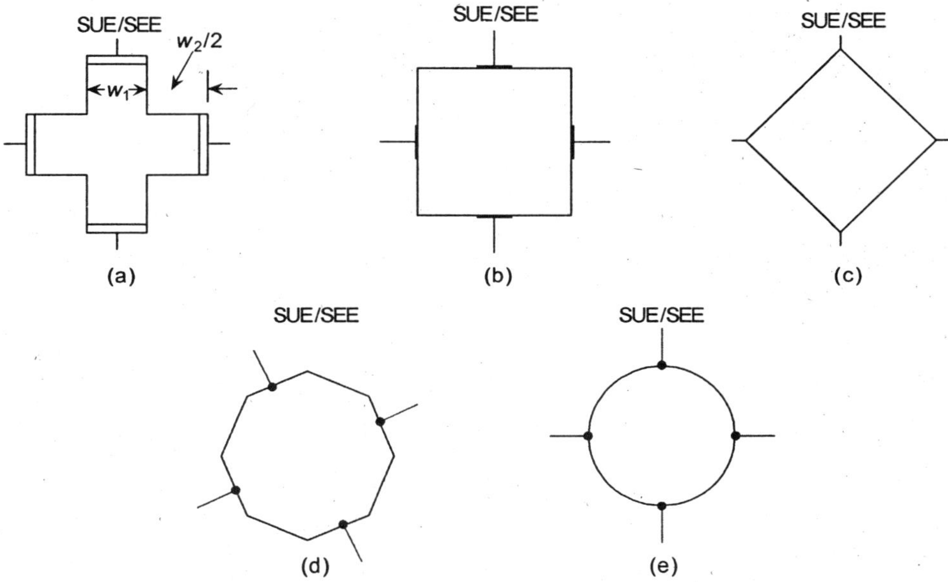

Fig. 4.30 Hall sensors of varying geometries.

Expressions for geometrical correction factors K_g's for different designs are tabulated in Table 4.2. Important parameters in this respect are:

For rectangular types—l/w ratio, θ_h, α_s/w where α_s is the SEE surface area,
for rotational invariant types—the ratio of the total electrodes length to the total plate periphery length λ, and so on.

Hall devices are fabricated by following the current trend in IC fabrication. The commonly available commercial sensors are developed using the bipolar IC fabrication process. An active Hall plate is usually an n-type epitaxial layer which is enclosed in a p-region with n^+-diffusion region covered by metal layers for contact. A depletion layer surrounds the n-type active region isolating the plate from the rest of the chip. Figure 4.31 gives a diagrammatic representation of the

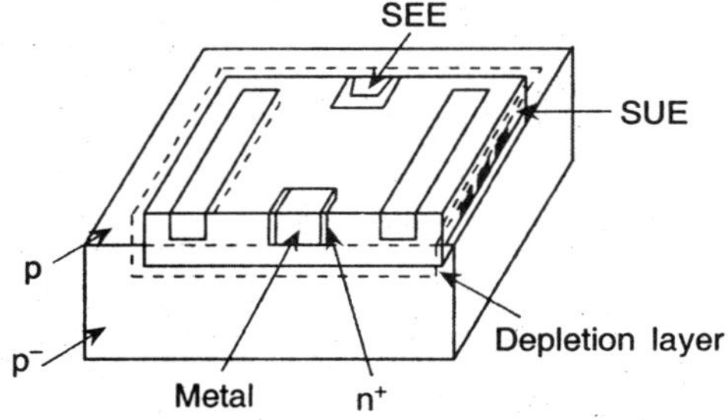

Fig. 4.31 A hall sensor produced through IC process.

same. Vertical Hall devices are also commercially produced with an advantage that the contacts are available from the top surface of the chip. In making these devices, n-type silicon, GaAs may be used. High mobility can be achieved using InSb and InAs.

Table 4.2   The geometrical correction factors

S. No.	Type and constraints on parameters	K_g
1.	Rectangular $l/w > 1.5$, $\alpha_s/w < 0.15$, θ_h small, < 0.45	$1 - \dfrac{16}{\pi}\left\{\exp\left(\dfrac{-\pi l}{2w}\right)\right\}\dfrac{\theta_h}{\tan\theta_h} \times \dfrac{1 - 2\theta_h}{\pi w \tan\theta_h}$
2.	Rectangular $l/w > 3$, $\alpha_s/w < 0.05$,	1
3.	Rectangular $l/w < 0.5$, $\alpha_s/w \to 0$ $\theta_h < 0.45$	$\dfrac{0.742 l}{w}\left(1 + \dfrac{\theta_h^2}{6}\right)\left(1 + 2.625 - \dfrac{3.257 l}{w}\right)$
4.	Rectangular $l/w < 0.3$, $\theta_h \to 0$ $\alpha_s/w \to 0$	$0.742\,\dfrac{l}{w}$
5.	*Rotational invariant $\lambda \leq 0.7$, $\theta_h \leq 0.015$	$1 - 1.94\left(\dfrac{\theta_h}{\tan\theta_h}\right)\left(\dfrac{\lambda}{1 + 0.414\lambda}\right)^2$ $\approx 1 - \dfrac{1.94\lambda^2}{(1 + 0.414\lambda)^2}\left(1 - \dfrac{\theta_h^2}{3}\right)$

*There is similarity between the relations for K_g in the third and the fifth rows.

Sensor Performance

The performance criteria for these sensors are specified as (i) sensitivity, (ii) noise, (iii) drift, (iv) nonlinearity, and so on, in the manner as for other sensors.

As the sensor is basically intended to measure magnetic field, sensitivity would be defined as the incremental value of V_H per incremental change in B_z but this requires I_x or V_c not to vary. This is called the *absolute sensitivity*

$$S_a = \left.\frac{\partial V_H}{\partial B_z}\right|_{I_x = \text{constant}} \tag{4.53a}$$

Another way of defining sensitivity is by including I_x and/or V_c in the expression which would then be called *relative sensitivity* S_{ri} or S_{rv}. Thus,

$$S_{ri} = \frac{1}{I_x}\left(\frac{\partial V_H}{\partial B_z}\right) \tag{4.53b}$$

and

$$S_{rv} = \frac{1}{V_c}\left(\frac{\partial V_H}{\partial B_z}\right) \tag{4.53c}$$

Combining Eqs. (4.53) with Eqs. (4.48) and (4.51), it is seen that S_{ri} is dependent on K_g and the plate surface carrier density $n_n t$; and S_{rv} is dependent on K_g and μ_{Hn}, K_g being a function of w and l as well. Larger value is obtained for shorter plates.

Sensitivity also arises due to environmental parameters such as temperature, light, pressure, and so on. Such a sensitivity is called *cross sensitivity* which is, obviously, different for different parameters. Cross sensitivity is often treated as a secondary sensitivity with respect to the concerned parameter. Thus, S_a and S_r are primary sensitivities and secondary sensitivity with a parameter p (temperature, pressure, and so on) is given as

$$S_{ssp} = \frac{1}{S_{a,r}} \left(\frac{\partial S_{a,r}}{\partial p} \right) \tag{4.54}$$

It has been seen that Hall voltage is a function of Hall coefficient h_c or a function of Hall scattering factor r_n for electrons as carriers. These quantities vary with pressure and temperature respectively and secondary sensitivity may be defined to cover these parameters as

$$S_{ssT} = \frac{1}{r_n} \left(\frac{\partial r_n}{\partial T} \right)$$

and

$$S_{ssp} = \frac{1}{h_c} \left(\frac{\partial h_c}{\partial p} \right)$$

It is known that Hall device is susceptible to two types of noise—that due to thermal variation and that due to frequency variation in the low range which becomes dominant and is known as '$1/f$ noise'. The actual figures are obtained experimentally by biasing the device at 100 kHz for thermal noise and 100 Hz for $1/f$ noise.

It is sometimes noticed that even if no field is applied, a static or a slow-varying output signal is obtained from the device which is attributed to the defects in fabrication or piezoresistive effects producing asymmetry voltages at SEE's This is known as *offset* and is characterized by an equivalent value of B given by B_o for an offset voltage V_o in terms of the absolute sensitivity as

$$B_o = \frac{V_o}{S_a} \tag{4.55}$$

One very convenient method of eliminating the offset is to bias two matched devices orthogonally and connect their outputs in parallel.

Nonlinearity in Hall device may be observed if h_c, K_g, or effective thickness of the plate vary with field or current and the nonlinearities are then called 'material', 'geometrical', or 'junction field effect' types. The first two types—Material and geometrical nonlinearities, can be made to cancel each other in the design and technology process, for example, GaAs Hall device shaped as a cross shows very low nonlinearity. Since the last one arises due to a feedback modulation of the plate thickness t, by the generated voltage V_H, it may be reduced by or compensated for using the effect in the feedback circuit. If in the usual calibration curve and the best fit curve, voltage is denoted as V_H and V_{Hb}, percentage nonlinearity is defined as

$$\% \text{ nonlinearity} = 100 \frac{V_H - V_{Hb}}{V_{Hb}} \tag{4.56}$$

4.5 INDUCTANCE AND EDDY CURRENT SENSORS

Both inductance and eddy current sensors follow the Faraday's law of induction which is mathematically stated as

$$\oint \mathbf{E}\,d\mathbf{l} = -\frac{d}{dt}\iint_A \mathbf{B}\cdot d\mathbf{A} \tag{4.57}$$

to mean that the voltage induced in closed turns of a coil is proportional to the time rate of change of flux linkage with it.

The essential difference between the two, however, is that the inductance sensors use the effect of voltage induction whereas the eddy current types use the current induced due to alternating magnetic field. Both these variaties are, perhaps, industrially the most useful ones as they are easily adapted to measure displacement, rpm, proximity, force, weight, acceleration, torque, pressure, and so on.

A voltage, in proportion to a variable to be measured, can be induced in a number of ways such as

  (i)   by varying the coupling between the two coils,
 (ii)   by changing inductance of two coils when a soft magnetic core is displaced inside them,
(iii)   by varying magnetic flux linkage when an air gap is varied or when the direction of magnetic polarization is changed, and so on.

Such voltage is easily picked up for indication. In contrast, eddy current is produced by moving an electrical conductor in an alternating magnetic field and its effect is utilized in the same way.

4.5.1 Variable Inductance Sensors

One of the simplest of the induction type sensors is the variable inductance sensor where a permanent magnet is placed in a magnetic circuit and a coil around a soft core. An air gap in the circuit varies in relation to the variable to be measured which changes the flux linkage. A typical scheme is shown in Fig. 4.32. Often the circuit is closed by a ferromagnetic yoke. Figure shows the technique for measurement of rpm with a moving gear.

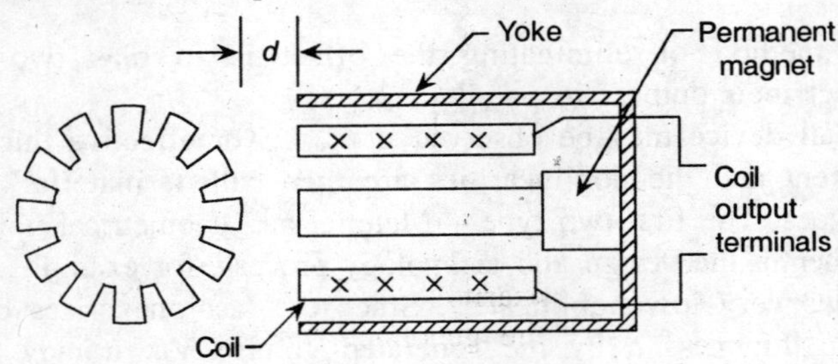

Fig. 4.32 Variable reluctance sensor.

The changing flux induces a voltage V according to Eq. (4.57) as

$$V = -\frac{N\,d\phi}{dt} \tag{4.58}$$

where

N is the number of turns in the coil and

ϕ is the flux through the coil.

Such sensors give high resolution, can measure with an accuracy of ±0.1%, and provide an output voltage in the range 0.25–75 V. Frequency range can vary between 1 Hz and 1 MHz. Limitations of the sensor exist in its dimensions and eddy effects—the latter increases the pick-up impedance by reducing the magnetic cross-section. Output voltage becomes a function of the rpm as well as the air gap. Obviously, with less air gap, voltage for same rpm is more because of more flux linkage or variation of it in the moving condition, as is evident from the graphs of Fig. 4.33.

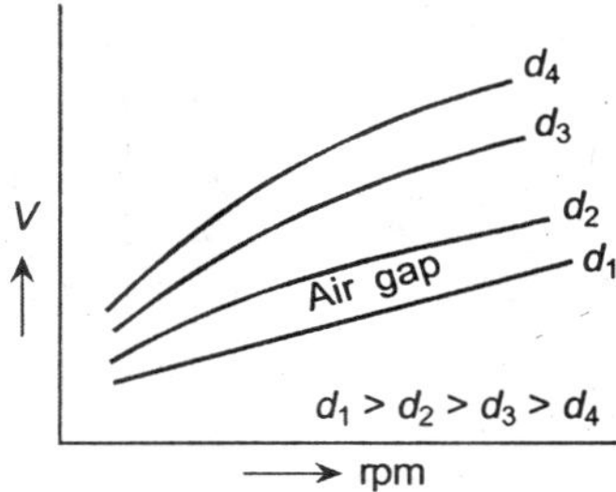

Fig. 4.33   Voltage–rpm characteristics with gap variation.

4.5.2   The Plunger Type Sensors

This is a variety that measures the change in inductance of a coil or a pair of coils, produced due to the displacement of a ferromagnetic core or plunger inside the coils. One such typical sensor assembly with two coils is shown in Fig. 4.34. The sensor can also be designed with single coil, in which case, the characteristic of the sensor itself becomes highly nonlinear. The type described in Fig. 4.34 is also known as the 'variable reluctance type sensor'. The two coils are identically designed and wound on a hollow cylindrical bobbin. A cylindrical armature is moved inside with a magnetically insulated rod. The inductance of a single layer single coil is given by

$$L = \frac{4\pi^2 n^2}{l^2}[lr^2 + (\mu - 1)r_c^2 l_c] \times 10^{-9}, \; l > l_c \tag{4.59}$$

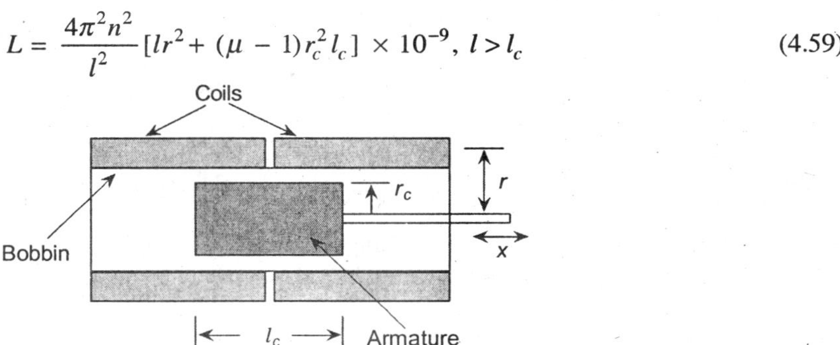

Fig. 4.34   The plunger type displacement transducer.

where μ = permeability of the core material,

n = number of turns in the coil,

r = radius of the coil,
r_c = radius of the core,
l = length of the coil, and
l_c = length of the core.

The coils are arranged in a bridge circuit with two equal resistances R's in the other two arms as shown in Fig. 4.35. When the core is centrally disposed, that is, $L_1 = L_2 = L$, the outpput V_o is then zero. For non-zero position of the armature, V_o also is non-zero because now $L_1 = L + \Delta L$ and $L_2 = L - \Delta L$ (say), so that

$$V_o = V_s \left[\frac{L + \Delta L}{(L + \Delta L) + (L - \Delta L)} - \frac{1}{2} \right] = \frac{V_s}{2} \frac{\Delta L}{L} \tag{4.60}$$

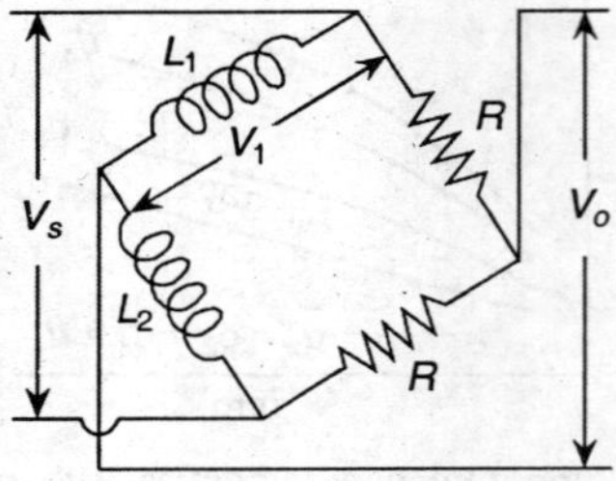

Fig. 4.35   The bridge circuit with coils.

with the movement of the core dx, change in inductance of the coil is dL (say), so that

$$\frac{V_o}{V_s} = \frac{1}{2} \frac{dL}{dx} \frac{\Delta x}{L} \tag{4.61}$$

since $\Delta L = (dL/dx)\, \Delta x$

The term dL/dx of the coil assembly is generally known and for a given small range, it is constant, so that V_o/V_s is proportional to Δx. Actually, for a movement Δx of the core, change in inductance of the coil is

$$\Delta L = \left(\frac{4\pi^2 n^2}{l^2} \right) r_c^2 (\mu - 1)\, \Delta x \tag{4.62a}$$

which gives the fractional change in inductance representing the sensitivity as

$$\frac{\Delta L}{L} = \frac{\Delta x}{l_c} \left(\frac{1}{1 + (l/l_c)(r/r_c)^2 (1/(\mu - 1))} \right) \tag{4.62b}$$

Similarly for the other coil, the same change (ΔL) in L, would occur but with an opposite sign, so that the overall change in inductance ΔL_t is given by

$$\frac{\Delta L_t}{L} = \frac{\Delta x}{l_c} \left(\frac{2}{1 + (l/l_c)(r/r_c)^2 (1/(\mu - 1))} \right)$$

$$= k \frac{\Delta x}{l_c}, \quad k < 1, \tag{4.63}$$

For $l_c \to l$, $r_c \to r$, and $\mu \to 1$, k approaches 1.
This calculation has neglected the leakage flux distribution.

4.5.3   Variable Gap Sensors

Variable gap sensors also utilize the variation in inductance. As the name indicates, these make use of the variation of air gap within a magnetic circuit. For a larger linear range, a differential arrangement is more appropriate as shown in Fig. 4.36. A single unit consists of a C-shaped

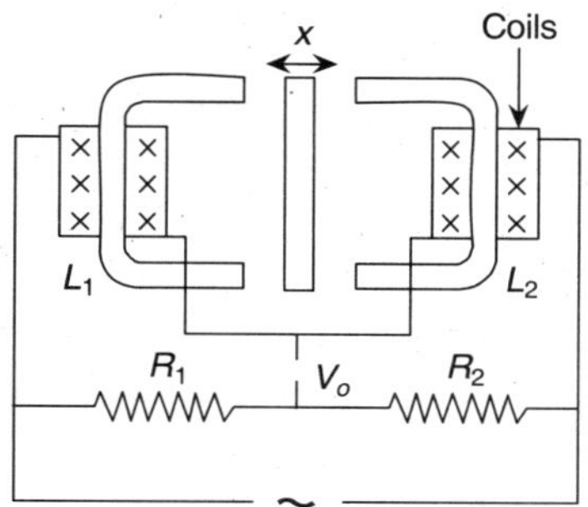

Fig. 4.36   Differential gap type sensor.

ferromagnetic core around which a coil is wound. A ferrous armature is made to approach or move away from the ends of the C-shaped core causing variation in the ratio between the air gap and the core length. Assuming permeability of the air gap as 1, and of core as μ, length of air gap to be l_a, length of core be l_c, core cross-sectional area as α, and number of turns in the coil to be n, the inductance of one coil is given by

$$L = \frac{n^2 \alpha}{\dfrac{l_c}{\mu} + l_a} \tag{4.64}$$

which becomes maximum for $l_a = 0$. The transducer, in the differential mode, is arranged for a voltage output V_o with an input V_s. With an air gap variation of x by only small amount either way, V_o/V_s linearly changes with x. This is graphically depicted in Fig. 4.37. The range is limited to $\pm x_p$. With proper choice of the material of the core (high μ), which is built with laminated sheets to avoid eddy effects, sensitivity could be made quite high so that a resolution as low as 1 nm can be obtained. The nonlinearity within the stipulated range is within $\pm 0.5\%$, but accuracy is comparatively poor, less than the type of Fig. 4.34.

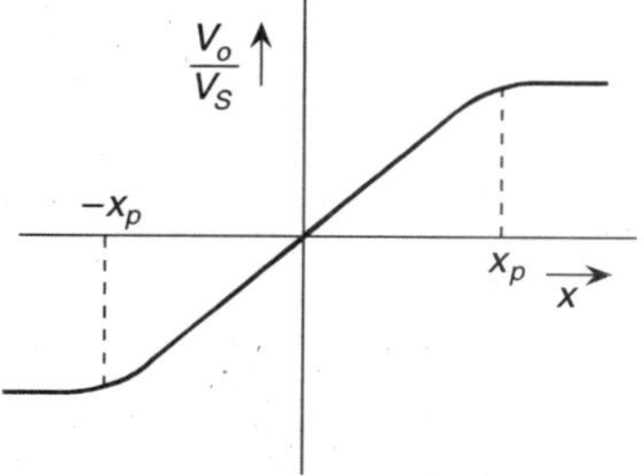

Fig. 4.37   Response characteristics.

4.5.4 Linear Variable Differential Transformer (LVDT)

Some aspects of an LVDT have already been described in Chapter 2 while this subsection incorporates some other aspects.

The linear variable differential transformer is a modified version of the plunger type sensors but arranged with two sets of coil, one as the primary and the other set as the secondary having two coils connected differentially for providing the output. Thus, it is a differential transformer. The coupling between the primary and the secondaries varies with the core plunger moving linearly as in the case of the plunger type. The secondaries are located on the two sides of the primary coil on the bobbin or sleeve. Figure 4.38 presents the arrangement of LVDT. An alternating supply of appropriate voltage V_i and frequency f is impressed across the primary coil and depending on the position of the core with respect to the primary and the two secondaries, an output voltage V_o is obtained from the secondaries as shown in Fig. 4.39. The induction in one secondary coil, according to the law is

$$V_{os} = -\frac{nd\phi}{dt} = -M\frac{di_p}{dt} \tag{4.65}$$

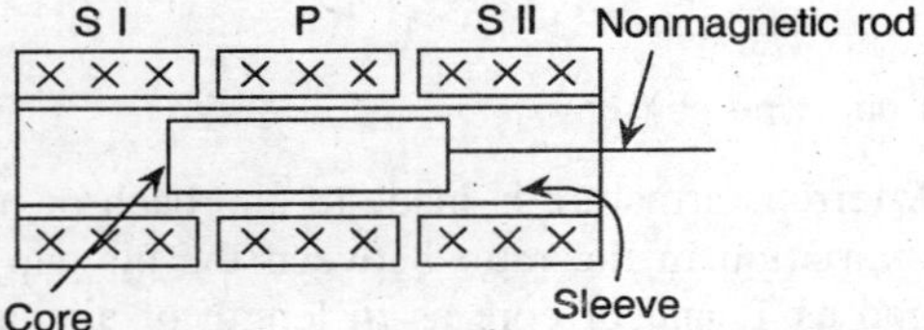

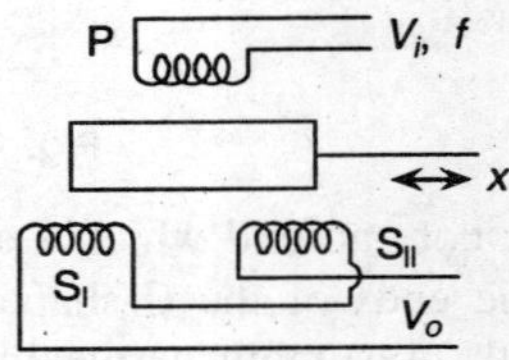

Fig. 4.38 Scheme of an LVDT.

Fig. 4.39 The equivalent model of LVDT.

where n = number of turns in the coil of the secondary,
 ϕ = magnetic flux,
 M = mutual inductance between primary and the concerned secondary, and
 i_p = primary current.

For the two coils differentially connected,

$$V_o = V_{os1} \sim V_{os2} = (M_1 - M_2)\frac{di_p}{dt} \tag{4.66}$$

Both M_1 and M_2 being functions of x, $M_1 - M_2 = M(x)$. If the function is linear over a certain range, $M(x) = kx$, so that

$$x = \frac{V_o}{k(di_p/dt)} \tag{4.67}$$

As the sensor basically forms a transformer, the loss components are also to be considered for obtaining the output V_o per unit displacement of the core. The loss components are to be considered for all the transducers of the inductive type. When arranged in a bridge in a differential manner, the loss components can, however, be compensated by appropriate circuit components. The equivalent circuit of the LVDT, in this concern, is shown in Fig. 4.40. Solving for the magnitude ratio per unit displacement $|V_o/V_i|/x$, angle by which the output voltage V_o lags the input voltage V_i at a frequency $f = \omega/2\pi$ and, if the meter load is R_m, we get

$$\left|\frac{V_o}{V_i}\right|\frac{1}{x} = \frac{k\omega R_m/\{(R_s + R_m)R_p\}}{\sqrt{[\{1 - \omega^2(\tau_m^2 + \tau_p\tau_s)\}^2 + \omega^2(\tau_p + \tau_s)^2]}} \tag{4.68}$$

and

$$\phi = 90° - \tan^{-1}\frac{\omega(\tau_p + \tau_s)}{1 - \omega^2(\tau_m^2 + \tau_p\tau_s)} \tag{4.69}$$

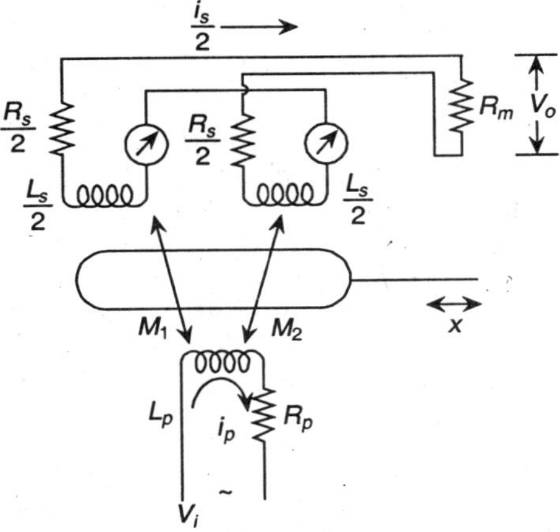

Fig. 4.40 The equivalent circuit of LVDT.

where

$$\tau_m = \frac{M_1 - M_2}{\sqrt{(R_s + R_m)R_p}}$$

$$\tau_p = \frac{L_p}{R_p}, \text{ and}$$

$$\tau_s = \frac{L_s}{R_s + R_m}$$

The phase-rectified secondary output voltage V_o with x is shown in Fig. 4.41 for a given V_i where the linear range limits are indicated by $\pm x_m$. This limitation is inherent in all differential

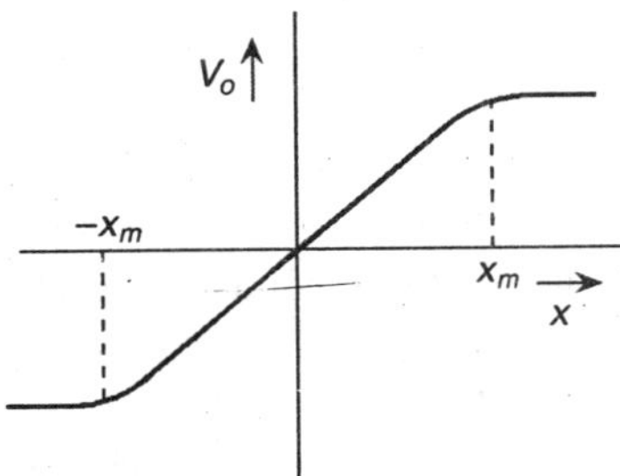

Fig. 4.41 Phase-rectified output characteristics.

systems and methods of extending the range have been proposed mainly by appropriate design and arrangement of the coils. Some of these are:

(i) balanced linear tapered secondaries (not much improvement is obtained in this case),

(ii) overwound linear tapered secondaries (linearity improves in this case to a certain extent, although range does not),

(iii) balanced overwound linear tapered secondaries (a little unbalance detected in the case (ii) with overwound linear tapered secondaries) is eliminated here, range specification is also similar to case (ii),

(iv) balanced profile secondaries (linearity range is extended by proper profiling of the secondary coils), and

(v) complementary tapered windings (linearity range is extended in this case as well but the winding is quite complicated as sectionalized winding is done).

Some of the winding arrangements are shown in Fig. 4.42. Improvement is obtained to a certain extent by providing a magnetic keeper which is nothing but a magnetic tube that surrounds the sensor. It can be noticed from Eqs. (4.68) and (4.69) that there is a frequency at which the output can be maximized and phase for the output and input are the same. But this condition is, obviously, load-dependent.

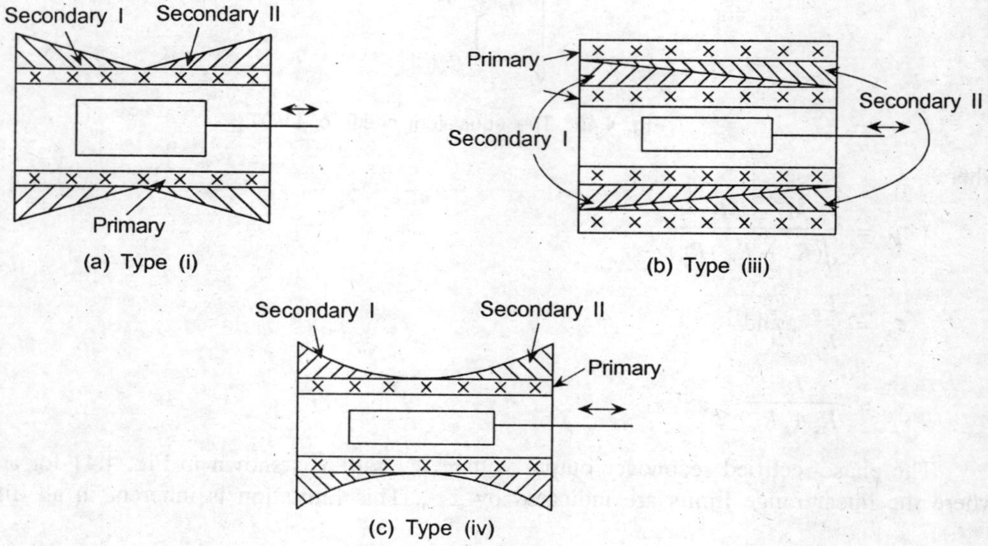

Fig. 4.42 Different secondary coil structures for better linearity and low remanence.

The LVDT's can be designed in various sizes for various ranges from a few μm to even 1 m movement of the core. Linearity from ±0.25% to ±0.5% is obtainable over the ranges. Tolerance, however small it may be in the design of the secondaries, gives rise to non-zero null output, that is, $V_o \neq 0$, at $x = 0$. But careful design can give a repeatable value ±0.05% *FS*. Depending on the choice of the materials, the LVDT can be used in a temperature range of −50–500°C. For avoiding eddy current losses, the core is made of stamped sheet by stacking and is also provided with a slot or notch all along the length. Using ferrite core, the supply frequency may be increased which increases the sensitivity to a certain extent.

4.6 ANGULAR/ROTARY MOVEMENT TRANSDUCERS

The principles discussed till now can be utilized to sense angular or rotary movements also. For an angular movement such as twist or torsion of a shaft, a section of the same may be made with half of it made of a ferromagnetic material, the other half being made of non-magnetic material. Two coils are arranged so that the inductance of the two parts would depend on the amount of magnetic material in the linkage path of the coil fluxes. The scheme is shown in Fig. 4.43.

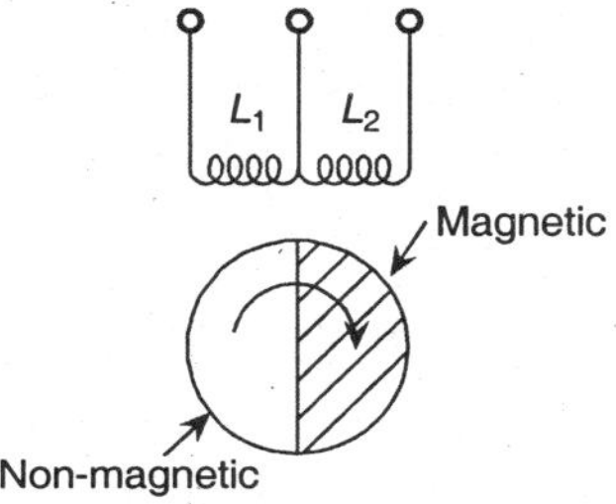

Fig. 4.43 A split-coil rotary transducer.

Another technique for rotary motion measurement is a modified version of the LVDT as depicted in Fig. 4.44.

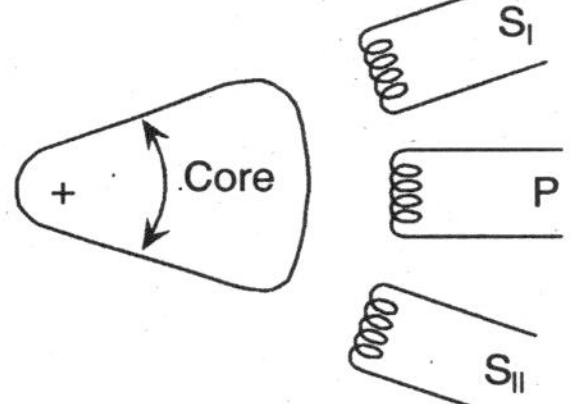

Fig. 4.44 Modified version of LVDT for angular movement measurement.

4.6.1 Synchros

By changing the magnetic coupling between coils, ac-excited electromechanical sensors have been developed. For measurement at distant points, such devices are adopted and are known as *synchros.*

Synchros, as sensors, are of two types, namely (i) torque type and (ii) control type. The general constructional features of a synchro are represented in Fig. 4.45. It consists of a stator with three windings S1, S2, and S3 separated by 120° in space and a rotor R, which is supplied with an ac voltage.

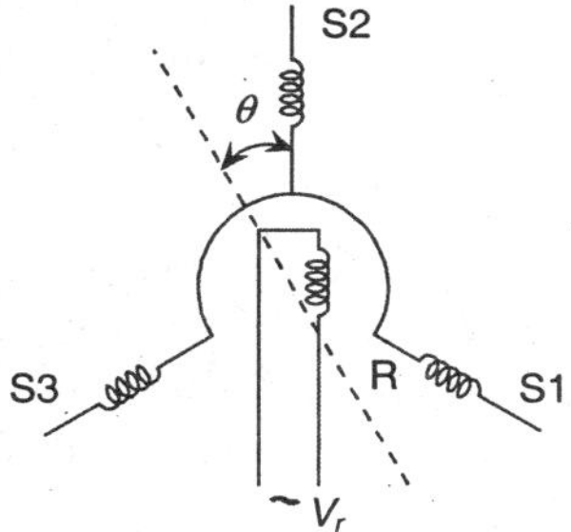

Fig. 4.45 General constructional features of a synchro.

In a 'torque type sensor', two such units are coupled as shown in Fig. 4.46. A rotation of the rotor R_1 by an angle θ changes the voltages induced into the stator windings S11, S21, and S31

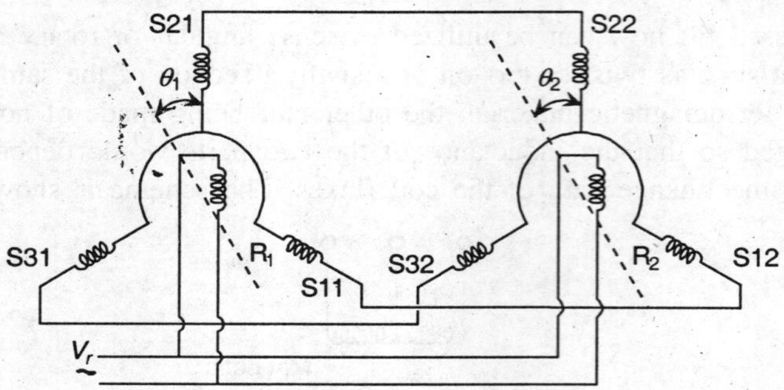

Fig. 4.46 Torque type synchro sensor.

in magnitude and phase. And since these windings are connected electrically to S12, S22, and S32, same voltages with phases as in those of windings of stator 1 produce a field so that rotor R_2, if not oriented as rotor R_1, would receive a torque and rotation till it attains the same rotational position as that of R_1. With a scale arrangement, the rotated angle, thus produced, may be measured. For a single synchro unit such as that of Fig. 4.45 with the rotor angle θ for an input sinusoidal voltage $V_r = V_{ro} \sin \omega t$, the voltages induced in windings S1, S2, and S3 are

$$V_{s1} = KV_{ro} \sin\omega t \cos(\theta + 120°) \tag{4.70a}$$

$$V_{s2} = KV_{ro} \sin\omega t \cos\theta \tag{4.70b}$$

and

$$V_{s3} = KV_{ro} \sin\omega t \cos(\theta + 240°) \tag{4.70c}$$

where K is a constant, such as the ratio of the rotor to the stator turns. From Eqs. (4.70), the line voltages are

$$V_{s12} = K\sqrt{3}\, V_{ro} \sin\omega t \sin(\theta + 240°) \tag{4.71a}$$

$$V_{s23} = K\sqrt{3}\, V_{ro} \sin\omega t \sin(\theta + 120°) \tag{4.71b}$$

and

$$V_{s31} = K\sqrt{3}\, V_{ro} \sin\omega t \sin\theta \tag{4.71c}$$

If $\theta = 0°$, $V_{s2} = KV_{ro} \sin\omega t$, that is, a maximum and $V_{s31} = 0$. This position of the rotor is marked as the zero position or the reference position.

In torque type sensors, it is tacitly assumed that $\theta_1 = \theta_2$. In such a situation, there is no compensation current because of any unbalanced terminal voltages. If, however, $\theta_1 \neq \theta_2$, a torque would be produced on the receiver synchro rotor till the equality is achieved. The torque is approximately sinusoidal in form

$$T \approx K_t \sin(\theta_1 - \theta_2) \tag{4.72}$$

It is shown in Fig. 4.47 where maximum torque occurs at $\theta_1 \sim \theta_2 = 90°$. For $\theta_2 \to \theta_1$, that is, $\theta_1 - \theta_2$ being small, torque versus $(\theta_1 - \theta_2)$ curve is approximately linear. The torque sensors are usually designed with a resolution of $5 \times 10^{-4} - 10^{-2}$ Nm/degree with an angular error of $\pm 0.5' - \pm 1.5'$. A rotational speed of 300 rpm is quite common for this case.

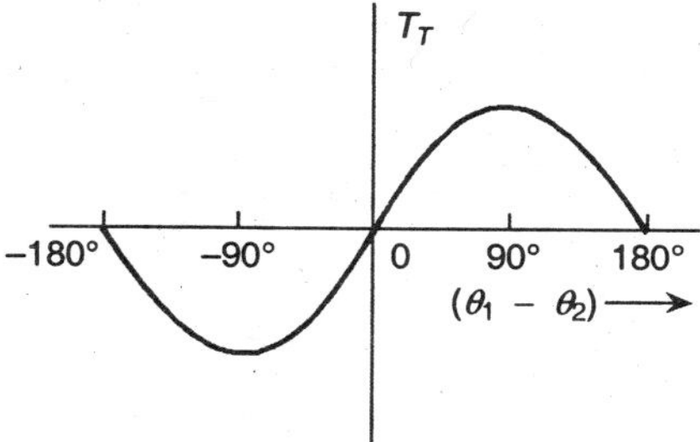

Fig. 4.47 The torque-angle characteristics.

If the torque becomes large and the error is to be kept small, the receiver synchro rotor is not connected to the line, instead an error voltage is taken out from the rotor R_2, amplified and a servo motor is driven to bring the rotor R_2 in position. Even otherwise, without this feedback arrangement, as shown in Fig. 4.48 with the servo motor, an error voltage may be obtained across the winding of rotor R_2 which is now held fixed. This voltage gives a direct indication of the rotation of the rotor R_1. This arrangement is known as the *synchro control transformer*. The error voltage $V_e(t)$ is proportional to the cosine of the angle between the two rotors $\cos(\theta_1 - \theta_2)$. Thus,

$$V_e(t) = K'V_{ro}\, \sin\omega t\, \cos(\theta_1 - \theta_2) \qquad (4.73)$$

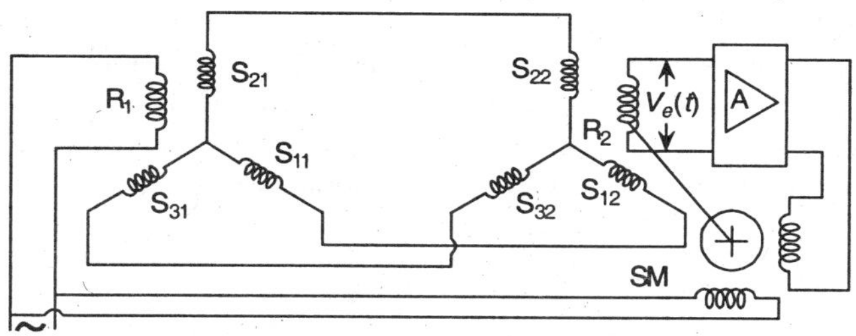

Fig. 4.48 Synchro-positioner without feedback.

If the two rotors are oriented at right angles, $V_e(t) = 0$, and with $\theta_1 = 0$ and $\theta_2 = 90°$, the transmitter and receiver rotors are said to be at electrical zero position. If $\theta_1 - \theta_2$ is close to $90°$, then

$$V_e(t) = K'V_{ro}\, \sin\omega t\, \sin\{90° - (\theta_1 - \theta_2)\}$$

$$\propto (\theta_1 - \theta_2) \qquad (4.74)$$

which means that the error voltage is proportional to the angular rotational difference of the rotors. In case of synchro control transformer, the rotor of the receiver unit is usually made cylindrical to make the air gap practically uniform. The accuracy is of the order of $\pm5'-\pm15'$.

To measure sums or differences of angles, a synchro differential unit is inserted in between the synchro torque type systems. The differential unit has star connected windings on both the stator and the rotor. The rotor, however, has a cylindrical structure. The scheme of connection is shown in Fig. 4.49 where both synchro torque units are used as transmitters S_{T1} and S_{T2} and the differential unit is used as synchro differential receiver (SDR). If S_{T1} rotor rotates by θ_1 and S_{T2} by θ_2, the free SDR rotor would rotate by $(\theta_1 - \theta_2)$ for balancing. Synchro differential unit may also be used as a transmitter when its rotor is a driven one.

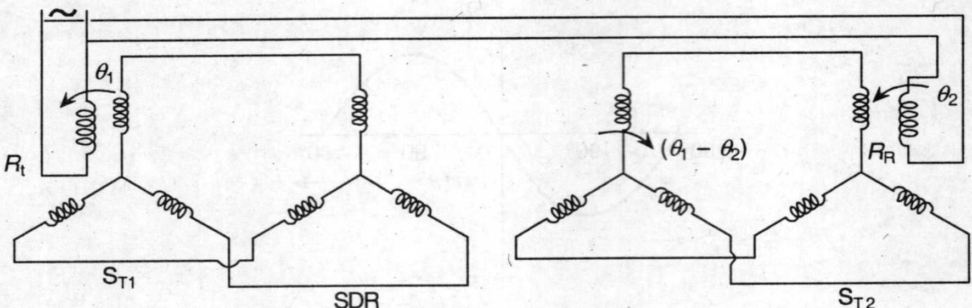

Fig. 4.49 Synchro differential unit introduced between synchro torque systems.

4.6.2 Synchro-resolvers

These are induction type devices designed to transform angular rotor position into a set of signals which vary with sine and cosine of the rotor position with respect to the position of a set of stators. Thus, resolving essentially means conversion from one coordinate system to another. The most common type of synchro resolver has two stator windings with their axes at 90° and two rotor windings with their axes also at 90°. The stator windings act as transformer primary and the rotor windings as transformer secondary. The operation of the resolver is akin to the operation of a synchro control transformer.

The zero position synchro resolver with two stators and two rotors is shown in Fig. 4.50(a) while Fig. 4.50(b) depicts the system with rotor cross coil rotated by 60° from zero position. Assuming V_{p1} is a voltage that is 1.5 times V_{p2}, the voltages induced can be obtained as V_{s1} and V_{s2} for these new rotor positions. Using the vector diagram shown in Fig. 4.50(c),

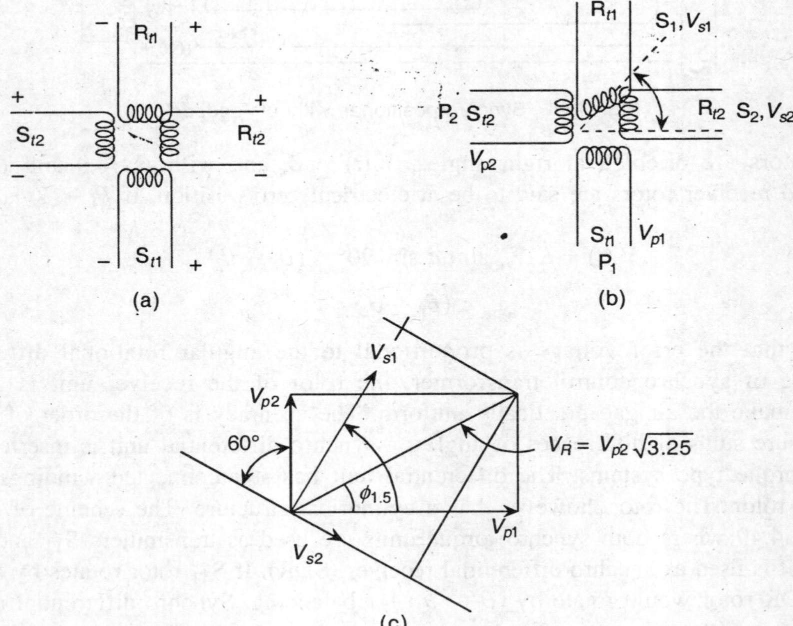

Fig. 4.50 (a) Scheme of a synchro-resolver, (b) resolver with rotor cross-coil rotated by an angle, and (c) resolver vector diagram.

$$\tan\phi = \frac{1}{1.5} = 0.667$$

so that $\phi = \tan^{-1}(0.667)$.
Resultant voltage,

$$V_R = V_{p2}(1 + 1.5^2)^{1/2} = \sqrt{3.25}\ V_{p2}$$

And hence,

$$V_{s1} = \sqrt{3.25}\ V_{p2o}\ \sin(60° - \tan^{-1}(0.667)\ \sin\omega t \tag{4.75a}$$

and

$$V_{s2} = \sqrt{3.25}\ V_{p2o}\ \cos(60° - \tan^{-1}(0.667)\ \sin\omega t \tag{4.75b}$$

where V_{p2} is effectively $V_{p2o} \sin\omega t$.

It must be pointed out that the rotors can be made primary while the stators are secondary. 'Supply to' or 'output from' the rotors is made by a commutator and brushes. Brushless resolvers also are available where an internal rotating transformer is used.

Another variation of resolvers is the *inductosyn* which is a sort of multipole resolver and used mainly in machine tool control. It can be constructed to measure either rotary or linear movements. In either type, two magnetically coupled parts—the scale and the slider are there. The scale consists of a circular/linear fret type winding usually designed as printed circuit track and bonded to a disc. The scale disc is fixed to the body and acts as the stator. The slider is the rotor part positioned opposite to the scale and is allowed to turn around more linearly along the scale. It consists of two separate fret type windings printed like the scale. The two windings are separated by a small gap.

4.7 EDDY CURRENT SENSORS

Eddy current is induced in an electrically conducting material due to the magnetic field. This current, however, varies with time; also the fields generated by these eddies interact with the excitation fields causing variation in the eddies.

The eddy current phenomena due to the magnetic fields can be described using the requisite Maxwell's equations. But the analysis of the operation of the sensors using eddy current phenomena is extremely difficult. No attempt is, therefore, made to do the same.

There are two types of such sensors, namely

(i)   the eddy current tachometer and
(ii) the eddy current proximity sensor.

Without going into the intricate constructional details, the basic scheme of the tachometer is shown in Fig. 4.51 which transforms a rotary movement in terms of the angular speed ω into a pointer deflection ϕ. There is a permanent magnet (PM) of at least two poles fixed to a shaft which rotates with angular speed ω. An eddy current cup made of aluminium or some such electrically conducting material is held closed to the magnet carried by a spindle and restrained by a torsion spring. When the multipole magnet rotates, magnetic flux passes radially through the conducting material but changes because of the rotation of the magnet. Thus, an alternating flux density $\mathbf{B}$ is produced in the cup whose rate of change produces an electric field $\mathbf{E}$ so that

$$\text{curl } \mathbf{E} = -\frac{d\mathbf{B}}{dt} \tag{4.76a}$$

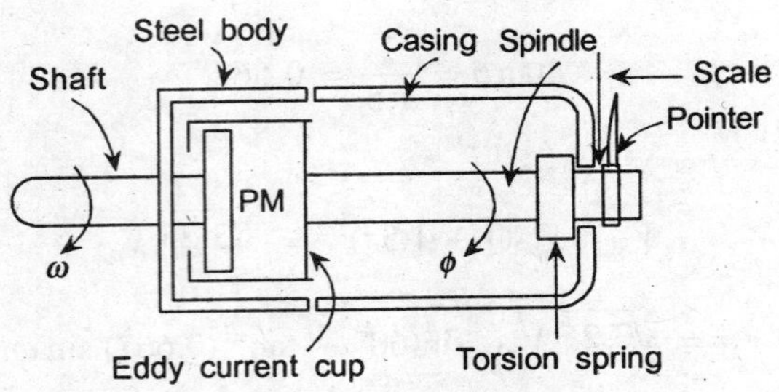

Fig. 4.51 Eddy current type tachometer.

This electric field produces the eddy current of density **J** in the shell of the cup which in turn generates a secondary magnetic field of intensity **H** given by the equation

$$\text{curl } \mathbf{H} = \mathbf{J} \tag{4.76b}$$

The two magnetic fields now interact with each other and a torque is produced in the cup which tends to follow the rotation of the magnet. This torque T is proportional to ω and if the cup is restrained by a torsion spring, an angular deflection of the spindle is indicated by a pointer mounted at the end of the spindle, when the magnetic torque is balanced by the spring torque.

In an eddy current proximity sensor, a high frequency LC oscillator is designed using a coil with air or ferrite core which is positioned so that the magnetic field produced by the coil spreads around the axis of the coil in all directions. Without any foreign material in this field close to the inductance, the circuit oscillates at its resonance frequency and constant amplitude. If, however, a conductor appears in the field, eddy currents are induced in it, the density of which is given by

$$\mathbf{J} = \sigma\left(-\frac{d\mathbf{V}_p}{dt}\right) \tag{4.77a}$$

where σ is the electrical conductivity and

$\mathbf{V}_p$ is the vector potential given by

$$\mathbf{B} = \text{curl } \mathbf{V}_p \tag{4.77b}$$

As a result, the generated magnetic field **H** reacts with **B** and the oscillation is damped, reducing amplitude and even stopping it. With and without a conductor in the coil field, the oscillator circuit impedance becomes high and low, and correspondingly the current becomes low and high. Not that a quantitative study can be made with such a sensor to show how close the conductor is, but the oscillator current versus distance of the conductor from the coil shows a characteristic curve which is indicative of high and low values only, as shown in Fig. 4.52(b) while Fig. 4.52(a) shows the transducer scheme.

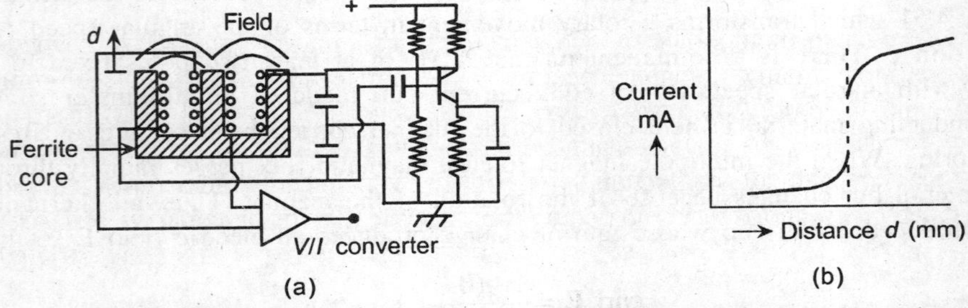

Fig. 4.52 (a) Eddy current proximity sensor, (b) current versus distance plot.

The size, shape, and material of the conductor are important for eddy current generation. The current generation depends on the parameters σ and μ of the material as also on the oscillator frequency ω. The size of the conductor should be large enough at the recommended operating distance so that it does not alter the output current by producing variations. Thickness of the conductor should be larger than the skin depth ds which is a function of ω, μ, and σ and is given by (approximately)

$$ds = \frac{0.14}{(\omega\mu\sigma)^{1/2}} \tag{4.78}$$

Normally the frequency is between 0.1 and 1 MHz; and thin conductors of ferromagnetic materials are good enough. These have a range from 1 to 50 mm with a resolution of 0.1 mm and a linearity within 0.1%. These can work under temperature limits of -40–$200°C$.

4.8　ELECTROMAGNETIC FLOWMETER

This is another commercially very important transducer that utilizes the law of electromagnetic induction.

A nonmagnetic pipe arranged with a pair of electrodes on the opposite faces, is provided with a magnetic field **B** orthogonal to the line connecting electrodes, that is, E_1 and E_2 (Fig. 4.53). A conducting liquid passing through the pipeline in full flow, behaves as a conductor as if it is moving in a magnetic field so that the charge carriers in the field get deflected by the Lorentz force

$$\mathbf{F} = e(\mathbf{v} \times \mathbf{B}) \tag{4.79}$$

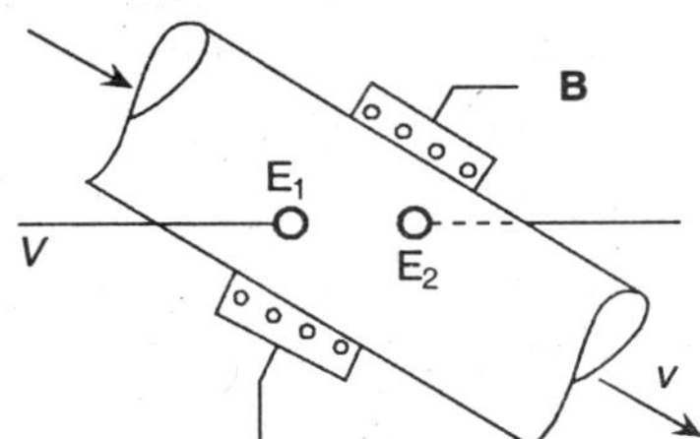

Fig. 4.53　Basic scheme of an electromagnetic flowmeter.

where **v** is the liquid velocity and e is the electron charge. In electric field of strength **E**, the force is

$$\mathbf{F} = e \cdot \mathbf{E} \tag{4.80}$$

and

$$\mathbf{E} = \mathbf{v} \times \mathbf{B} \tag{4.81}$$

so that the voltage induced between the electrodes separated by a distance d, the pipe diameter, is

$$\mathbf{V} = \mathbf{d} \cdot \mathbf{E} = \mathbf{d}(\mathbf{v} \times \mathbf{B}) \tag{4.82}$$

when d, v and B all are at right angles, we can write

$$V = k\mathbf{d} \cdot \mathbf{v} \cdot \mathbf{B} \tag{4.83}$$

where k is a constant ideally unity but dependent on, to a certain extent, the conductivity of the liquid. Such a sensor is used for measuring flow rates of conducting fluids from 0.25 m/s upto 20 m/s with a resolution of 0.1 m/s. A minimum conductivity of 1 µs/cm is necessary for the sensor to be effectively operative. Usually ac field is used to avoid polarization. Pipe diameters used lie in the range of 0.2–200 cm.

4.9 SWITCHING MAGNETIC SENSORS

In usual remagnetization in ferromagnetic materials, flux changes continuously with external magnetic field. In such materials, domain wall displacements and rotation occur. If, however, well-defined anisotropy is introduced through mechanical stresses or by tempering in magnetic fields, as in the case of hard-drawn Ni (10%–20%) wires, magnetization under tension would form a single domain with the saturation magnetization along the axis. As a consequence, this one domain magnetization can be either parallel or anti-parallel to the saturation direction also called the 'preferred direction'. Wires with this preferred direction of magnetization along the axis have their anisotropy enhanced with the application of tensile stress and the stress anisotropy energy density E_ρ is then given by

$$E_\rho = \frac{3}{2} \lambda_s S \tag{4.84}$$

where λ_s and S are the same as in Eq. (4.5b). When λ_s is positive, with tensile stress S, magnetization and length of the wire in the direction of magnetization, increase. E_ρ can be made very high with appropriate materials such as cobalt–iron and nickel–iron alloy.

With such materials when E_ρ is quite large and the applied magnetic field is sufficiently high, the wire becomes a single-domain magnetized specimen. On reversing the field of strength H_r above the coercivity H_c, switching in remagnetization takes place which is determined by $(H_r - H_c)$ and not by dH/dt. But, as mentioned, the wire must be under stress with the surface energy of domain wall being atleast

$$E_{DS} = 4k_D \sqrt{E_\rho} \tag{4.85}$$

where k_D is a constant so that $(H_r - H_c)$ can be made high as it is proportional to $\sqrt{E_{DS}}$, that is $\sqrt{S}$. Also, $(H_r - H_c)$ is inversely proportional to the grain diameter (d) which should obviously be made small, that is, the grains are made fine. In fact, it is shown that

$$H_r - H_c \approx \frac{\sqrt{S}}{d} \tag{4.86}$$

A typical sensor with tensile stress consists of a field coil through which a stressed wire is extended. A pick-up (search) coil for picking up the transitory pulse voltage during switching is shown in Fig. 4.54. Figure 4.55(a) shows the hysteresis loop indicating H_c and H_s and Fig. 4.55(b) shows the pulse voltage during transition. It must be mentioned here that for a single domain remagnetization, voltage pulse magnitude is nearly equal to the magnitude of $(H_r - H_c)$ but the pulse has a building switching time which is proportional to the square of the wire diameter and B and inversely proportional to its resistivity and the quantity $(H_r - H_c)$. Thus,

$$V_p \approx (H_r - H_c) \tag{4.87a}$$

and

$$T_p = \frac{B\,d_w^2}{\rho(H_r - H_c)} \tag{4.87b}$$

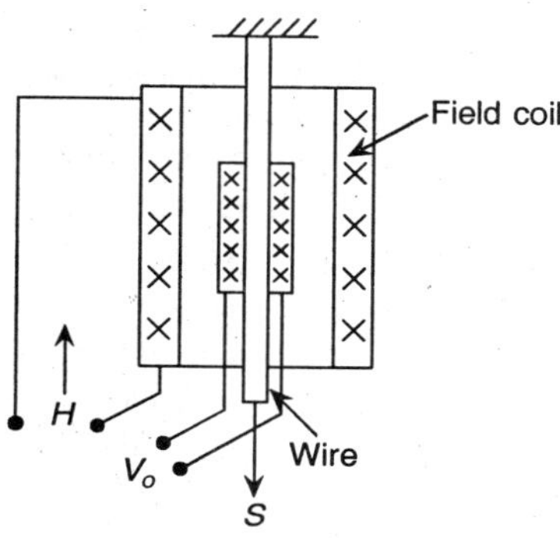

Fig. 4.54 Search coil type pick-up for transitory pulse voltage.

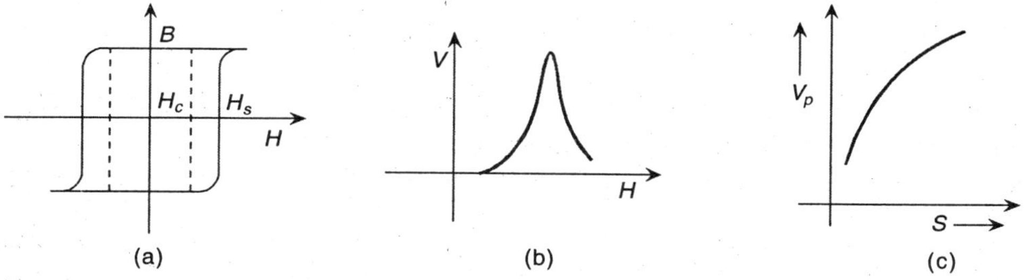

Fig. 4.55 (a) The hysteresis loop indicating coercive and saturation forces, (b) the pulse voltage in transition, and (c) pick-up voltage–stress characteristics.

Combining Eqs. (4.86) and (4.87a), one can get the relation between the pick-up voltage amplitude and the tensile stress which is shown in Fig. 4.55(c).

Depending on the way the wires are prepared for use as sensors, two types of sensors are specified: (i) the Wiegand sensor and (ii) the pulse wire sensor.

4.9.1 The Wiegand Sensor

The Wiegand sensor uses a wire manufactured from an alloy consisting usually of Co(52), V(10), and Fe(38) and of a diameter varying from 0.2 to 0.3 mm. This is done by cold working process under tension to cause plastic deformation in the outer shell of the wire while the inner shell or the core is under elastic stress. The coercivity for the inner and outer shells are 10–20 A/cm and around 30 A/cm respectively. The wire actually is a twisted set of wires having a soft magnetic core and a hard magnetic outer shell. The wire length varies from 8–30 mm. The wire, initially, is magnetized to saturation in such a way that the inner core and the outer core are oppositely magnetized, as shown in Fig. 4.56.

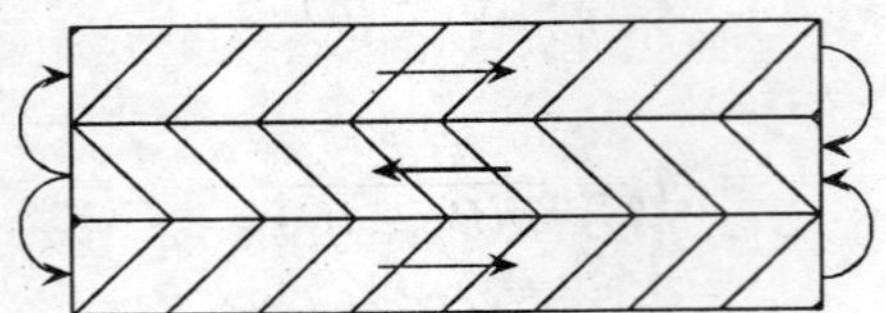

Fig. 4.56 A Wiegand sensor.

Wiegand wires are seen to produce high pulse voltages (2–2.5 V) of constant magnitude from a field frequency of 0.1 kHz–0.001 Hz, even dc while the ordinary wires show almost zero output below 50 Hz or so. Besides, Wiegand wires have a wider working temperature range, −180°–180°C. However, with increasing wire length, pulse amplitude increases.

By applying a proper field, even a permanent magnet is suitable for the purpose, the inner core induces a voltage at the search coil that appears as a pulse. In the process, both the core and the outer shell are magnetized in the same direction. Also, the pulse can be in one direction only and when sharp positive increase of B occurs in the $B–H$ curve, the Barkhausen jump takes place. The magnetic status of the wire needs be restored by a magnetic field of the correct magnitude for next pulse to be obtained.

4.9.2 The Pulse Wire Sensor

In pulse wire sensors, the inherent stress is actually generated by a composite wire made of two different alloys with varying thermal expansion coefficient unlike Wiegand sensors where torsion is the cause of this stress. In this composite form, permanent magnetic materials are not found to be satisfactory and hence, an additional permanent magnet wire is attached parallel to the switching wire. The schematic diagram of the sensor is shown in Fig. 4.57. While the core is made of the material as discussed earlier, the outer shell comprises of Ni(28), Co(18), and Fe(54). The pulse voltage and its duration are similar to those of the Wiegand wire, but the wire can now be driven symmetrically. The tensile strength required for switching core is given by stressing. For introducing tensile stresses by difference of coefficients of thermal expansion, a Ni–Fe core and permanent magnetic shell of Cr(28), Co(10), and Fe(62) are used.

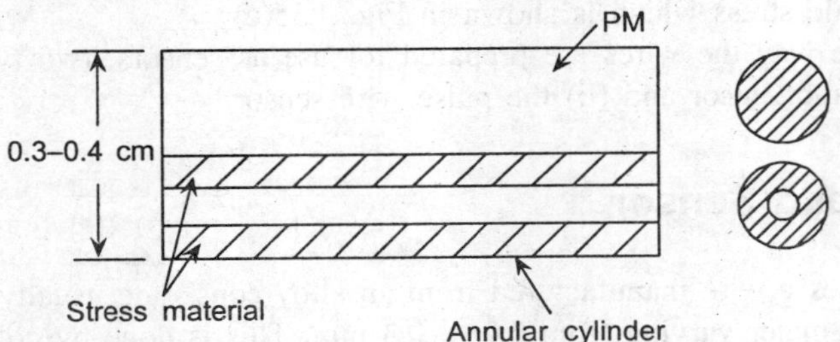

Fig. 4.57 The pulse wire sensor.

4.10 SQUID SENSORS

Superconducting quantum interference device (SQUID) sensor operates on one of two effects, (i) flux quantization or (ii) Josephson effect—both of which are observed in the presence of superconductivity. Superconductivity, in turn, is characterized by the condensation of conduction electrons into pairs that have opposite momentum and spin.

Obviously, **SQUID** has low operating temperature. This limits its commercial application to a great extent. Basically, it is a magnetic sensor of very high resolution but requires practically useful sophisticated infrastructure.

However, one of the outcomes of the Josephson effect is that if a radio frequency energy of frequency f falls on a very thin insulating junction between two superconductors and a current I_i is allowed to flow through this junction as shown in Fig. 4.58(a), the voltage V_o available across the junction changes in steps (Fig. 4.58(b)) with current, each step being described by

$$\Delta V_o = \left(\frac{h}{2e}\right) f \tag{4.88}$$

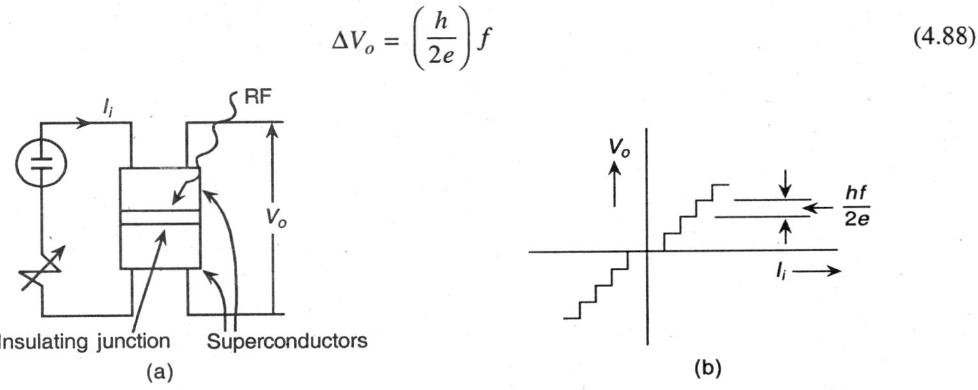

Fig. 4.58 (a) Scheme of a sensing device using the Josephson effect, (b) The voltage–current characteristics with steps.

where h is Planck's constant and e is the electron charge. The quantity $2e/h$ is known as the *Josephson's constant*. This technique is used to maintain voltage standard in terms of frequency. Insulation junction is quite effectively made by producing a dry solder between the superconductors.

SQUID detectors are now quite efficiently used for **comparison of the Josephson junction** *emf* with the standard cell *emf* as it provides an accuracy of the order of 1 in 10^7.

REVIEW QUESTIONS

1. (a) What are the different types of magnetic sensors? **On what principles do they work?** Outline briefly.

    (b) What is ΔY-effect? How is it used in practice for magnetic field sensing? What materials are specifically suitable for the purpose?

2. What are the differences between Matteucci effect, Villari effect, and Wiedemann effect? How are these three effects used in developing magneto-elastic sensors?

3. Distinguish between the operations of cross transductor type and the four-branch type yoke coil designs for torque sensing. Can they be used in any type of shafts? How are magnetic inhomogeneities in the shaft surfaces accounted for?

4. Describe with diagrams, the principle of operation of a coaxial type torque sensor. What is an inactive zone in such a sensor? Why is it provided?

5. Sketch the scheme of a transformer type magnetoelastic force transducer. What is pressductor? How does it work and where is it used? What materials are used for such sensors and what is the range of such sensors?

6. Develop the principle of anisotropic magnetoresistive sensors. How is it used in sensing magnetic field?
   A metallic magnetoresistor is placed in a magnetic field with its length perpendicular to the field. How does the resistance vary with this field?

7. How does a semiconductor type magnetoresistor differ from a metallic one? How is the former type dependent on its l/w ratio? What type of design can take care of this influence?
   What are the different materials used for the same and what is the effect of change in doping level?

8. (a) Describe the basic principle of a Hall device and show how can it be used as a magnetic field sensor?
   On what factors and parameters of the sensor does the Hall voltage output depend for a given field condition?

   (b) How is performance of a Hall sensor evaluated? What are its primary and secondary sensitivities?

9. How does an LVDT measure displacement? How is the nonlinearity attempted to be eliminated? Discuss coil structure design with reference to the above.
   Deduce the condition under which the output voltage of the LDVT is in phase with the input.

10. Briefly discuss about the operation of synchro-sensors in (i) torque type and (ii) control type modes.
    What is the order of resolution of a torque type synchro-sensor?

11. What is a synchro-resolver? Explain the operation of a synchro-resolver for measuring angular rotation. Append diagrams as needed.
    What are inductosyns? Where and how are they used?

12. How does an eddy current proximity sensor operate? How is the induced eddy current density related to the conductivity and vector potential? How do the permeability of the material and the supply frequency influence the measurement?

13. (a) How can switching magnetic sensors be used for stress measurement? Can they be made with any ferromagnetic material? What other factors are important in the design of such a sensor?

    (b) Explain the operation of 'pulse wire' sensor with diagrams.

14. (a) What is the major advantage with SQUID sensor that makes it popular in scientific work but why then, it has limited commercial popularity?

    (b) What is a Josephson constant? Discuss a SQUID that is used for maintaining voltage standard in terms of frequency?

Chapter 5

Radiation Sensors

5.1 INTRODUCTION

Gradual development of instrumentation and measurement systems for industry and scientific work has widened the use of sensors of a special type which, in general, can be classified as radiation sensors. Earlier, it was photosensors or photosensistors, but sensing radiation other than optical or photonic has also acquired an important place now. Of commercial importance, are the infrared, ultraviolet, visible, photon-type, x-rays, and nuclear radiations such as β- and γ-rays. Their detection or sensing has also undergone large scale change. Earlier classification of radiation sensors or sensistors can be put as follows:

(i)   the photoelectric cell such as photoemissive cell,
(ii) the photoemf cell such as photovoltaic, barrier layer, boundary layer type, and
(iii) the photoconductive cell such as light sensitive resistors.

Of these, the photoconductive cell was described earliest, far back in 1873, then came the photoemf cell in 1876, and the photoelectric cell in 1889. The photosensistor was considered as a combination of two electrodes in an electrolyte. The concept has not changed much but with the advent of material science and technology, scope has widened and newer sensors other than those specified previously joined the family and the radiation was also not limited to be optical alone, as mentioned. According to the radiation ranges, frequency or wavelength, the electrodes change in size and shape and the electrolyte also changes, that is, becomes gas, liquid, or solid.

One of the fundamental laws on which some of these sensors are based is the photoelectric effect. Radiation energy propagating through space in quanta when collides with matter, certain integral number of quanta called photons are emitted, reflected, and others absorbed depending on the material characteristics. If the incident photon energy $h\nu$ (h is Planck's constant and ν is the frequency of radiation), then

$$h\nu = \frac{1}{2}mv^2 + \phi e \tag{5.1}$$

where

$mv^2/2$ is the energy of the electron emitted from the atom of the matter by the impact of the photon and

ϕe is the energy required to release the electron (e = charge of electron, ϕ = work function) from the material.

ϕ, the work function is a characteristic of the material. As seen from Eq. (5.1), the kinetic energy of the photoelectron is dependent only on the incident photon energy which is transmitted to the electron. The intensity of incident radiation determines the number of electrons released. The mechanism is explained by the band structure. If the photon energy is sufficient to raise an electron in the material to a vacant conductivity band level, the electrical conductivity of the material increases. Two situations may now arise:

(a) The incident radiation energy $h\nu$ is just sufficient to transfer an electron into a vacant conductivity level and not beyond. This process leads to increased photoelectric conductance of a substance and the effect is sometimes referred to as the 'inner photoelectric effect'.

(b) $h\nu$ is high enough, it then causes the electron to be detached and emitted from the material. This is known as the 'outer photoelectric effect', and is effective in gaseous systems.

5.2 BASIC CHARACTERISTICS

The important characteristics that need to be considered for the photodetectors are (i) work function, (ii) spectral sensitivity and spectral threshold, (iii) quantum yield and quantum voltage, (iv) time lag, (v) drift, fatigue and so on, (vi) static and dynamic responses, and (vii) linearity. This section discusses these characteristics in detail.

(a) *The work function:* The energy E, which is spent in overcoming the surface attractive forces is given by

$$E = \phi e \tag{5.2}$$

where e is the electronic charge.

Work function ϕ is a physical constant for a given material and is usually expressed in electron volts. For the metallic elements, work function is observed to be smaller for higher atomic number. Alkali metals have smaller work functions of which caesium has the smallest value, 1.54 eV. So, for photodetectors, alkali metals make a good choice. However, alkali metals are electropositive and loose electrons easily (small E and hence, small ϕ). This makes them vulnerable to atmospheric condition that contain electronegative oxygen or hydroxyl ions. If such metals are to be used, they are encapsulated and protected. They are, however, used only as a surface layer on metal plates of higher work functions. Thus, Na, K, Rb, or Cs layers are laid on Ag, Be, Ta, Ni, Al, Cu, Ca, Zr, or W plates. Oxides of alkali metals have still smaller work function. There are special compounds such as BiS_3 or $Ag–Cs_2O–Cs$ where work function can be drastically reduced.

The alkali metals have a single electron (valence electron) on the outermost orbit so that the force keeping this electron in its orbit is smallest and a low work function is enough to dislodge it completely from the atom. When the number of electrons on the outermost orbit increases as in the case of platinum (for example) where it is 4, they are bound with higher energy and hence, the metal correspondingly has a higher work function.

(b) *Spectral sensitivity and spectral threshold:*   When the electron velocity is zero, $v = 0$ in Eq. (5.1), which occurs at absolute zero, then, electron escape from metal surface is possible with radiation if

$$h v > \phi e \qquad (5.3a)$$

and a critical or a threshold frequency is obtained as

$$v_0 \geq \frac{\phi e}{h} \qquad (5.3b)$$

the corresponding threshold wavelength being given by

$$\lambda_0 \leq \frac{hc}{\phi e} \qquad (5.3c)$$

where c is the velocity of light. Substituting the values of h and c, and expressing ϕ in volts as it is actually a volt equivalent, we obtain

$$\lambda_0 = \frac{1.2395}{\phi} \ \mu m \qquad (5.4)$$

For caesium, ϕ being 1.54, $\lambda_0 = 0.8045$ μm.

Table 5.1 gives some idea about λ_0 and ϕ for different useful metal elements and some double metal films used for photoemission.

Table 5.1 λ_0 and ϕ for metals and double metal films

Surface	λ_0(μm)	ϕ(eV)	Surface	λ_0(μm)	ϕ(eV)
Li	0.5580	2.20(2.28)	Na–Pt	0.5900	2.08
Na	0.6470	1.90(2.46)	K–Pt	0.7700	1.60
K	0.6820	1.80(2.24)	Rb–Pt	0.7950	1.56
Rb	0.7300	1.69(2.18)	Cs–Pt	0.8900	1.38
Cs	0.8045	1.54(1.91)	Ni–Pt	0.3318	—
Ca	—	2.51(2.70)	BaO–Pt	0.9200	1.34
Ba	—	2.29(2.51)	Li–W	0.6700	1.83
W	0.2620	4.58	Th–W	0.4900	2.52
Ni	0.3050	5.01	Ba–Ag	0.7900	1.56
Pt	0.1960	6.30	Cs–Cs_2O–Ag	1.0000	1.23
			Pt–W	0.2804	—

Work function may be 'extended' because of the presence of the 'space charge' which requires to be overcome by providing additional 'energy'. This may occur if different frequency radiations are incident on the surface as the low energy (low $h v$) electrons cannot reach the range of the high energy electrons and form the space charge. Also, during the process of electron escape, energy loss may occur which is proportional to the change in the surface temperature that becomes less during escape of electrons. In Table 5.1, two values of ϕ are given for a few useful metals. The values within the brackets are the ones found to be more useful for practical purposes.

The work functions of the metals change when oxygen reacts with them. The work function increases for Fe, Ti, Ni, Zr, Ag so that the threshold wavelength decreases whereas it decreases for Th, U, Ca, Ba, Cs, shifting the threshold wavelength towards the longer side.

Thus, photoelectric emission from metal surface occurs only if the wavelength of the incident radiation is less than the threshold wavelength or in other words, frequency is greater than the threshold frequency. Further, the amount of emission is proportional to the intensity of the incident radiation, that is, the number of photons, although the proportionality is not linear for a fixed frequency distribution of the radiant power. The spectral sensitivity of 'photosensistors' higher than the threshold frequency is, in fact, a function of frequency of the incident radiation. However, it has been shown that for alkali metals with increasing atomic weight, the sensitivity peak decreases and the wavelength for peak sensitivity increases. Figure 5.1 shows the relative spectral sensitivity curves for photosensistors of the alkali metal group along with that of a human eye. Table 5.2 shows the list of peak wavelengths, atomic weights, and relative responses with the response approaching 1 for the eye.

Table 5.2 List of atomic weights, peak wavelength, and relative response for some metals

Element	Atomic weights	$\lambda_p(\mu m)$	Relative response (Eye → 1) at 0.5580 µm
Lithium	6.94	0.4050	1.50
Sodium	22.997	0.4190	1.35
Potassium	39.096	0.4400	1.15
Rubidium	85.48	0.4730	0.75
Caesium	132.91	0.5390	0.50

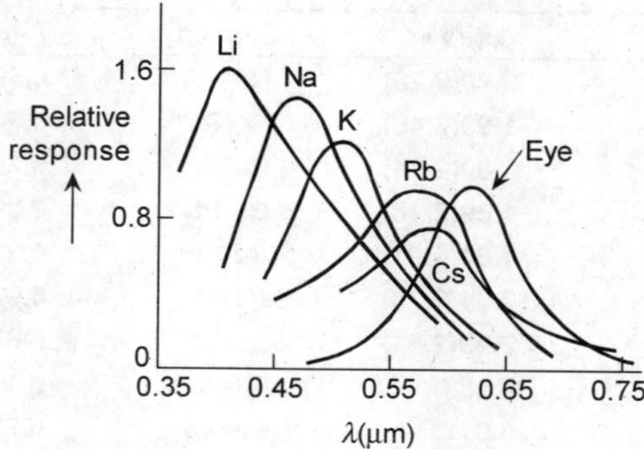

Fig. 5.1 Relative spectral responses of some useful metals as compared with human eye.

The absolute spectral response is often called the *quantum efficiency* and is expressed as a percentage of the maximum attainable for that particular wavelength. For a radiation source, the total power corresponds to the total area under the power distribution curve such as the one shown in Fig. 5.2 for a tungsten lamp and for a photosensistor. The total response to such a distribution curve corresponds to the integral of the product of the spectral response function and the power distribution function, which is actually the area under a newly found curve obtained by the product of the ordinates at each abscissa. The ordinates of this new curve represent the spectral sensitivity. This is often expressed in microampere per lumen or more generally, by microampere per microwatt.

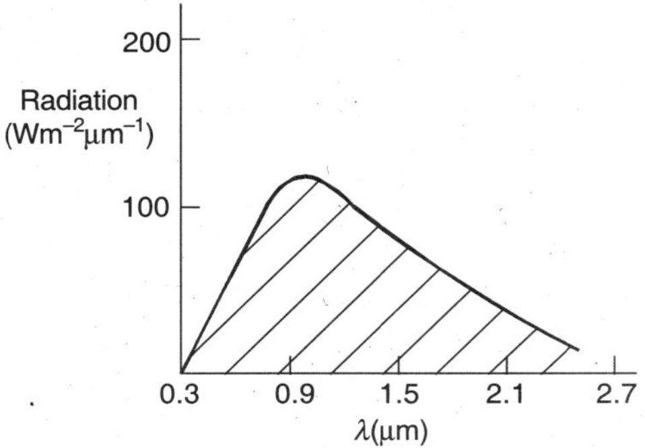

Fig. 5.2 Power distribution of radiation.

Practical use of this property requires to obtain photoemissive surfaces that operate in a frequency range extending from near UV to near IR, also covering the visible range. Good results have been obtained by making photocathodes consisting of alloys of alkali metals and Bi or Sb which covers the entire visible spectrum. However, other cathodes are also in commercial use. Figure 5.3 shows three absolute spectral response curves for materials made by alloying.

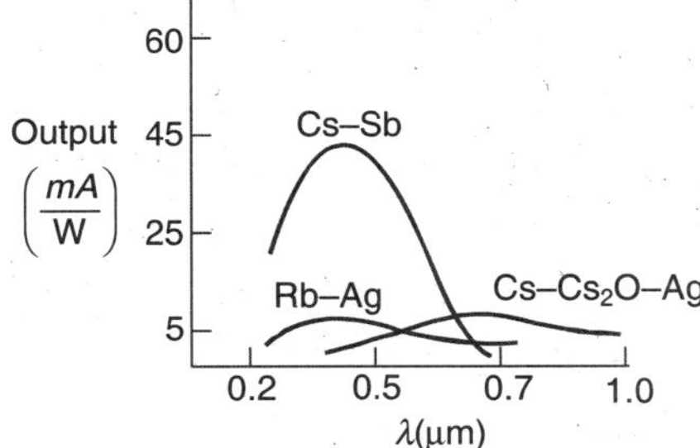

Fig. 5.3 Spectral response curves of some alloyed materials.

(c) *Quantum yield:* It is a ratio of the number of electrons emitted by sensistor cathode to the number of photons it receives for the purpose. The number of electrons emitted constitutes the current. At any wavelength λ, the number of electrons emitted can be given by a number $6.242 \times 10^{18}\zeta$ where ζ is called the sensitivity and a flow of 6.242×10^{18} electrons is required to produce 1 A current.

Also, to free one electron an energy

$$E_\lambda = \frac{12395}{\lambda}\,\mathrm{eV}$$

$$= \frac{1.9857 \times 10^{-8}}{\lambda}\,\mathrm{ergs}$$

is required. Therefore, one watt (= 10^7 erg/s) of power would release

$$\frac{10^7 \times 10^8}{1.9857}\,\lambda = 5.036 \times 10^{14}\lambda \text{ electrons/s}$$

producing a current of $\dfrac{(5.036 \times 10^{14})\lambda}{(6.242 \times 10^{18})}\,\mathrm{A/W}$

The quantum yield is, however,

$$Q_y = \frac{6.242 \times 10^{18}\, \zeta}{5.036 \times 10^{14}\, \lambda}$$

$$= \frac{12395\zeta}{\lambda} = E_\lambda \zeta \tag{5.5}$$

At peak wavelength,

$$\zeta = \zeta_p \quad \text{and} \quad \lambda = \lambda_p$$

So that

$$Q_{yp} = E_{\lambda p}\, \zeta_p \tag{5.6}$$

The energy that a photoelectron acquires by the impact of a photon is expressed as *quantum voltage*. The maximum kinetic energy it can have after escape from the surface is

$$E_m = E_\lambda - \phi \tag{5.7}$$

This energy gives the velocity of the electron with which the photoelectron travels perpendicular to the surface. The threshold wavelength is then defined as the one for which $E_m = 0$. The value of the quantum voltage is, expressed in electron volts, and is given by

$$E_\lambda = \frac{h\nu}{e} = \frac{hc}{\lambda e} \tag{5.8}$$

$$= \frac{1.2395}{\lambda}\ \text{eV} \tag{5.9}$$

where λ is in μm.
Basically therefore,

$$\lambda E_\lambda = \text{constant} \tag{5.10}$$

For photoelectric effect to be actually occurring $E_\lambda > \phi$. Also, Eqs. (5.9) and (5.10) indicate 'colour' of radiation, for visible case for example. Thus, for green radiation, $\lambda = 0.5461$ μm, and

$$E_{\lambda g} = \frac{1.2395}{0.5461} = 2.26974\ \text{eV}$$

The quantity termed work function as used for metals is called *ionization potential* for gases and vapours. This potential is higher for the same metal in its vapour state. For examples, for Cs vapour ionization potential is 3.87 but its work function is 1.54. Table 5.3 shows the differences in the quantity for some other metals in solid and vapour states.

Table 5.3 Work functions and ionization potentials

Element	Work function (eV)	Ionization potential (eV)
Cs	1.54	3.87
K	1.80	4.32
Na	1.94	5.12
Rb	2.15	4.16
Li	2.21	5.37
Sr	2.30	5.67
Ca	2.51	6.09
Hg	4.53	10.38

(d) *Time lag:* Time lag for photosensistors obviously varies over a wide range. It is very small, of the order of 10^{-8} s in photoemissive cells and quite large, of the order of 5×10^{-2} s in light sensitive resistors. In gas-filled photocells, the time lag is of the order of 10^{-5} s. Flashing light with intervals less than 5×10^{-2} s cannot be properly detected by LDR's and only a low average value is indicated. The time response characteristics of the photosensistor become explicit by the general relation

$$y = y_0(1 - e^{-\alpha t}) \tag{5.11}$$

where y_0 and α are constants, α being very large for the photoemissive cells, medium for a class of photovoltaic cells and gas-filled photoemissive cells, and small for the photoresistor types.

(e) When the incident radiation is fluctuating at a frequency larger than 100 Hz, the response of the detector in many cases is not what it should be and, in fact, it does not follow the fluctuations faithfully. This is known as *dynamic fatigue* and is prominent in LDR's. Even otherwise, for a steady high energy incidence, the photodetector output is not always in conformity with the input. Such a discrepancy is often known as *static fatigue* which is predominant in photovoltaic cells, for example, selenium cells show more static fatigue than the others.

(f) *Drift:* Drift or what is conveniently referred to as the 'transient change' in response during a short period after the cell is irradiated, is more common in photovoltaic cells. The amount of change as well as the period depend on the cell storage time in darkness prior to exposure and wavelength range of the source, and it is explained by suggesting that a space charge tends to develop hampering the 'response progress'.

Another kind, often termed as *ageing*, is also the loss of sensitivity with time and is often explained by the fact that there occurs an accumulation of electrons on the inside wall of the sensor.

(g) *Static and dynamic response:* 'Static response' is defined as the ratio of the output to the input for steady illumination. This is basically the *static sensitivity*. Thus,

$$S_{\text{st}} = \frac{i_a}{\phi_l} \tag{5.12}$$

where i_a is the anode current, say, for a photoemissive cell, and ϕ_l is the light flux incident on the cathode.

The 'dynamic response' is similarly given by the dynamic sensitivity

$$S_{\text{dy}} = \frac{\partial i_a}{\partial \phi_l} = \frac{\partial i_a}{\partial t} \bigg/ \frac{\partial \phi_l}{\partial t} \tag{5.13}$$

A photosensor can be represented by its equivalent circuit consisting of a diode in parallel with a capacitance and a high resistance. It is this capacitance that changes the dynamic response compared with the static one.

(h) *Linearity of response:* None of the photosensistors show ideal linearity of response, particularly in the loaded condition. Of these, the photovoltaic cell is said to produce a voltage due to 'inner' photoelectric effect and linearity between this voltage V and the incident light flux ϕ_l is ideal only for zero external load and the corresponding current is ideally linear for zero value of load resistor. For photoresistors or LDR's, the intensity versus resistance approximately exhibit the usual $R = R_o e^{-\beta I}$ characteristics. For photoemissive cells, however, the linearity is much better.

5.3 TYPES OF PHOTOSENSISTORS/PHOTODETECTORS

Of the various types of photodetectors, the commonly used ones are: (i) photoemissive cells and photomultipliers, (ii) photovoltaic cells including photodiodes, and (iii) photoconductive cells and light detecting resistors.

5.3.1 The Photoemissive Cell and the Photomultiplier

This type of radiation sensor shows an 'external' effect which is possible when a photoelectric cell consists of a pair of electrodes separated either by a rarefied rare gas or vacuum. A typical scheme is shown in Fig. 5.4. Light is made to fall on a properly coated surface (called photocathode) to have very low work function resulting in release of electrons which are then attracted towards the

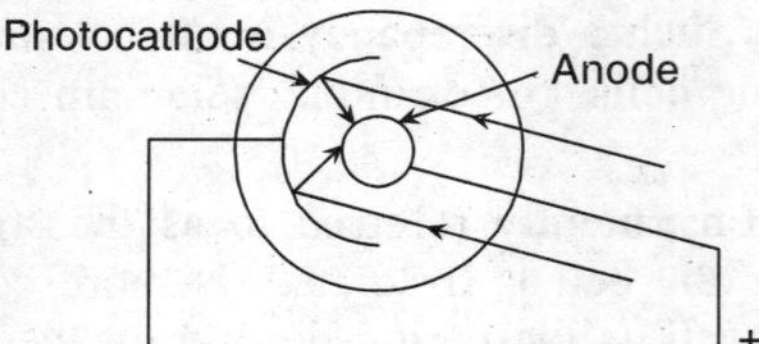

Fig. 5.4 A basic photoemissive cell.

positively charged electrode called the anode. The external circuit can be connected with a resistance so that the change in the current through this indicates the intensity of the optical radiation, falling on the cathode. The current with a single pair of electrodes is very small and photomultiplication process is incorporated in most cases for large current output. The technique takes advantage of the secondary emission of electrons and for this, a number of electrodes called 'dynodes', are used that are basically secondary emitters of electrons. Each successive electrode is kept at a higher potential for the electrons to be attracted by it. The use of nine to eleven such dynodes is a very common practice. A schematic representation of mounting solid surface dynodes is shown in Fig. 5.5.

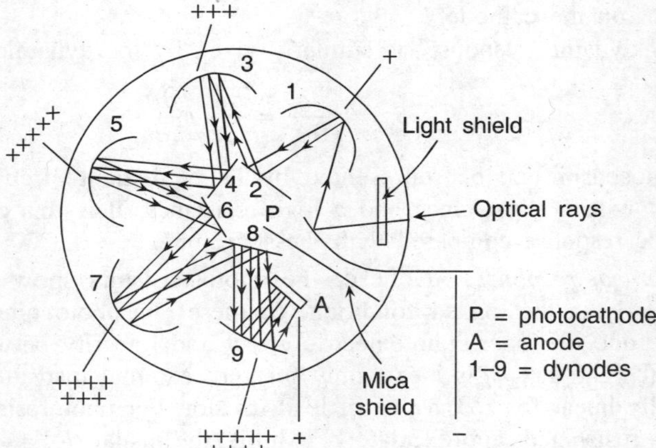

Fig. 5.5 A photomultiplier.

Light shield is actually a grill connected to the photocathode and in this way the electrode assembly is electrostatically completed. Optical radiation reaches the photocathode P through this shield and electrons liberated from the photocathode are first attracted towards dynode 1. These electrons, by impact on the surface of the dynode 1, release a number of secondary electrons whose number depends on the energy of the photoelectrons. As a result, successive impacts occur on the other dynodes increasing electrons exponentially and at the end a 'copious' stream of electrons is collected by the anode A and an external load may now be connected to it to produce the output current I given by

$$I = I_p K^n \tag{5.14}$$

where

I_p is the initial primary photoelectric current constituted by the electrons leaving the photocathode,

K is a constant dependent on the cathode (dynode) materials and is often known as dynode emission coefficient, and

n is the number of stages.

The potential difference applied between successive stages is about 100–130 volts. Such an arrangement can produce a current amplification of 10^6–10^7 or even higher. A mica shield is provided between the photocathode and subsequent multiplying stages for isolation, to prevent spurious electron emission.

If the voltage per dynode pair is V and n is the number of dynode stages, sensitivity S of the detector is observed to be proportional to $V^{n/2}$. Thus,

$$S = k_s V^{n/2} \tag{5.15}$$

where k_s is a constant that depends on the materials chosen. Cathode areas are to be optimized since larger areas contribute to noise by spurious or uncontrolled release of particles. Typical current sensitivity/amplification characteristics of a photomultiplier are presented in Fig. 5.6 for different dynode stage voltage supplies.

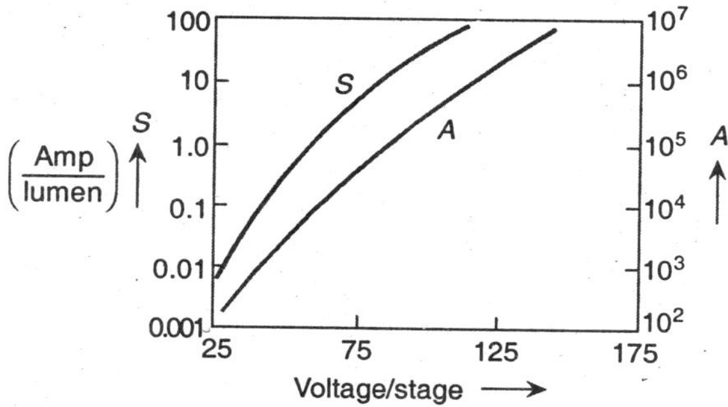

Fig. 5.6 Current–voltage and amplification–voltage characteristics of a photomultiplier.

There is thermionic emission from the photocathode also and a permanent 'dark current' is produced. This contributes to noise due to its random nature and ultimately sensitivity gets limited because of this noise. The guiding equation for such an emission is the 'Richardson's equation' given as

$$J = AT^2 \, e^{-b/T} \tag{5.16}$$

where

A is a constant,

b is a constant but slightly dependent on temperature T, and

J is the emission current density determined by the rate of electron escape through the surface.

At room temperature, number of electrons per second per cm^2 area may become as large as 10^{10} obliterating the primary emission due to a single photon incidence. In this way, the sensitivity becomes limited.

Varieties of dynode arrangement are available and have been developed. One such serial arrangement of dynodes in the pattern of venetian blinds is shown in Fig. 5.7. Figure 5.8 shows the voltage V_{da} versus I plots for different lumen incidence where per stage voltage is fixed. Emission from dynodes depends on the voltage per dynode pair and the energy of electrons received by it for impact. For a good vacuum, secondary emission is not otherwise decreased. Low work function photocathode coating, as has already been mentioned, is that of Cs–Cs$_2$O–Ag and for this coating, secondary emission also increases. The velocity v, a primary electron attains, is given by

$$v = 5.95 \times 10^4 V^{1/2} \tag{5.17}$$

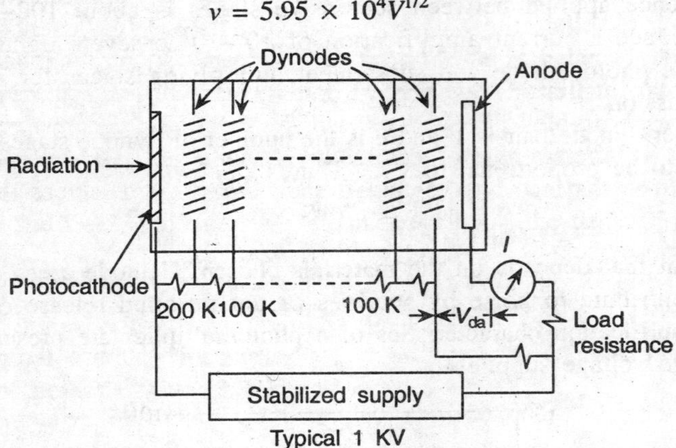

Fig. 5.7 Photomultiplier with serially arranged dynodes.

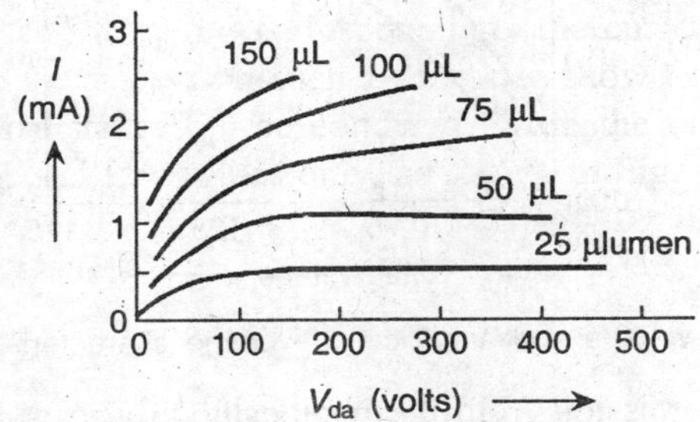

Fig. 5.8 Photomultiplier characteristics with radiation as a parameter.

where V is the electrostatic potential so that the kinetic energy acquired by the primary electron

$$E = \frac{mv^2}{2} = \frac{(5.95)^2 \times 10^8}{2} Vm \tag{5.18}$$

Table 5.4 shows the dynode emission efficiency for some commonly used metallic and alloy surfaces with specific dynode voltages.

Table 5.4 Emission efficiency with voltage for different surfaces

Surfaces	Voltage	Efficiency (maximum) (K)
Lithium	100	0.56
Beryllium	150	0.90
Magnesium	300	0.95
Aluminium	300	0.97
Iron	300	1.0
Copper	240	1.32
Molybdenum	375	1.25
Silver	250	0.93
Caesium	400	0.72
Platinum	250	1.01
$Cs–Cs_2O$ on Ag	600	10.0
$Cs–Cs_2O$ on Ni	550	5.0
$Cs–Cs_2O$ on Mo	500	3.0
$Cs–Cs_2O$ on Fe	500	2.0
Cu–Be (2%) alloy	400	1.95
Ag–Al (2%) alloy	560	2.8
Ag–Ca (2%) alloy	360	4.5

Depending on the cathode coating material, the spectral response or the spectral sensitivity of the cell changes. Manufacturers provide the response charts for the cells they produce. Some typical special response curves are shown in Figs. 5.9(a), (b), (c), and (d).

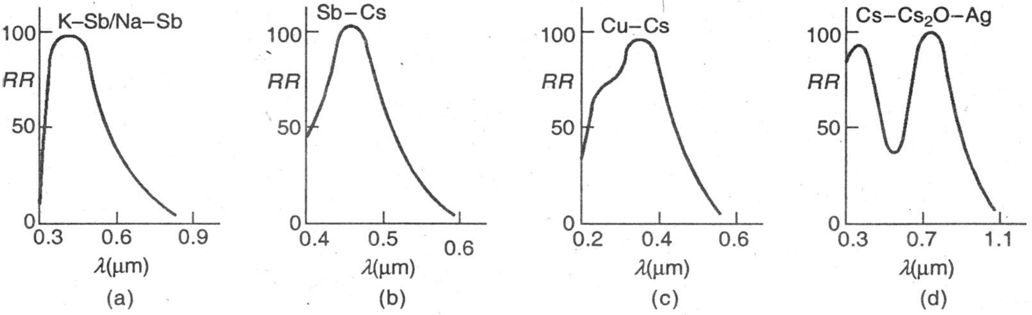

Fig. 5.9 Relative spectral response characteristics of some photoemissive surface materials.

Attempts are made to utilize the visible and IR range as far as possible. Cathode materials for spectral range are chosen with appropriate work function. Table 5.5 shows the list of cathode materials for three distinct spectral ranges.

Table 5.5   Spectral ranges and cathode materials

Spectral range	Cathode materials	Work function category
Infrared	Cs–Cs$_2$O–Ag, Ba–Ag, BaO–Pt, Cs–Pt.	Low
Visible	Cs–Cs$_2$O–Ag, Cs–Mg, Li–W, Se–Fe, Sr,	
	Ba, Cu$_2$O, K, Rb, Li, Na–Pt, K–Pt and so on.	Low–medium
Ultraviolet	Cd, U, Na, Ca, Ta, Ti–Ni, W–Ni,	High
	Zr, Ni, Mo, Zn, Pt, Th–Ni, etc	

The vacuum photoemissive cell has a variation when it is filled with an inert gas at a very low pressure. With a potential between the cathode and the anode exceeding a certain critical value for the filling gas, the photoelectron emitted, gets accelerated and ionizes a gas atom into another electron and a positive ion. The positive space charge close to the photocathode may induce secondary electron emission from it which partly neutralizes the positive ions. A part of these secondary electrons with some photoelectrons and 'gas-ionized' electrons move towards the anode to form photoelectric current, which, obviously, is much more compared to that produced in a vacuum photocell. Depending on the gas and pressure, there is a multiplying factor called gas factor G which is rarely higher than 10. As the gas ionization potential is high, photon cannot ionize a gas atom. Figure 5.10 shows the technique of generating a 'photocurrent' with gas tube type photocell. Table 5.6 shows the ionization potentials (V_{ip}) of commonly used gases.

Table 5.6   Ionization potentials

Gas	Ionization potential, V_{ip}
Xenon	12.0
Krypton	13.9
Argon	15.7
Neon	21.5
Helium	24.5

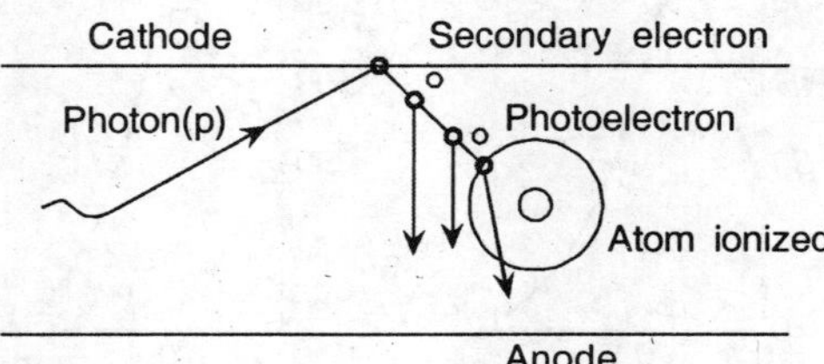

Fig. 5.10   Photocurrent generation mechanism.

With increasing anode voltage, the gas factor tends to increase but anode voltage is to be kept limited to a value to avoid breakdown or gaseous discharge. The gas factor G versus relative potential V, a ratio of applied potential to breakdown potential as a percentage, is given in Fig. 5.11. The actual value of the breakdown potential varies from 200–300 volts. Because the positive ions also move, they also contribute to photocurrent indirectly, although they move with much smaller speed. This increases the cell response time. However, proper geometrical design can reduce this time to a certain extent.

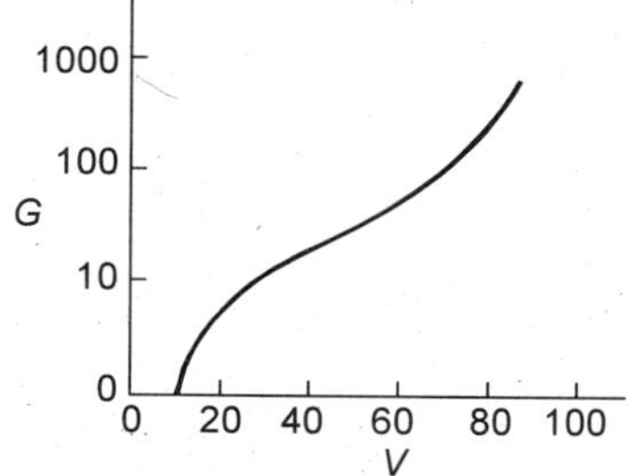

Fig. 5.11   Gas factor, *G*, versus relative potential, *V*.

5.3.2   The Photoconductive Cell

It has earlier been discussed that in intrinsic semiconductors thermal energy can cause a small proportion of the valence electrons to jump to the first conduction band. The holes produced in the process and the free electrons so produced move in the crystal lattice resulting in a kind of conduction. Interestingly, energy can be in any form of radiation including photoradiation. If this phenomenon occurs due to photons, it is known as *photoconductivity,* and occurs when sufficient number of electrons shift into the conduction band after being irradiated by photons changing the conductivity of the material. These devices include the photoresistors or light dependent resistors (LDR).

The LDR

The phenomenon of photoconductivity as in Sec. 5.3.2 is an oversimplified presentation. As specified in the beginning of this chapter, only when the photon energy is sufficient to shift the electrons does the conductivity appear to change implying that only below a certain wavelength, it is possible and a sharp cut-off is likely to be observed when the photonic energy is equal to the semiconductor energy gap. But this is not wholly true and as shown in Fig. 5.12, a gradual transition does occur. This has been explained using the probability that a photon ($\lambda > \lambda_{\text{threshold}}$) gives its energy to an electron which already is excited to a certain extent by thermal agitation and the sum total of this energy raises the electron to the conduction level. Also, the width of the energy gap fluctuates as a result of thermal vibrations within the lattice. Generally, however, photons of lower energy, in the far infrared region for example, pass through the material without having to part with their energy and semiconductors such as germanium are being used to make windows or lenses for such radiations.

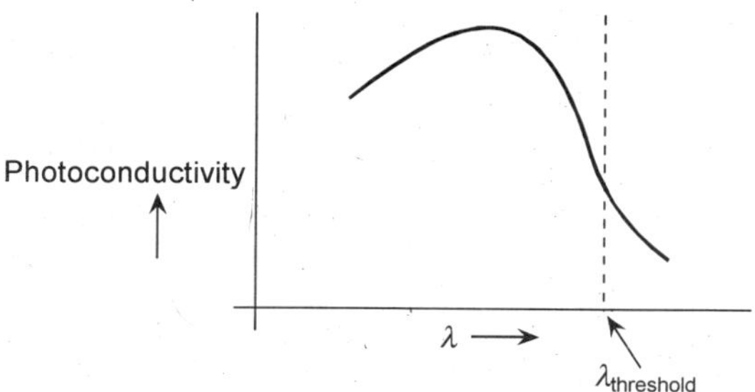

Fig. 5.12   Photoconductivity versus λ plot for photoconductors.

Presence of impurities or imperfections in the semiconductor material alters (enhances) recombination of holes and electrons and affects the carrier life time. Also, imperfections in the form of trapping centres can trap a carrier and subsequently release it, affecting the speed of movement of carrier and correspondingly changes in resistivity after any change in incident radiant energy can occur.

Earlier works have shown that the resistance of a photoconductor changes with incident photoradiation intensity but in a complex manner. Materials such as sulphides of lead, thallium, and selenium have shown an approximate relation between the resistance r of the sensor, its dark resistance r_d, and intensity of illumination I, as

$$r = \left(\frac{1}{r_d} + \frac{I^{\beta^{-1}}}{\alpha} \right)^{-1} \tag{5.19}$$

where α and β are constants that are, however, material dependent. The nature of the r–I relation is approximately given in Figs. 5.13(a) and (b).

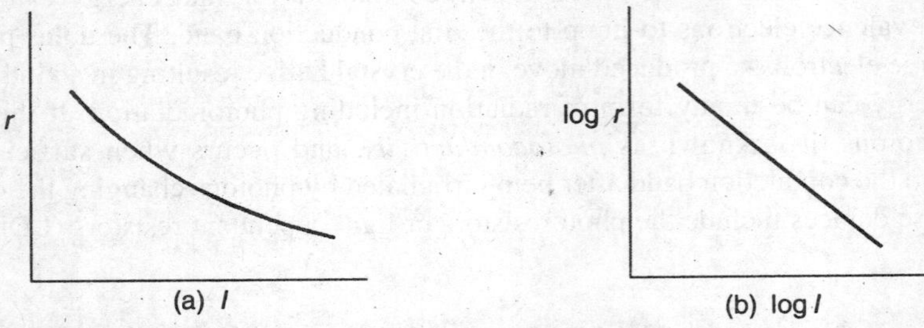

Fig. 5.13 Resistance versus intensity of illumination plots (a) linear scale, (b) log scale.

It has subsequently been investigated and found out that the electrical and spectral properties of semiconductors, specifically the intrinsic type, are dependent on many factors some of which can be brought under control during the fabrication process by which the spectral response and speed of response may be controlled mainly by regulated impurity level. The materials for making such resistors are presently chosen from the sulphides and selenides of Cd and Pb. Appropriate impurity regulation can be made in CdS and CdSe with Ga, Cu, and Ag, particularly for peak response but in a spectral range of 0.53–0.70 μm. CdSe, however, is more temperature-sensitive than CdS. The electrode connections for CdS and CdSe are made with indium (In) and/or tin (Sn) to avoid nonlinearity in V–I characteristics of these contacts which has been seen to occur with gold terminals.

For low energy photon detection, say in the far infrared region, the semiconductor is heavily doped with an 'activator' impurity. Gold is a good doping material for germanium. Table 5.7 shows a list of doping materials for the type of semiconductors with the usable range of wavelengths on the high side. But, for proper photo-activation, they must be cooled to very low temperatures when their response time becomes small, approximately 1 μs.

Table 5.7 Doping materials for different semiconductors

Semiconductor	Doping material	Maximum value of λ (μm)
Germanium	gold	9
Germanium	mercury	15
Germanium	copper	29
Germanium–silicon	gold	14
Germanium–silicon	zinc	16

Metallic films of selenium, thallium, and lead change resistance with photo-irradiation; so also does a film of tellurium, but it recovers only after a very long time after the exposure is withdrawn. Selenium resistance cell at one time was quite in use. Thin film of Se, of about 0.003 mm deposited on an insulating base encapsulated in an inert atmosphere or evacuated atmosphere was used. The latter processing is necessary as oxygen and water vapour/moisture in air cause deterioration of the sensitive selenium layer. Its response characteristics are shown in Fig. 5.14. But irradiation must not be modulated since increasing modulation frequency decreases the relative sensitivity almost exponentially.

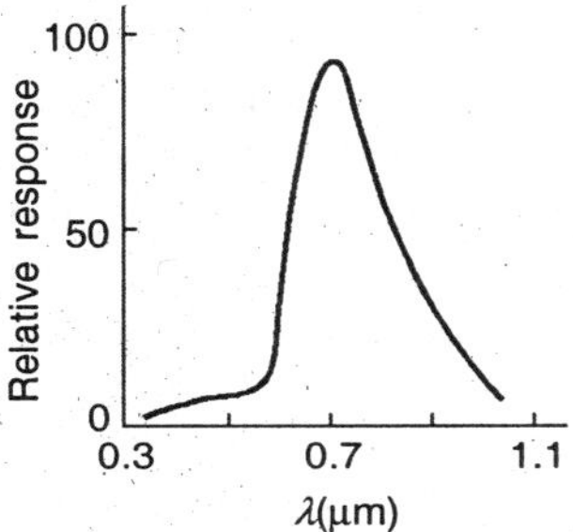

Fig. 5.14 Relative spectral characteristics of selenium cell.

Metal sulphide resistors are by far the most important variety. As has already been mentioned, PbS and CdS are very common kinds and are available in variety of shapes and sizes. Table 5.8 shows some characteristics of a few photoresistors and Fig. 5.15 shows the responsivity to wavelength for PbS, CdS and CdSe for fixed illumination.

Table 5.8 Photoresistor characteristics

Material	*Usable resistance change ratio*	λ *range* (μm)	λ_{peak} (μm)
Thallium sulphide	$10^4 : 1$	0.6–1.8	1.1
Lead sulphide	$10^3 : 1$	0.3–3.5	2.0–2.7
Lead selenide		0.3–4.0	3.0
Lead telluride			3.5
Cadmium sulphide	150 : 1	0.25–0.9	0.52
Cadmium selenide		0.3–1.1	0.74
Cadmium telluride		0.3–1.2	0.85
Bismuth sulphide	$10^2 : 1$	0.4–4.0	0.7
Bismuth telluride			1.0

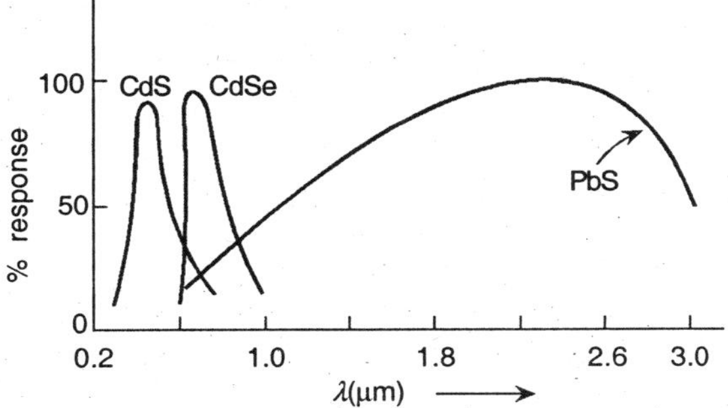

Fig. 5.15 Spectral response characteristics of certain useful photoconductor materials.

Resistance ratio change is dependent on the illumination in lux and also on the supply given to the conductive cell. It is assumed above that the illumination level changes in a ratio 1:1000 and a voltage of ·100 V is impressed.

The material is usually deposited on quartz or ceramic strip. Specially made ceramic materials alone also show photoresistive property. A combination of 90 parts by weight of TiO_2 with oxides and/or titanates of bivalent elements such as Sr, Pb, Mg, Zn, Ba, and Cd shows good photoconductivity with dark to light resistance varying in a ratio of about 15:1. Titanium oxide has an absorption band at around 0.4 µm and it shows peak spectral response in the red region, around 0.65–0.7 µm.

When a photoconductor is to be used only in the visible region, specification or characterization with reference to illumination level may be made while for IR region, reference about the radiation units or colour temperature must be made. With reference to Fig. 5.15, it can be seen that a photoconductive cell has to be characterized with reference to the operating waveband, such as 0.3–3 µm for PbS. However, manufacturers often characterize the sensitivity of their products by stating a condition in which this is tested. One such specification is that a cell when illuminated by tungsten lamp at colour temperature of T_c K whose radiation is chopped at f_c Hz, the cell area receives a lumen of the sensitivity which is given as $S = I_p/l$ in mA(peak)/lumen. Companies give numerical values for this relation. The testing arrangement is shown in Fig. 5.16(a). For obtaining the effective value T_c, a practical idea is to use a lamp of specific wattage. Thus, for $T_c = 2700$ K, a tungsten filament lamp of 75 W running at normal voltage is accepted.

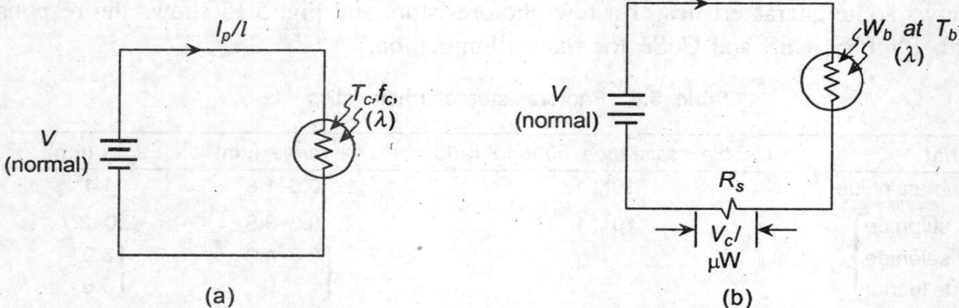

Fig. 5.16   (a) Photoresistor testing arrangement and (b) the arrangement with standard resistance R_s.

For infrared radiation, the characterization is through a blackbody radiation emitted at T_b (= 200°C say). If the emitted radiation received by the cell is W_b, the cell running with a standard voltage and a standard resistance (R_s) would produce (across R_s) a specified rms voltage V_c per µW of radiation. Figure 5.16(b) shows this arrangement with R_s. The load resistor R_s is arbitrarily found but was initially chosen for maximum power transfer which is not necessary here. It is, instead, to be selected on the basis of the following 'electronic' stage such as an amplifier. Radiation is chopped to obtain alternating signal for convenience of amplification and specification where I_{peak} and V_{rms} are arbitrary. The chopping frequency f_c must be well within the device cut-off frequency f_α. Supply is chosen such that the cell can withstand it conveniently.

Photocurrent

Photoconducting materials have been discussed earlier. Their structural features are nonetheless significant. Figures 5.17(a) and (b) show a simple and planar photoconductor structure. In the former, that is Fig. 5.17(a), the entire slab of the semiconductor material acts as the photoconductor which must have a depth d, with $d > d_p$. The term d_p denotes the penetration depth which is the reciprocal of the optical absorption coefficient α. This ensures maximum light absorption. If the thickness d becomes very large, a smaller thickness part of it only serves as the photoconductor, and the rest becomes a shunt resistance path. For minimizing loss component, often an anti-reflection coating is provided on the radiation receiving surface of the material. Figure 5.18 gives an idea about the values of penetration depth for different wavelengths. Radiation incident on the materials generates holes and electrons which must have sufficient mobility and life-time to be collected by the ohmic contacts. To help faster movement, external field is applied as has already been discussed. Figure 5.17(b) shows a structure where the active photoconductive region is formed near the surface of the semiconductor and close to the ohmic contact zones.

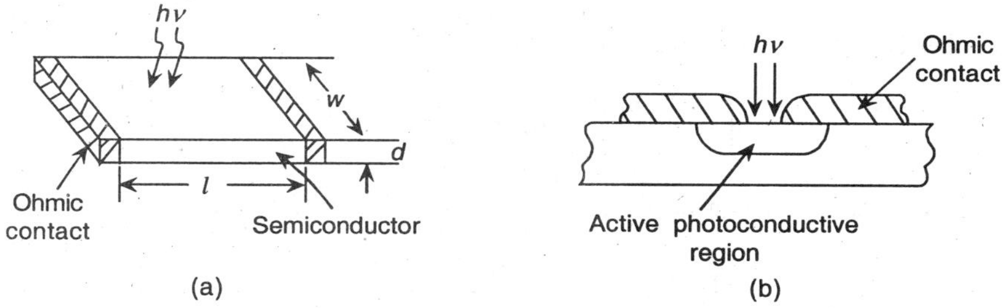

Fig. 5.17   Semiconductor-based photoconductor: (a) simple, (b) planar.

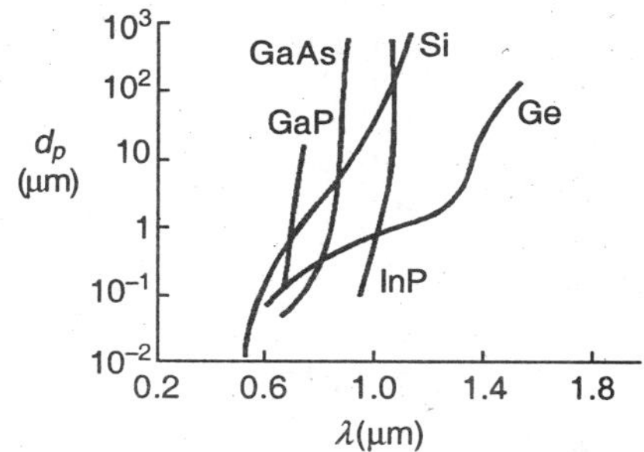

Fig. 5.18   Penetration depth versus wavelength plots for semiconductor sensors.

As has been mentioned, some of the charge carriers generated are lost due to (i) recombination, (ii) traps, and (iii) 'sweep out'. The first two phenomena are well understood while the sweep out occurs when the carriers recombine at the external ohmic contacts.

Referring to Fig. 5.17(a), if the photoconductor is n-type and of the dimensions as shown, electron–hole pair generation rate per unit volume with incident radiation power P, is given by

$$n_0 = \int_0^d n_0(x)\, dx \qquad (5.20\text{a})$$

$$= \int_0^d \frac{\eta \alpha P}{h v w l d} \exp(-\alpha x)\, dx$$

$$= \frac{\eta P \alpha}{h \nu w l d} \left[1 - \exp(-\alpha d) \right] \tag{5.20b}$$

In photogeneration process, the number of positive and negative charge carriers is same so that $\Delta n = \Delta p$ and hence, when an electric field E is applied, the current through the material can be calculated, for a charge unit e, as

$$I = eE(\mu_n \Delta n + \mu_p \Delta p) w d$$

$$= eE \Delta p (\mu_n + \mu_p) w d \tag{5.21}$$

If the concentration of the optically generated holes is p_1, the concentration of the holes in equilibrium state is p_2, and τ_p is the minority (hole) carrier life-time, then $(p_1 - p_2)/\tau_p = \Delta p / \tau_p = n_0$. Hence

$$I = eE n_0 \tau_p (\mu_n + \mu_p) w d \tag{5.22}$$

Using Eq. (5.20b), Eq. (5.22) can be derived as

$$I = \left(eE\tau_p \frac{\eta P \alpha}{h \nu l} \right) (\mu_n + \mu_p)(1 - e^{-\alpha d}) \tag{5.23}$$

Also, velocities v_p and v_n can be given by $v_{p,n} = \mu_{p,n} E$ and transit times $t_{n,p}$ by $l/v_{n,p}$, so that

$$I = \left(\frac{e \eta P \tau_p}{h \nu} \right) \left(\frac{1}{t_n} + \frac{1}{t_p} \right) (1 - e^{-\alpha d}) \tag{5.24}$$

gives the expression for the photocurrent.

The device gain defined as the ratio of the carrier collection rate to the carrier generation rate G, is given by

$$G = \left(\frac{I/(w l d e)}{n_0} \right) \tag{5.25}$$

Combining Eqs. (5.22) and (5.25), we get

$$G = \tau_p (\mu_n + \mu_p) \frac{V}{l^2} \tag{5.26}$$

where V is the applied voltage, equal to El.

Minority carrier life-time determines the device response time. This time also limits the bandwidth, as has already been mentioned. The high frequency limit is, thus,

$$f_h = \frac{1}{2 \pi \tau_p}$$

The 'idealized' response characteristics of the photoconductor with λ, shown in Fig. 5.12 are rarely met in practice. The curves of Fig. 5.18 show that with decreasing wavelength, carrier generation is only at the surface or near-surface depths. Such a condition is more defect-oriented like having more trap densities resulting in loss of generated carriers, at least temporarily, and response characteristics shown in Fig. 5.19 are more appropriate.

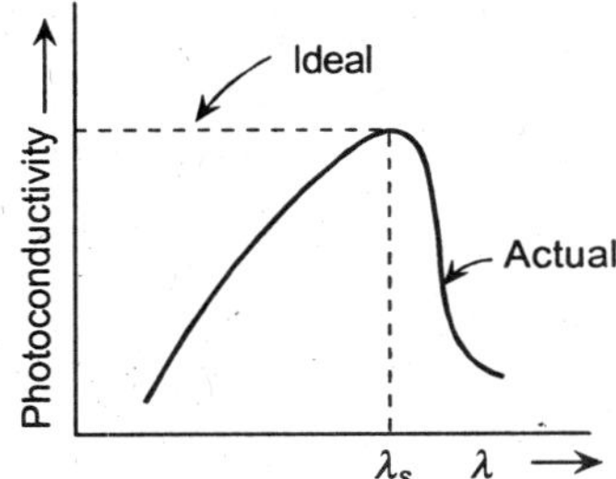

Fig. 5.19 Comparing ideal and actual spectral photoconductivity.

Noise

There is always some inherent noise in the photoconductors. A rigorous analysis of all such noise sources and their effects is very complex. In general, there are four common sources of noise that lead to (i) Johnson's or thermal noise, (ii) shot noise due to recombination, (iii) flicker noise, also known as contact noise, current noise, excess noise or $1/f$-noise, and (iv) interference or intersymbol noise.

(i) Johnson noise: It is a well understood phenomenon and is known to be generated due to the random motion of carriers within a resistive material. The motion of the carriers, however, is induced by thermal kinetic energy. This random motion produces a current or a voltage that fluctuates about a zero mean value. For a resistance R, temperature T, electrical output modulation frequency f and frequency interval Δf, the average value of the squared voltage is given by

$$\langle v^2 \rangle (f) = \frac{4Rhf\Delta f}{e^{hf/kT} - 1}$$

$$\approx \frac{4Rhf\Delta f}{hf/(kT)} \quad \text{for } kT \gg hf$$

$$= 4kTR\Delta f \tag{5.27}$$

Its circuit equivalent model has, therefore, a source of voltage $(\langle v^2 \rangle)^{1/2}$ with a noiseless resistance R. In series with a load resistance R_L, it would give a current (squared average) as

$$\langle i^2 \rangle = \frac{\langle v^2 \rangle (f)}{(R + R_L)^2} \tag{5.28}$$

Maximum power is dissipated for $R = R_L$, this power is called the *available noise power*, P_n and is given by

$$P_n(f) = \langle i^2 \rangle R_L = kT\Delta f \tag{5.29}$$

This is called *white noise,* since the power is independent of frequency spectrum, and, for the same resistance, it is same for equal intervals of frequency anywhere in the spectrum. If the spectrum bandwidth is BW, power is given by

$$P_n = kTBW \tag{5.30}$$

As power is directly proportional to BW, noise level increases with high frequency, or, for digital case, high bit rate.

(ii) Shot noise: This is generated due to the random fluctuations of the number of carriers of the charges themselves. As photons are absorbed in the photoconductor surface to generate carriers, this generation itself becomes random in nature. This phenomenon was first analyzed by Schottky and he obtained an expression for the average noise current due to this. Accordingly,

$$\langle i_s^2 \rangle (f) = 2eI\Delta f \tag{5.31}$$

I being the average dc current generated by photons.

In a ·photoconductor, energy received is in discrete steps and the corresponding pulses generated follow Poisson's distribution. But the recombination being a statistical phenomenon, the pulse widths are not the same. Considering these, the shot noise current can be obtained as

$$\langle i_{sp}^2 \rangle (f) = \frac{4GeI}{1 + 4\pi^2 f^2 \tau_p^2} \tag{5.32}$$

where G is the gain and I is the current given by Eq. (5.24). It can be shown that $G = \tau_p/t_n$, also $t_n \ll t_p$ and $d \gg 1/\alpha$, so that I is simplified to

$$I = \frac{e\eta PG}{h\nu} \tag{5.33}$$

Using Eq. (5.33) in Eq. (5.32),

$$\langle i_{sp}^2 \rangle (f) = \left(\frac{4e^2 \eta P \tau_p^2}{h\nu t_n^2} \right) \left(1 + \frac{1}{4\pi^2 f^2 \tau_p^2} \right) \tag{5.34}$$

For a bandwidth BW, Eq. (5.34) can be integrated to obtain

$$\langle i_{sp}^2 \rangle = \left(\frac{2e^2 \eta P \tau_p}{\pi h \nu t_n^2} \right) \tan^{-1}(2\pi BW\, \tau_P) \tag{5.35}$$

Usually, $BW\tau_P \ll 1$, so that from Eq. (5.32) and (5.35) one can simplify

$$\langle i_{sp}^2 \rangle = 4eGIBW \tag{5.36}$$

(iii) Flicker noise: This noise is observed in almost all measurement systems. Its power spectral density varies as k/f^n, $0.8 \le n \le 1.4$ where k is a constant. The spectral density has been found to exist in a frequency range 10^{-6}–10^6 Hz. The phenomenon is not well explained.

(iv) Interference or intersymbol noise: To detect high bit rate discrete signals, a slow photoconductor faces overlapping of the bits which results in a background current that persists like a noise current. This is termed as *intersymbol noise* or *interference*.

For low bit rate signals, however, Johnson noise turns out to be most predominant and it is with this noise that the noise equivalent power is considered. In fact, the characterization of a sensor of this type is rationalized in terms of two parameters, namely specific responsivity S_R and the noise equivalent power (NEP). The first one is defined as the rms microvolts obtained across a matched resistor (R_m) with the cell irradiated with 1 μW/cm^2 radiation and the combination of R_m and R_{cell} is supplied by 1 V. Obviously, S_R is wavelength-dependent and a set of S_R–λ curves for a cell is supplied ·by the manufacturer. The NEP is often defined as the radiation incident on the cell to produce a signal voltage equal to the noise voltage. If W is the radiation energy incident per unit time, a is the cell area, V_s is the signal voltage, and V_n is the noise voltage then *NEP* may be given by the relation

$$NEP = \left(\frac{Wa}{V_s}\right)V_n \tag{5.37}$$

However, the signal-to-noise ratio in terms of squares of currents is

$$\frac{S}{N} = \frac{I^2}{\langle i^2 \rangle} \tag{5.38}$$

Often, detectivity D is specified, which is defined as the inverse of *NEP* for

   (i)   One Hz bandwidth at a given wavelength,
  (ii)   Given modulation frequency, and
 (iii)   A given source radiation temperature.

Another more commonly used term, specific detectivity $D*$ is the normalized D to 1 cm^2 detector area. This is mathematically expressed as

$$D* = 1.3 \times 10^{-11}\left(\frac{30}{T}\right)^{5/2}\left[\frac{\exp(h\nu_0/(2kT))}{\dfrac{h\nu_0}{kT}\left(\dfrac{h^2\nu_0^2}{k^2T^2} + \dfrac{2h\nu_0}{kT} + 2\right)^{1/2}}\right] \text{cmHz}^{1/2}/\text{W} \tag{5.39}$$

where ν_0 is the minimum detectable frequency.

Although digital signals are two level signals (0 or 1), often a spread is encountered around these ideal values and for analysis purposes, the distribution is considered as Gaussian. Keeping this under consideration, the error probability in total is obtained by summing the contributions of the individual ones. For two level signals, the error probability is

$$P(e) = \frac{1}{2}\left[1 - (erf)\left(\frac{1}{2}\sqrt{\frac{S/N}{2}}\right)\right] \tag{5.40}$$

where the error function $erf(y)$ is given as

$$erf(y) = \frac{2}{\sqrt{\pi}}\int_0^y e^{-x^2}dx \tag{5.41}$$

with Fig. 5.20 showing the plot.

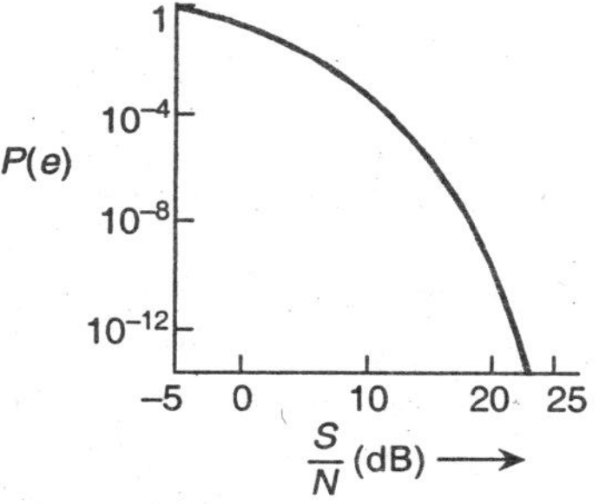

Fig. 5.20 Error probability versus *S/N* ratio.

It must be stressed at this point that only the average power in a pulse is determined and not the true power in a pulse. This is all the more true if the incoming power is low. However, the number of photons in pulse stream is more easily found as their occurrence follows Poisson's distribution. If on an average, N photons are considered, their number in a pulse, n, is given by the probability

$$P(n) = \frac{N^n e^{-N}}{n!} \tag{5.42}$$

For two level signals, with no photons in the pulse, ($n = 0$), the probability error of having a '1' is given 2×10^{-9} as the standard, from which N is obtained as 20 and hence, the average number of photons per bit in the minimum detectable signal is 10. Therefore, the average power in the pulse is

$$P_{av} = \left(\frac{h\nu}{\eta}\right) \times 10 \times \text{bit rate} \tag{5.43}$$

where $h\nu/\eta$ is the average energy per absorbed photon.

For practical applications, manufacturers provide a set of curves containing informations relating to cell voltage versus cell current at various illuminance levels and the corresponding dissipation levels of the cell, response time for different illumination levels. Figures 5.21(a) and (b) show the nature of these curves.

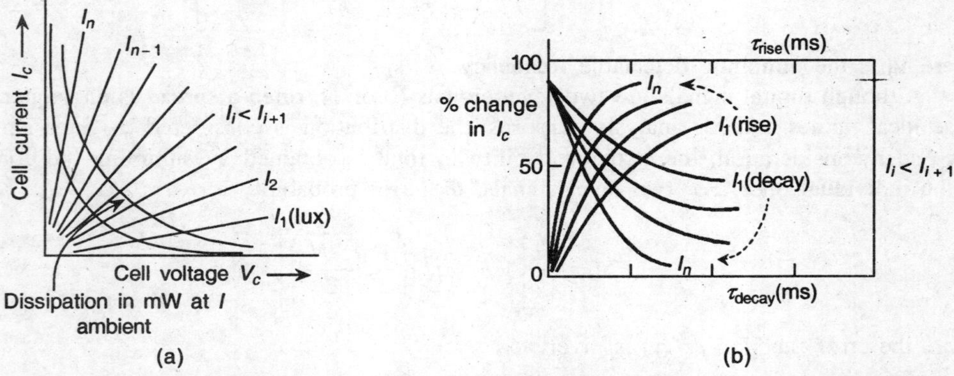

Fig. 5.21   (a) Cell current versus cell voltage for different illumination levels and (b) Cell current versus rise or decay time for different illumination levels.

Storing of photoconductors is an important aspect. Sensitivity is higher when it is exposed to radiation after storing in a 'dark' condition than in an ambient condition. For repeatability, the storing should be at the same condition all the time for a reasonable period.

5.3.3   Photovoltaic and Photojunction Cells

The photoemf cells are seen to generate emf when irradiated with radiation of proper range and illuminance. Earlier theories of photoemf generation such as photochemical reaction, the current generation at interface layer called the barrier layer or boundary layer, and electron diffusion theory of the semiconductor photoeffect have all been subsequently replaced or modified. Two kinds of photocells have, however, long been known—the backwall type in which the electrode exposed to the incident radiation becomes positive, and the frontwall type where it becomes negative as shown in Figs. 5.22(a) and (b). The barrier layer, as already mentioned, is the interface of electrodes at which the charges develop and move.

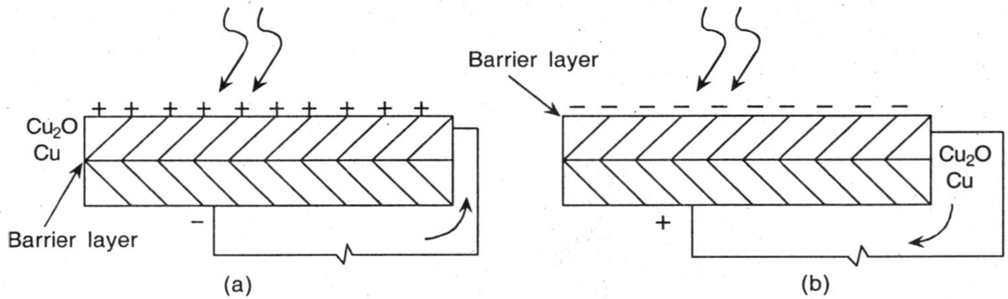

Fig. 5.22 Barrier layer cell: (a) backwall type and (b) frontwall type.

It is now known that the emf generates following the same electronic process as in p-n junction of a semiconductor.

The conventional photoemf cell, called the photovoltaic cell, consists of a layer of semiconductor on a metal disc. A thin translucent or transparent layer of a precious metal is sputtered on top. The incident radiation on the top layer passes through it and is absorbed by the upper surface of the semiconductor and electrons are freed. These electrons flow away from or towards the incident radiation depending on the type of cell, as already discussed in Figs. 5.22.

Commercial photocells are of two different varieties as far as material choice is concerned. One uses copper oxide on copper while the other uses selenium on metal base as shown in Fig. 5.23.

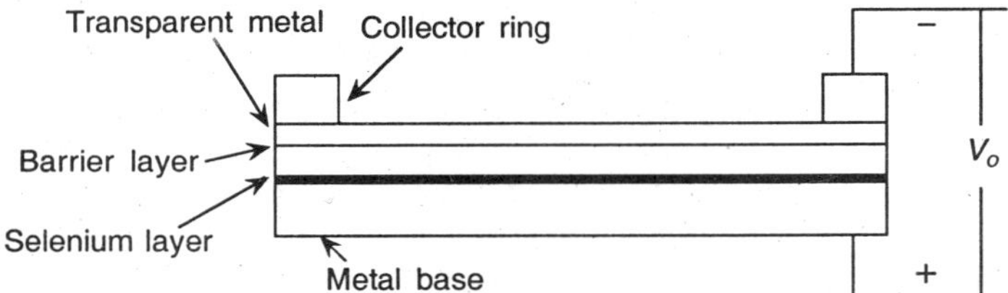

Fig. 5.23 Structure of a selenium photovoltaic cell.

The selenium cell characteristics are very close to that of human eye. Modern selenium cell is a little different. Pure selenium is a p-type semiconductor, the metal base aluminium or brass. At the first stage of manufacturing, selenium film coating on the base is heated converting it to crystalline state. Cadmium is then diffused in selenium to form p-n junction as cadmium oxide forms the n-type layer. This layer was earlier called the barrier layer. A very thin layer of gold forms the transparent metal layer and it also acts as the alternate electrode. The cell, thus formed, is encapsulated in suitable resin with leads/terminals properly taken out.

More recently, it has been found that silicon is more efficient than selenium but its peak response occurs in the IR region and silicon photovoltaic cell has been very successfully adopted as the solar cell for conversion of sunlight into emf or current. It may be either p-on-n type or n-on-p type. The latter is more resistant to degeneration by high energy radiation and is used in satellites more effectively than the other.

Across a p-n junction, carrier migration occurs and hence, a potential difference is developed. The operation of a p-n junction with or without an impressed electric voltage across it can have three distinct modes of operation shown in Figs. 5.24(a), (b), and (c). If the p-n junction is forward biased, it is used basically as an electric circuit element for rectification purposes and photo-irradiation has no control in its current. The V–I characteristics are shown in the first quadrant of Fig. 5.25.

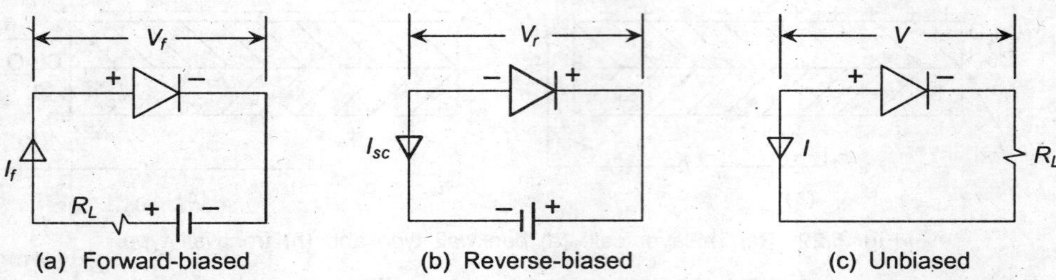

Fig. 5.24   Operation scheme of a p-n junction diode: (a) forward biased, (b) reverse biased, and (c) unbiased.

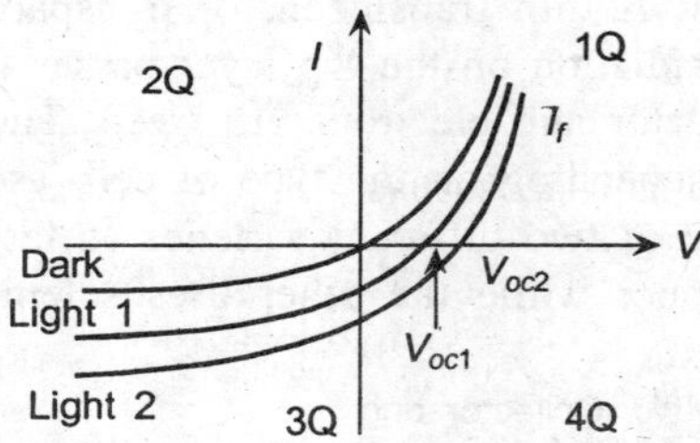

Fig. 5.25 V–I plots of the p-n junction diode in three quadrants.

If the diode is reverse biased, small leakage current due to movements of intrinsically generated carriers would flow. Now with photo-irradiation, the rate of generation of intrinsic carriers (electrons in p-region, holes in n-region) is enhanced and a set of curve in the third quadrant of Fig. 5.25 are obtained with increasing illumination. This mode of operation is commonly encountered in photodiodes and they can be called photojunction cells. Photodiodes are discussed latter in a little more detail.

In the unbiased condition, when the cell is irradiated with photons, a current passes through a load resistance R_L (Fig. 5.24(c)) or in the open circuit condition, an open circuit voltage V_{oc} is obtained across the junction diode. This is the photovoltaic mode of operation and the corresponding characteristic set of curves is shown in the fourth quadrant of Fig. 5.25.

While photovoltaic cells have been in the run for over seven decades, the junction cell or 'photodiodes', as are commonly known, do not have a long history behind them and have undergone structural changes considerably to suit the operational aspects.

For the photovoltaic cell, the family of characteristic curves as depicted in the fourth quadrant of Fig. 5.25 is dependent both on illuminance and load resistance. It is to be noted that unfortunately, the illuminance versus open circuit voltage is nonlinear but is linear for short-circuit current, as shown in Fig. 5.26. With increasing load resistance, the current illuminance curves tend to become more and more nonlinear as shown in the figure.

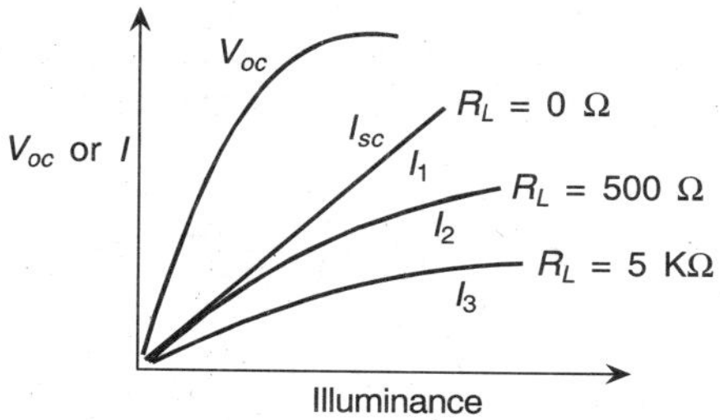

Fig. 5.26   Photovoltaic cell characteristics for varying load resistance.

An equivalent circuit model of the photoemf cell has been suggested as shown in Fig. 5.27 so that the load current I is given by

$$I^{-1} = \frac{k}{V\phi}\left(1 + \frac{R_L}{r}\right) + \frac{R_L}{V} \tag{5.44}$$

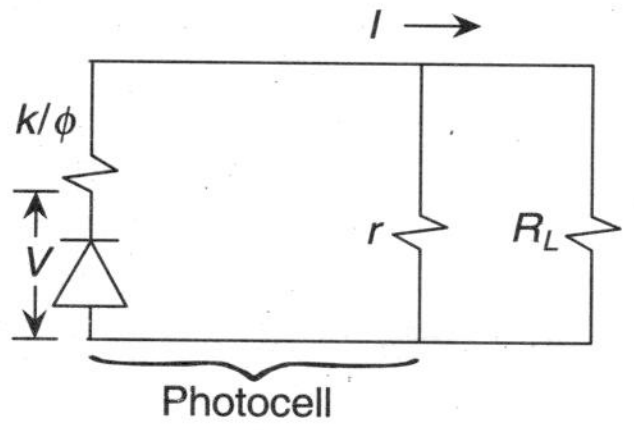

Fig. 5.27   Equivalent circuit model of a photovoltaic cell with steady illumination.

where ϕ is the intensity of illumination and k is a constant. If any externally applied voltage V_E is applied as in the case of Figs. 5.24(a) and (b), the relation can be modified to

$$\frac{V_E}{R_L} - I = \frac{1}{R_L}\left[\frac{kI/\phi - V}{1 + k/(r\phi)}\right] \tag{5.45}$$

An increase in the temperature reduces the output of the photocell. This is attributed to the fact that the temperature changes the internal leakage resistance. For frequency response studies, the effect of the internal capacitance of the cell should be considered and the equivalent circuit modified to the one given in Fig. 5.28. Neglect for present, k/ϕ for maximum power output, which states that $r = R_L$. In practice, this ideal condition does not hold good and the manufacturers recommend a relation

$$0.625 \le \frac{R_L}{r} \le 0.750 \tag{5.46}$$

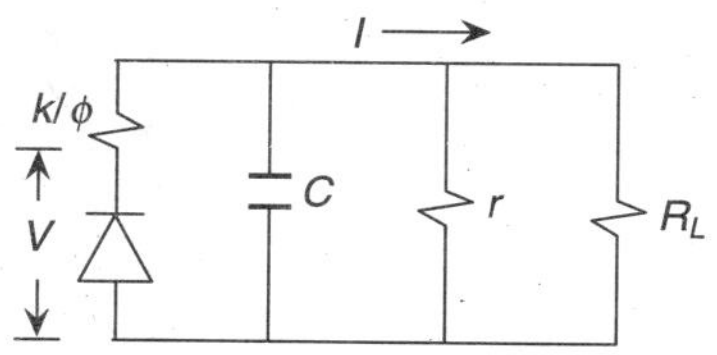

Fig. 5.28   Equivalent circuit model with varying radiation intensity.

Spectral sensitivity of photocells as has been discussed, depends on materials and type of cell. Selenium cell has response characteristics close to the human eye. Figure 5.29 shows this comparison graphically.

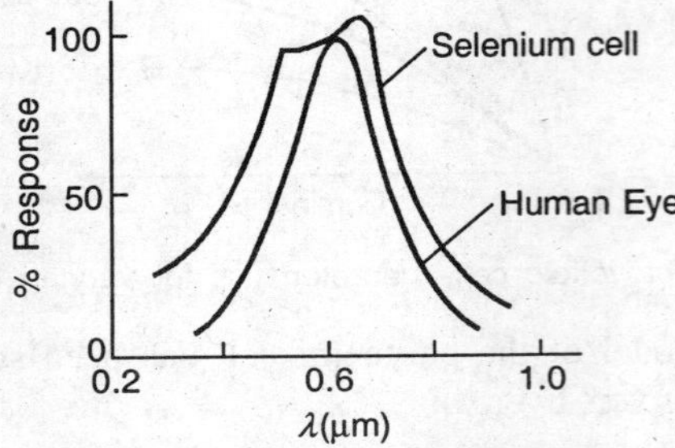

Fig. 5.29 Selenium cell spectral response compared with that of human eye.

Discussions have so long been centred around Se, Cu_2O, and Si photocells. Photovoltaic cells with Ge have also been produced extending the λ-range upto 2 μm. Cooled InSb cells operate upto 5 μm. Metal sulphide photovoltaic cells have also been in use. Thallium sulphide is one that produces high output of the order of 6 mA/lumen and response extends upto 1.5 μm, the peak being around 0.9–1.0 μm. Mixing metallic thallium with thallium sulphide in a proportion 2 : 1 or 3 : 1, the sensitivity as high as 10 mA/lumen has been attained. Molybdenum sulphide gives peak output as that of Tl_2S and response easily extends upto 2 μm.

Reverse biased junction photocell

This type of photocells deserves special mention because of their wider range of applications. Considering the band structure of a reverse biased p-n junction as shown in Fig. 5.30 (subscript c indicates conduction, v denotes valence, and f denotes Fermi level), it is specified that V is the reverse bias voltage, E is energy level, e is electron charge, and V_b is the potential difference between the conduction bands of the n and p type materials, often referred to as built-in, contact or diffusion potential. It is basically a potential barrier restricting electron transport from n to p regions or hole transport from p to n regions. This potential difference may also develop between the valence bands of the same types of materials. As is known, the total current density under the reverse biased condition is

$$J = J_s(e^{-eV/kT} - 1) \tag{5.47}$$

where

$$J_s = \left(\frac{eD_p}{l_p}\right)p_{n0} + \left(\frac{eD_n}{l_n}\right)n_{p0} \tag{5.48}$$

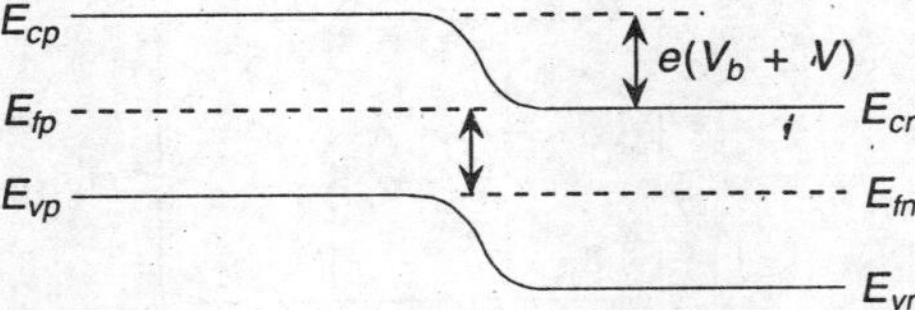

Fig. 5.30 The band structure of a reverse biased p-n junction.

where

D's are diffusion coefficients,

l's are diffusion lengths which are the lengths travelled by electrons or holes before recombination covering a transit time τ_n or τ_p,

n_{p0} and p_{n0} are electron or hole concentrations at the junctions. If $|V|$ is large, the first term within brackets in Eq. (5.47) vanishes and then

$$J = -J_s \qquad (5.49)$$

If the photon energy is greater than the band gap, then on incidence on a reverse biased p-n junction, it is absorbed and converted into an electron–hole pair. If this pair is created far away from the depletion region edges, its contribution in photocurrent is negligible. If, however, the generation is within a diffusion length from the depletion region edge on either side of the junction, the electric field within the depletion region (for n-type region) sweeps the holes across the junction to the other side and aids the current in this way as a generated hole effectively contributes an additional charge e to the existing current in the reverse biased p-n junction. On the p side, electrons are swept across, contribution again is a charge by one electron. The net flow of charge across the junction for a pair is e as long as the generation is within the depletion region. Thus, for a photocurrent density $-J_0$, Eq. (5.49) is modified to

$$J = -J_s - J_0 \qquad (5.50)$$

Obviously, J_0 depends on the optical power level at any point in the device. When a power P_0 is incident on the surface, a part of it is reflected back and the rest passes through the semiconductor material to be absorbed for this generation mentioned in precedence. If the air-to-semiconductor reflection coefficient is denoted as ρ_{as}, then at any depth x, the power level is given by the relation

$$P(x) = P_0(1 - \rho_{as})\, e^{-\alpha x} \qquad (5.51)$$

where α is the absorption coefficient. Figure 5.31 shows the plot between $P(x)$ and x. Referring to this figure, the power absorbed in the depletion region $(x_1 - x_2)$ is

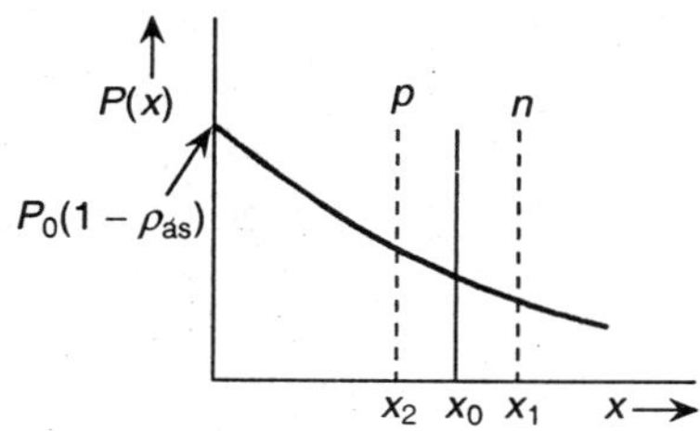

Fig. 5.31 $P(x)$ versus x plot for the reverse biased p-n junction.

$$P(x_1) - P(x_2) = P_0(1 - \rho_{as})(e^{-\alpha x_1} - e^{-\alpha x_2}) \qquad (5.52)$$

If now, the internal efficiency in the depletion region η_i is given by

$$\eta_i = \frac{\text{number of electron–hole pairs generated in depletion region}}{\text{number of photons absorbed in the region}}$$

then, the current density in the region is given by

$$J_0 = \left(\frac{e\eta_i}{h\nu}\right)\phi_o(1 - \rho_{as})(e^{-\alpha x_1} - e^{-\alpha x_2}) \tag{5.53}$$

Here, ϕ_o is the optical power flux. If anti-reflection coating is provided over the material, ρ_{as} becomes small. Also, for large J_0, $x_1 - x_2$ should be large with x_2 being small implying a wide depletion region starting very close to the air–device interface. In fact, PIN diodes have been developed following this philosophy and maximizing the depletion region.

For a constant optical power incidence, depletion region increases roughly as the square root of the voltage applied. However, for a fixed bias current, density is proportional to the incidence flux and this linear proportionality is useful in practice. In the fourth quadrant where current density J is negative and voltage is positive, the device acts as a generator of power. The solar cell is developed on this principle as has already been explained earlier in brief.

The external efficiency of a photodiode η_e, which is defined as

$$\eta_e = \frac{\text{number of carriers contributing in current generation}}{\text{number of incident photons}}$$

is given as

$$\eta_e = (1 - \rho_{as})\,\eta_i\,(e^{-\alpha x_1} - e^{-\alpha x_2}) \tag{5.54}$$

Expressing this in terms of current and power,

$$\eta_e = \frac{I_0/e}{P_0/h\nu} \tag{5.55a}$$

$$= R_s\,\frac{hc}{e\lambda} \tag{5.55b}$$

where R_s is the photodiode responsivity that equals I_0/P_0. Equation (5.55b) represents a relation that is ideal to some extent. The coefficient α is a function of λ and the ideal situation is required to be modified accordingly.

Usually PIN diodes are more common which have their current density versus optical flux relationship same as that given by Eq. (5.53). The depletion region in such a diode is around 1 μm wide, or even less which can be increased by incorporating a lightly doped material between the p and n regions. Figure 5.32 shows the arrangement. The increase in the width of depletion region so obtained enhances the efficiency as well as responsivity. This increase of the depletion region is from 10 to 100 μm but is associated with an increase in the transit time, thereby affecting the frequency response. Figure 5.33 shows plots of η_e versus width w and carrier transit time τ for radiations of different wavelengths.

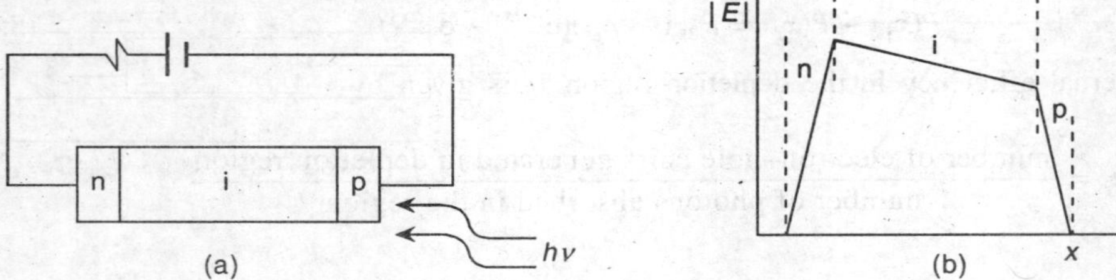

Fig. 5.32 (a) Structure of a diode with enhanced depletion region, (b) gradient of field with x.

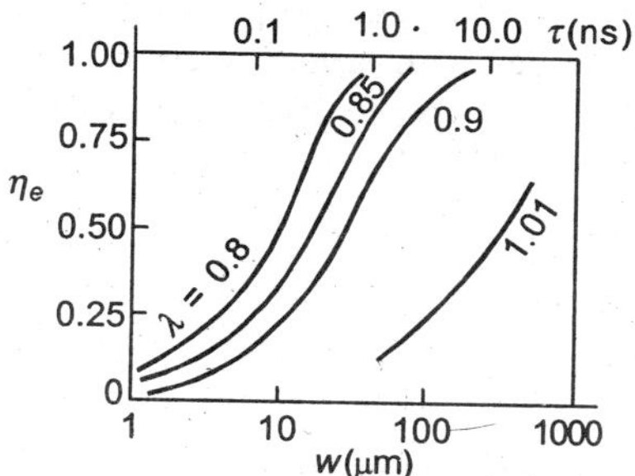

Fig. 5.33 Curves of absorption with increasing depletion layer width *w* for specific wavelength.

The structure of Fig. 5.17(b) describes a PIN diode which is more explicitly depicted in Figs. 5.34(a) and (b). The diode is developed over an n-substrate on which an intrinsic epitaxy is grown and then a thin layer of p-type material is diffused. SiO_2 windows are also produced because of the requirement of a subsequent high temperature diffusion of a dopant (p-type).

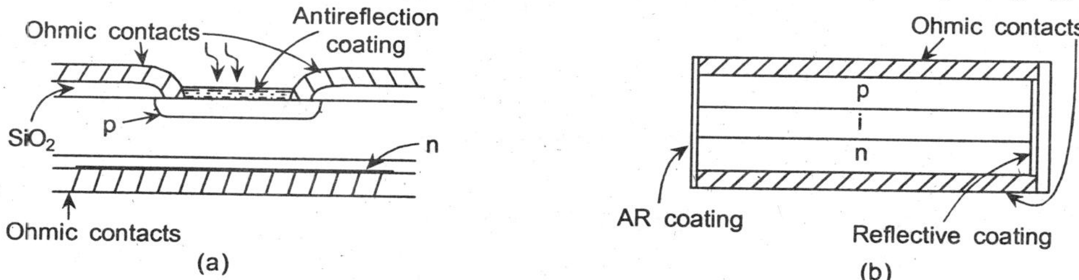

Fig. 5.34 The schematic structure of a PIN diode: (a) elevation and (b) side view.

Example

Calculate the external efficiency, responsivity, and the photocurrent in a PIN diode with intrinsic region thickness of 10 μm. Given, $\eta_i = 0.8$, $\rho_{as} = 0.1$ and incident optical power $P_0 = 10$ μW.

Solution

$$\eta_e = 0.9 \times 0.8 \left(e^{-10^6 \times 10^{-6}} - e^{-10^6 \times 10.1 \times 10^{-6}} \right) = 0.246$$

and as $R_s = e\eta_e \lambda/(hc)$;   and $I_0 = R_s P_0$

Using, $e = 1.6 \times 10^{-19}$ coulombs, $h = 6.63 \times 10^{-34}$ js,   and $c = 3 \times 10^8$ m/s

$$R_s = 0.198 \text{ for } \lambda = 1 \text{ μm}$$

Hence, $I_0 = 0.198 \times (10 \times 10^{-6}\text{W}) = 1.98$ μA

It must be mentioned that the photodiode leakage current is slightly dependent on the applied voltage. Figure 5.35 shows the characteristic curves for different voltages.

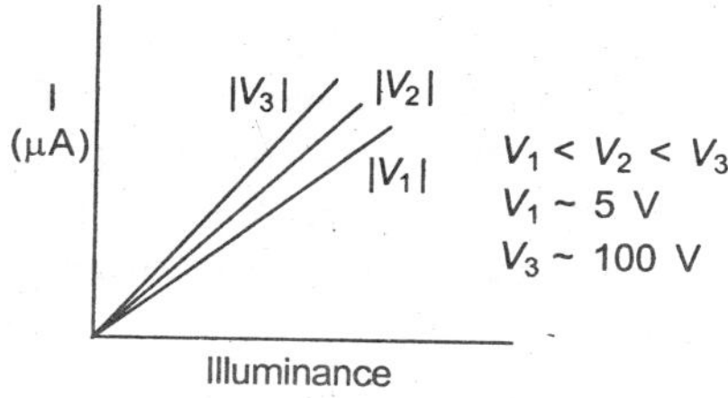

Fig. 5.35 Diode current verus illuminance for different bias conditions.

Small-sized photodiodes are used in switching operations, such as PIN diode with small junction area. On the other hand, the diodes with larger junction area can be used for illuminance measurement. The construction of an n-on-p type junction photocell is shown Fig. 5.36 and its equivalent circuit is shown in Fig. 5.37, when the device is illuminated. Here,

i_p is the photocurrent,

I_d is the dark current also called the leakage current,

r is the incremental resistance of the diode,

C is its capacitance (junction),

R_{sem} is the bulk resistance of the semiconductor material,

R_g is the resistance due to diffusion, that is, between the guard ring and the active region, and

R_L is the load resistance.

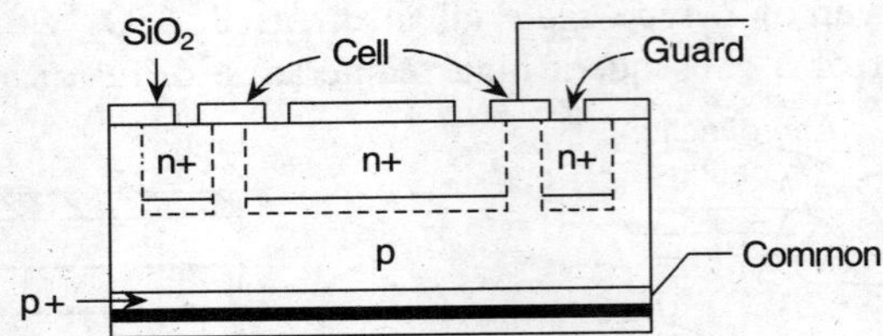

Fig. 5.36   Structure of a guard ring type n-on-p photodiode.

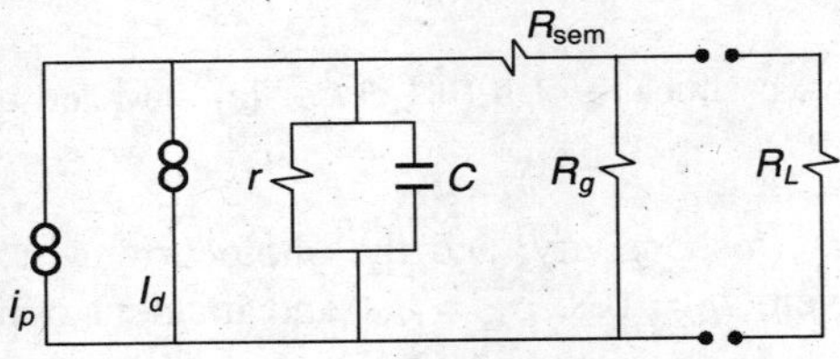

Fig. 5.37   Equivalent circuit of Fig. 5.36.

The effective time constant of the diode is given by the *RC* product as the cut-off frequency is given by

$$f = \frac{1}{2\pi RC} \tag{5.56}$$

where R is given by

$$R = \frac{R_g R_L}{R_g + R_L} + R_{sem} \tag{5.57}$$

In reverse biased operation, the values of R_{sem} and C are small compared with photovoltaic operation due to which the latter operation is slower than the former. From the depletion region point of view, as mentioned earlier, in the reverse biased operation, the applied voltage is fully impressed across the diode so that there is a steep potential gradient across the depletion region and the generated holes and electrons are fast removed to the electrodes. The rise time, thus, is less than 1 µs.

5.3.4 Position-sensitive Cell

Photovoltaic or photoconductive cells may be used in split-construction mode for position sensing as shown in Figs. 5.38(a) and (b) but the utility of such configurations is limited to the extent that these can be used for movement in one dimension only.

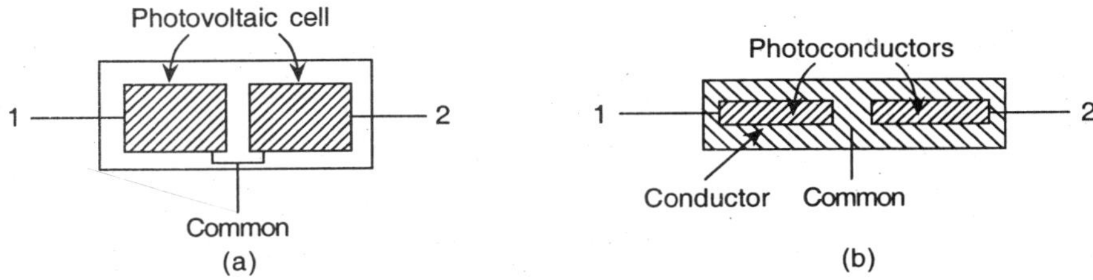

Fig. 5.38 Split construction type (a) photovoltaic cell and (b) photoconducting cell.

The structure of a cell shown in Fig. 5.39 is comprised of three layers: (a) highly resistive, (b) photoconductive, and (c) a conductive layer on an insulating substrate. Any bright spot, depending upon the position, would connect the resistive layer to the conductive layer through the photoconductor using the position of illumination. As a result, a 'differential' signal depending on the position is obtained.

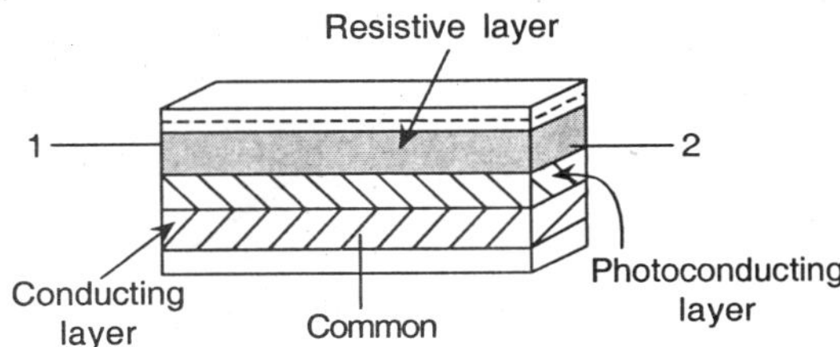

Fig. 5.39 Structure of a three-layer position sensitive cell.

A recent version of this cell consists of a p-n junction with ohmic contacts made in a transverse position. Figure 5.40 shows a typical structure. A base wafer has on one side an emitter of an alloy junction etched to expose the base for receiving image. Ohmic contacts have been made to the wafer as shown. Another modified version of this is shown in Fig. 5.41 which consists of a Schottky barrier structure with the ohmic contacts made at the bottom edges of the substrate. The circuit connections are also shown. Usually $R_1 = R_2$. Light beam falling on the device passes

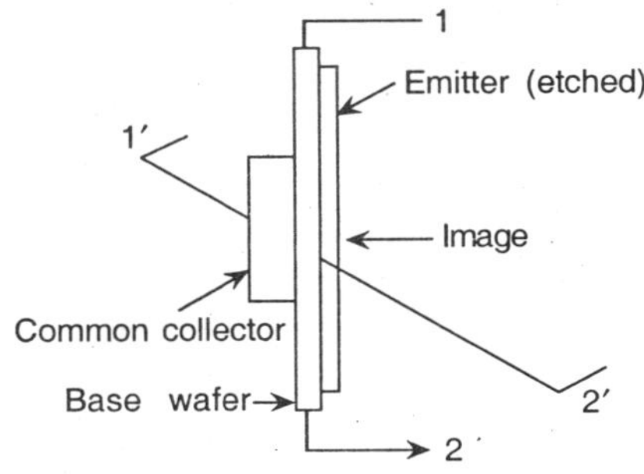

Fig. 5.40 A p-n junction type position sensitive cell.

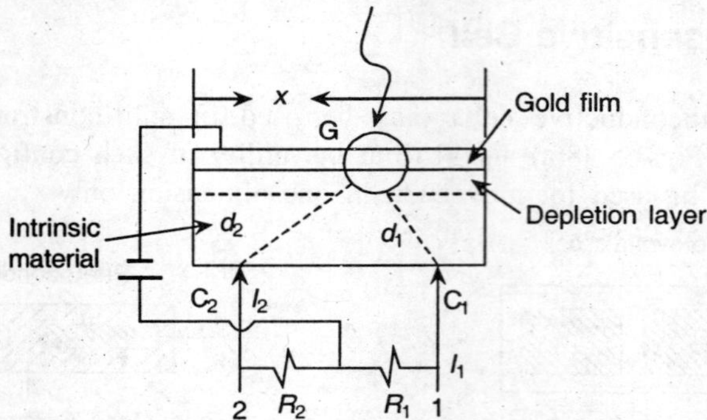

Fig. 5.41 A modified version of the scheme of Fig. 5.40.

through the gold film to release electron–hole pairs in the intrinsic region. The holes are collected by the gold film which acts as an electrode and the electrons are proportionally divided to form two currents I_1 and I_2 depending upon the resistance between the point of generation G and the ohmic contact points C_1 and C_2. These resistances are proportional to the distances d_1 and d_2. The differential current $I_2 - I_1$ with spot position x can be given by the graph shown in Fig. 5.42.

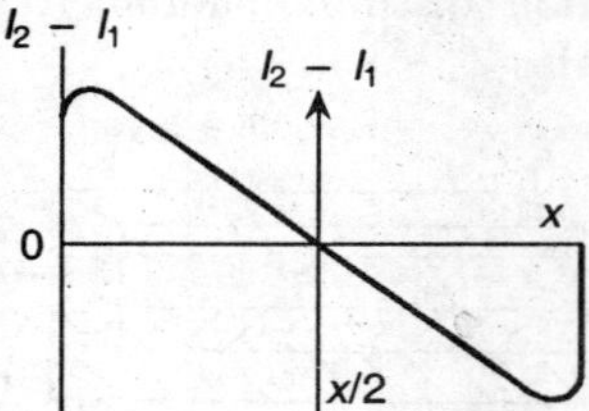

Fig. 5.42 Plot of differential current with spot position x.

5.3.5 Phototransistors and PhotoFETs and Other Devices

The phototransistors and photoFETs are made to amplify the photodiode current and consequently a large output is derived. For phototransistors, the base collector junction can be considered as a photodiode so that the leakage current I_{CBO}, which, now is a function of the illumination, is amplified by a factor $(1 + h_{FE})$.

In case of photoFETs, the irradiated junction produces a leakage current $I_G + I_P$, the sum of normal gate current and the photocurrent, so that we obtain a drop across a resistor R_g in the gate circuit of a value

$$V_G = (I_G + I_P)\, R_g \approx I_P R_g \tag{5.58}$$

This increases the drain current I_D and is obtained as usual with FETs.

Photocouplers used for electrical isolation by optical means often use other photodetectors such as photodarlington pair [Fig. 5.43(a)] and photothyristor [Fig. 5.43(b)] which basically consists of a photodiode and a thyristor in appropriate arrangement. A photocoupler consists of a light source usually an LED and a 'photocell' in a light-tight encapsulation. Opto-isolatorts or photocouplers are finding increasing use in measurement of current/transients in H.V. lines, patient isolation from mains-operated systems, and plenty of other control systems. Photocouplers are specified by the current transfer ratio (CTR), that is, the ratio of the output (I_2) to the input (I_1)

currents and is usually expressed in percentage. Photodarlington pairs of photoSCRs provide additional circuits to increase this ratio. For an LED and photodiode pair, this is a fraction increasing to more than 1000 per cent for an LED-photodarlington pair.

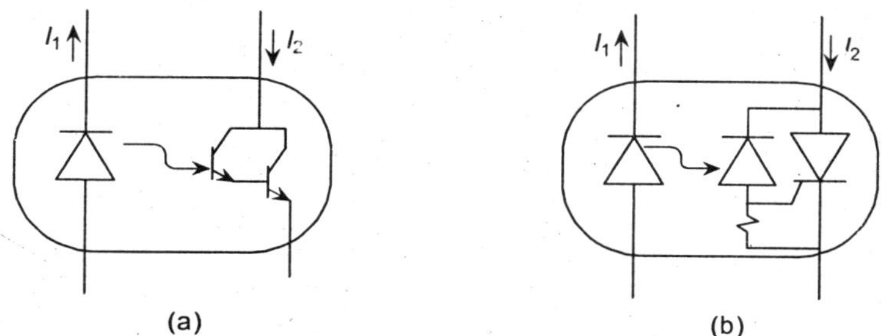

(a)   (b)

Fig. 5.43(a) A photodarlington pair and (b) a photothyristor.

Photocouplers have nonlinear characteristics and, therefore, require to be linearized by external circuits including photo-feedback. However, this is necessary for analog applications. Even then, over a small range it may be considered approximately linear as can be seen from the normalized curve of Fig. 5.44, for an LED–photodiode pair. Figure 5.44 shows two such ranges. For digital or switching operation, this nonlinearity however, is of no consequence. LED's are not optical or radiation sensors but they are quite often used in conjunction with many optoelectronic systems as sources and display units. The basic principle of an LED, that is, the light emitting diode, is that a forward biased p-n junction converts electrical energy into optical energy to emit light, for example, GaAs is one such junction. In fact, GaAs junctions have also been used in solid state lasers. When an electron falls from the bottom of a conduction band to the top of valence band and recombines with a hole there, it loses an amount of energy proportional to the width of the forbidden gap. This is emitted either as thermal energy or optical energy. If the band gap energy is E_g(in eV), the optical radiation wavelength λ is given by

$$\lambda = \frac{hc}{E_g} = \frac{1.237}{E_g}(\mu m) \tag{5.59}$$

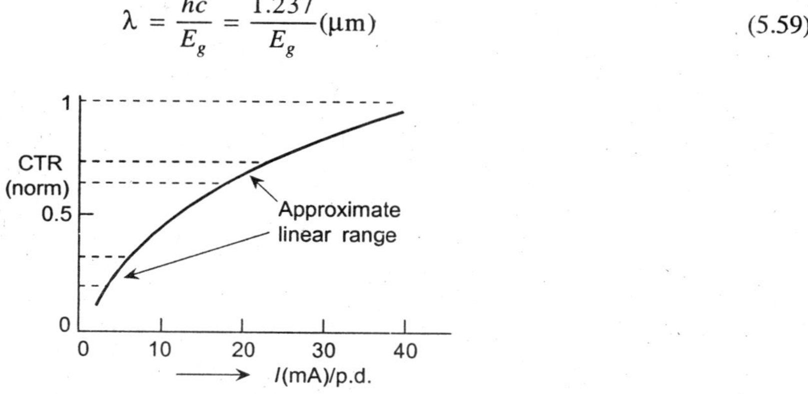

Fig. 5.44 Approximated small range linearization of photocouplers.

GaAs has $E_g = 1.37$ eV so that emitted wavelength is 0.903 μm in the infrared region. Other materials have also been known such as SiC, ZnS, GaP, Si, Ge. Figure 5.45 shows the emitted wavelengths versus band gap energy plot for some such materials.

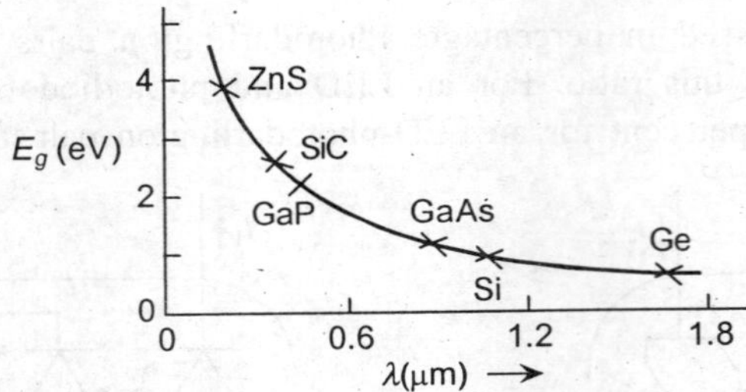

Fig. 5.45 Gap energy versus wavelength plot for different LED materials.

Commercial LEDs, in general, are made by epitaxial deposition of GaAsP on a GaAs substrate, the phosphorus to arsenic ratio is chosen for the appropriate colour of the emitted radiation which varies from green to red. Its quantum efficiency is higher in the red region but better relative sensitivity of human eye to green radiation is useful for that range. GaP is transparent and often a substrate of GaP is made and a P/As ratio of 0.49 with nitrogen impurity gives very stable green output LEDs. LEDs are used as forward biased diodes in a current range of 5–20 mA. It has a potential drop of about 1.6 V.

5.4 X-RAY AND NUCLEAR RADIATION SENSORS

There are a few common sensors for x-ray and nuclear radiation detection such as Geiger–Muller counter, proportional counter, and scintillation counter. Others are ionization chamber, electron multiplier tubes, and non-dispersive detectors such as cooled lithium drifted silicon detector.

Both x-rays and nuclear radiation are high energy radiations as compared to optical range of radiation. Different units correspond to their energy content or other relevant parameters. Roentgen, for example, is a measure of the intensity of the radiation in air and is defined as the quantity of x-ray or γ-radiation which produces 2.083×10^9 ion pairs, equivalent to 1 esu of charge, per cc of air at 0°C and at 1 atm pressure. It is denoted by R. Radiation damage that occurs due to nuclear radiation (γ-rays, for example) or x-rays is different with same number of Roentgens. This damage is often termed as *relative biological effectiveness* (RBE). When different types of radiations incur same extent of damage in humans, the amount of such ionizing radiation is called 1 REM (Roentgen equivalent man) which is equivalent to 1 R of 200 KV x-radiation and is sometimes known as 1 rad in tissue or 1 RBE. In terms of physical dose, it is called 1 REP (Roentgen equivalent physical) which means that in soft tissue an absorption of 93 ergs per gram of energy occurs. If absorption is 100 ergs/g, the corresponding ionization has an unit of 1 rad (in tissue) = 100/93 REP.

Curie, the popular SI radiation unit is the unit of the rate of radioactive decay and is defined as the quantity of radioactive material having 3.7×10^{10} disintegrations per second (dps). Microcurie or millicurie are often used in practice.

The principal nuclear radiations emitted by radioisotopes are (i) α-particles, (ii) β-particles and (iii) γ-rays. These are ionizing radiations. There is another kind of ionizing radiation, the neutrons. X-rays are ionizing radiations but are not nuclear by nature.

The non-ionizing radiations comprise of the (i) UV-visible-IR optical types and (ii) extremely low frequency, radio-frequency and microwave electromagnetic radiations, whose detectors sensors have already been discussed. It must be kept in mind here that the electromagnetic radiations are measured by thermocouples, thermopiles, diodes and bolometers.

Before proceeding to discuss the sensors in detail, a list of the characteristics of the ionizing radiations and the detectors usually used for them is presented here as Table 5.9.

Table 5.9   Radiation characteristics and detectors

Ionizing radiation	Characteristics	Detectors
α-particles (He^{++})	Positively charged, highly ionizing but low penetrating power, and discrete energy levels.	ionization chamber, proportional counter, scintillation counter (ZnS), semiconductors, plastic film detectors.
β-particles (e$^-$,e$^+$)	Electrons and positrons, more penetrating than α-particles, energy levels continuous.	Geiger–Muller counter, proportional counter, scintillation counter (solid/liquid), semiconductors, photo-films or plastic films.
γ-rays and x-rays (X-rays are non-ionizing.)	Penetrating electromagnetic types.	Geiger–Muller counter (mainly for X-rays), proportional counter (for X-rays), photon-spectrometer, scintillation counter (NaI), semiconductors (Ge, Li), thermoluminescent detectors.
Neutrons (n)	Indirectly ionizing.	p-n junction diode, etched track films, thermoluminescent detectors (Li-loaded), Counters filled with ^{3}He, BF$_3$.

Ionization chambers, proportional counters, and Geiger–Muller counters are basically gas-filled detectors. Gas-filled detectors can be grouped under two distinct classes. The one that indicates the arrival of a single particle by counting and is common in pulse-ionization chambers, proportional counters, and Geiger counters. The other one is the indicator of current ionization chambers.

The basic scheme of the gas-filled detector chamber is shown in Fig. 5.46. The central electrode is kept separated from the chamber, which also is an electrode, by an insulator. A supply voltage of V_s is impressed between the electrodes through a resistance R in parallel with a capacitance C. The time for an ion to be picked up by any of the electrodes, after production, is much smaller than this RC product. Collection of ions by the electrodes will produce a charge q across the capacitor C to produce a voltage pulse

$$V_o = \frac{q}{C} \tag{5.60}$$

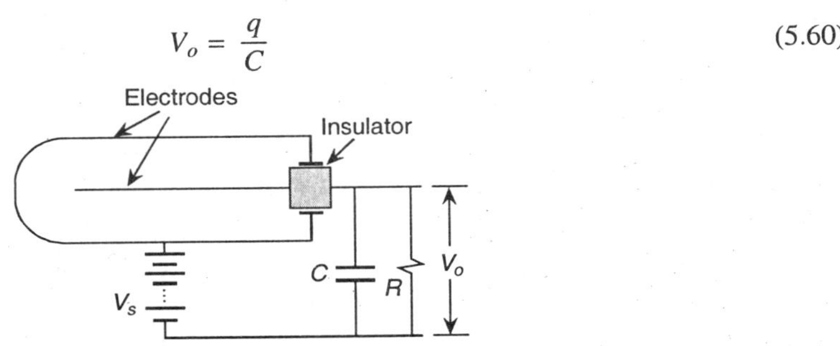

Fig. 5.46   Ionization chamber.

The ions are produced in pairs with the incoming radiation hitting the chamber gas. Depending on the value of V_s, the ions produced also vary. Figure 5.47 shows the V_s versus $\log_{10} n$ curve where n is the number of ions produced/collected by the electrodes. With respect to the voltage V_s, the curve is divisible into several regions. Five regions have been marked in Fig. 5.47. When voltage V_s is quite small, only a few volts, V_{s1}, as the maximum value, ion pairs are formed but the electrons cannot reach the anode as they are lost by recombination. Upto a voltage V_{s2}, of the order of tens or hundreds, the electrons gain sufficient energy and all the n pairs produced contribute n electrons to be collected by the anode, as in this region, recombination is negligibly small. Charge q across the capacitor would thus, be given by

$$q = ne \qquad (5.61)$$

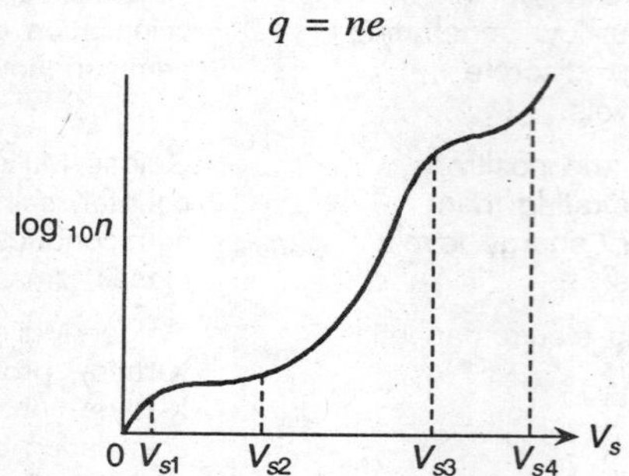

Fig. 5.47   The V_s versus $\log_{10}n$ curve.

where e is the charge of an electron, and the voltage pulse is

$$V_o = \frac{ne}{C} \qquad (5.62)$$

This region is known as the 'saturation region' or the 'ionization chamber region'.

When the voltage increases further, electrons are accelerated to acquire still higher energy and, in turn, produce secondary ionization and a multiplication in n occurs. This value of n increases almost proportional to the input, at least for over a major part of the section of the curve covering $V_{s2} - V_{s3}$. At the latter portion of the curve, ionization starts falling compared to the initial multiplication so that the last portion is known as the 'region of limited proportionality'. This region is, in fact, the 'proportional counter region'.

Over the region V_{s3} to V_{s4}, the increased voltage allows all the gas molecules close to anode to ionize so that a complete discharge of the counter can be generated. The output in this region is fairly constant and a plateau starts to form in the curve which is called the Geiger plateau and the region is known as Geiger–Muller region. Further, the increase in voltage initiates secondary ionization and also photon production starts which, further ionizes the gas molecules leading finally to a complete or continuous discharge. This is the region above V_{s4}.

5.4.1   The Ionization Chamber

There are various types of ionization chambers of which the cylindrical chambers are quite common though others such as parallel plate type and thimble type (parallel plate ionization chambers can be multiple-plate type) also find their use.

With the specified voltage applied between the electrodes, a field or potential gradient is set up in the chamber, as a result of which three types of ions are produced in the chamber transiting across. These are the electrons, the positive ions, and the negative ions produced by attachment of an electron to a neutral atom. The probability of this attachment is defined in terms of electron attachment coefficient α_c that depends on field strength, that is electron energy and the type of the gas. For most common gases, the value of α_c is around 10^{-6} whereas it has one of the highest values for halogens, around 10^{-3} and for oxygen or water vapour, α_c is about 10^{-4}. This coefficient can be made negligibly small by choosing the gas, field strength, and chamber geometry.

As has already been mentioned, in ionization chamber, recombination is very important. The rate of recombination is proportional to the product of the positive and negative charge densities. Recombination decreases the free charge densities, thus, if charge densities ar e_{d+} and e_{d-} for positive and negative ions respectively, then,

$$\frac{de_{d+}}{dt} = \frac{de_{d-}}{dt} = -\rho e_{d+}\, e_{d-} \tag{5.63}$$

where ρ is called the recombination coefficient and has a value around 10^{-6}–10^{-5} in many gases.

Because of ion movement in the chamber, a current is produced which consists of two parts, namely (i) the drift current due to imposition of a field and (ii) the diffusion current because of unequal charge distribution in the ionization chamber space and it is present even in absence of the field.

The drift current densities for positive and negative ions are given by the relations

$$J_{\mu\pm} = \pm e\, e_{d+}\, v_{\pm} \tag{5.64}$$

where

    e is the electron charge,

    $v_{\pm}$ is the drift velocity for the ions, and

    subscripts $+$ and $-$ stand for positive and negative ions respectively.

As electrons are faster than the other ions, they constitute the main current in the chamber. The drift velocity is a function of the field and of gas pressure and, of course, of the ion mobilities $\mu_{\pm}$. Thus,

$$v_{\pm} = (\mu_{\pm})\left(\frac{E_f}{p}\right) \tag{5.65}$$

The diffusion current density is a function of the charge diffusion coefficient D and the gradient of the charge density, grad e_d.
Thus,

$$J_{d\pm} = \pm e\, D_{\pm}(\operatorname{grad} e_{d\pm}) \tag{5.66}$$

The total current densities are thus, given by

$$J_t = J_{d\pm} + J_{\mu\pm} \tag{5.67}$$

If electron attachment, diffusion, and recombination are negligible with the entire current being only due to drift of charge produced at a constant rate for the potential gradient, then the ionization or the saturation current i_s is given as the integral of the pairs n_v produced per unit volume over the active volume V_a appreciated with charge e.
Thus,

$$i_s = e\int_{V_a} n_v(x, y, z)\, dx\, dy\, dz \tag{5.68}$$

Active volume is usually much less than the total volume of the chamber. Schematically it is shown in Fig. 5.48 by the dotted line for a parallel plate ionization chamber provided with guard rings for producing uniform field all through. However, i_s is reduced when (i) recombination occurs in the presence of negative ions, low potential gradient, and high production rate of ions; and (ii) diffusion occurs and opposes the drift (this opposition is induced by ion concentration in the chamber). Therefore, the larger the i_s, larger is its reduction as the ion concentration, in turn, is induced by drift.

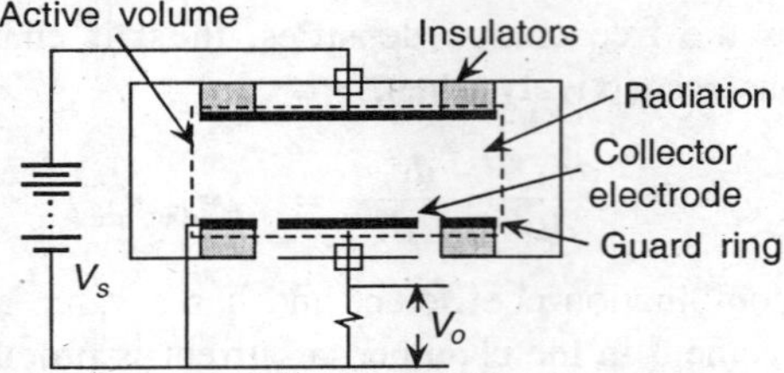

Fig. 5.48 Structure of an ionization chamber with guard ring.

Saturation current is also obtained in terms of specific ionization S, which is defined as the number of ions per unit length (cm) of path travelled by the particles per unit pressure. Thus,

$$i_s \propto SLP \tag{5.69}$$

where L is the path length traversed by the ionizing particles and P is the chamber pressure. Equation (5.69) is valid upto a definite upper range of pressure only, beyond which the losses increase.

In ionization chambers, specific ionization is not very large. When it is quite low, it is used in mean value measurement and the chamber is followed by a digital voltmeter or a digital current meter.

For large specific ionization, pulse type measurement is made with information on (i) number of ionizing particles, (ii) the time intervals between the incidences of ionizing particles, (iii) the energy distribution, and so on. A pulse amplifier is used in such a case with the choice of pulse height and pulse duration.

5.4.2 Proportional Counters

Proportional counters are more sensitive than ionization chambers and are often used for weak α- and β-particle sources or X-rays. These are gas-filled chambers and the gas multiplication increases the pulse size though this increase is independent of the primary ionization. In proportional counters, gas multiplication varies usually between 10^3 and 10^4 increasing to around 10^5–10^6 just below the Geiger–Muller region.

From the counter chamber design data, applied field; number of ion pairs produced, pressure of the gas, and mobility of ions; and the pulse size and the ion collection time, that is, pulse time can be determined. For a cylindrical design with a central collecting electrode of diameter d_c, outer electrode of diameter d_o, and distances, d_p and d_n, of the positive and negative ions respectively from the central axis, the pulse size is given by

$$V_p(t) = \frac{n_0 e}{C}\left[-\frac{\ln(d_o/d_n)}{\ln(d_o/d_c)} + \frac{\ln(d_o/d_p)}{\ln(d_o/d_c)} \right]$$

$$= -\frac{n_0 e}{C}\left[\frac{\ln(d_p/d_n)}{\ln(d_o/d_c)}\right] \tag{5.70}$$

where n_0 is the number of ion-pairs produced and C is the chamber capacitance.

Although electrons are the main charge carriers, or perhaps, the only charge carriers with the positive ions being very slow in movement, the collection of the positive ions determines the counting resolution. Using Eq. (5.65), the velocity of the positive ions is given by

$$v_+ = \frac{d}{dt}(d_p) = \mu_+\left(\frac{E_f}{p}\right) = \mu_+ \frac{V_s}{p\,d_p\,\ln(d_o/d_c)} \tag{5.71}$$

where V_s is the supply voltage to the electrodes. Integrating Eq. (5.71) w.r.t. time, we get

$$d_p = \left[\frac{2V_s\mu_+ t}{p\,\ln(d_o/d_c)} + d_c^2\right]^{1/2} \tag{5.72}$$

The initial condition is given by the term d_c^2. The maximum time of transit occurs for $d_p = d_o$, so that, from Eq. (5.72)

$$t = p(d_p^2 - d_c^2)\frac{\ln\left(\dfrac{d_o}{d_c}\right)}{2V_s\mu_+} \tag{5.73}$$

For improving the counting time, the output pulse is often differentiated.

5.4.3 Geiger Counters

The most popular and widely used of all the gas-filled counters is the Geiger/Geiger–Muller counter, in short the GM counter. It can measure all types of radiations, it has a high sensitivity and large output, is available at a cheap price in any shape and size required, and it can be made to have longer operating life-time by particularly using halogen gas filling. The commercially available varieties are (i) the end-window type (both with constant pressure filling or with gas flow), (ii) the cylindrical type, and (iii) the needle type.

In the end-window type, a metal-coated glass tube of cylindrical form has a thin tungsten wire of 0.002–0.01 cm diameter passing through the centre acting as the collector electrode with the body as the other. The end-window is usually made of mica sheet of a thickness less than 1 mg/cm^2. To avoid spark over the central electrode, it terminates into a glass bead as shown in Fig. 5.49. Radiation is received by the end-window.

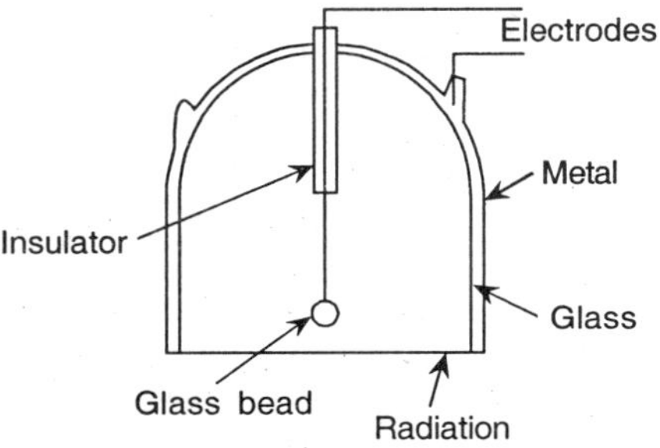

Fig. 5.49 A typical Geiger–Muller counter.

In the cylindrical GM counters, radiation is received by the side walls. This, also, is the case with the needle type GM counter which is used where insertion in a narrow channel is required. The schematic representation of the cylindrical GM counter is given in Fig. 5.50.

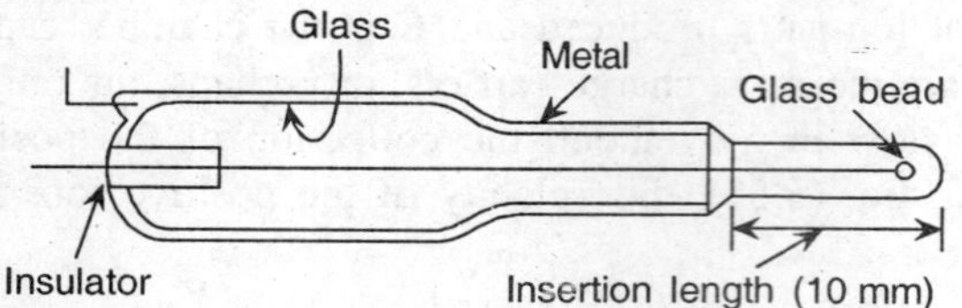

Fig. 5.50   Needle type design of a GM counter.

The GM counter chamber uses a gas at a low pressure of about 0.1–0.15 kg/cm^2 that consists of 90% inert gas such as Ar and Ne, and 10% ethyl alcohol or other organic vapours like methane. This mixture ensures charge transit through electrons only.

One important thing in gas-filled counters is the discharges mechanism. In the GM counter, specially, the Townsend discharge occurs and with the bulk of electrons in this discharge being collected by the anode, a positive ion sheath or cloud is left to reduce the field and stop the discharge. This is known as *quenching* of the discharge. With this cloud slowly moving towards the cathode, the field gradually gets recovered. However, with argon or ethyl alcohol filling, argon ions collide with alcohol molecules transferring their charges to alcohol ions. Alcohol ions, having less ionization potential, only reach cathode and get neutralized.

Because the potential gradient is less in this case, electrons may not escape from the electrode to initiate a secondary discharge. For ensuring complete quenching, external quenching mechanism is sometimes used. Such quenching also ensures that the counter dead time is its resolution time so that during recovery, no additional incoming radiation may initiate any secondary discharge.

5.4.4 Scintillation Detectors

Certain single crystals of organic or inorganic materials, activated glasses/liquids, or plastic fluors have the property that when they receive 'high energy' radiation, they produce very short duration light pulses or flashes called *scintillations*. These materials are known as scintillators. A scintillation counter basically consists of (i) a scintillator, (ii) a photomultiplier that converts light flashes into electrical pulses, and (iii) the associated electronic circuits consisting of amplifier, pulse shaper, scaler, discriminator, and so on. Figure 5.51 presents the block diagram of a scintillation counter.

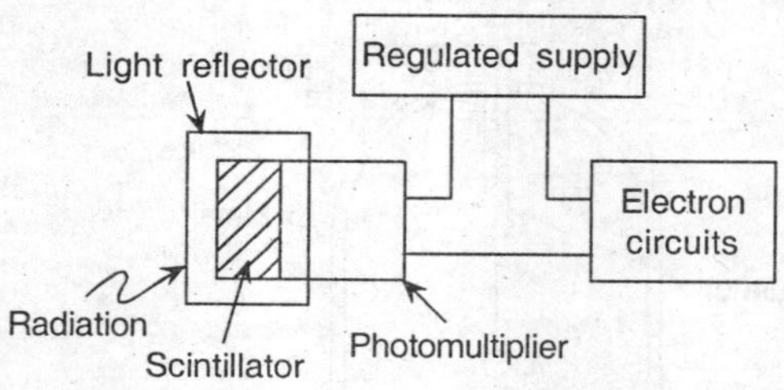

Fig. 5.51   Block diagram of a scintillation counter.

Depending on the type of radiation to be detected, different types of scintillators are used which have their own characteristic features. Only a fraction f_a of the radiation energy received by the scintillator is absorbed for conversion into light and a fraction f_l of this is converted into light. The fraction f_c of this light is then transmitted to the photocathode. The spectral matching between the crystal and photomultiplier and spectral response of the multiplier converts a fraction f_e into photoelectrons. Thus, if radiation energy received by the crystal is E, the number of photoelectrons produced is n and is given by

$$n = k_1 k_2 E \tag{5.74}$$

where k_1 is any dimensional constant and

$$k_2 = f_a f_l f_c f_e \tag{5.75}$$

For photomultiplier multiplication factor A_m, its output, in quantity of charge Q_m, is given by

$$Q_m = k_1 k_2 A_m e E \tag{5.76}$$

where e is the charge of an electron.

An ideal scintillator has every fraction, that is, f_a, f_l, f_c, and f_e as unity so that $k_2 = 1$. Unfortunately, most of the materials used are nontransparent to light and hence, f_l and f_c are never unity. Also, energy absorption by the scintillator is a typical of its properties but depends, to a certain extent, on its effective atomic number as it determines the energy absorbed by the material, or, f_a. Higher density materials are required for higher energy radiation such as γ-rays when absorption is larger. Absorbed energy is converted into light by luminescence. The excited crystal emits photoradiation and returns to normal condition again and in the process, a time lapse, known as the *decay time* occurs. Usually, decay time is quite small for most of the materials. Sometimes, however, phosphorescence occurs so that a metastable transition state is evolved and the decay time increases. A good portion of the absorbed energy which is not converted into light energy is transformed into heat raising the scintillator temperature and affecting the relative output of the scintillator. Table 5.10 shows features of a few commonly used scintillators which may be organic (Org) or inorganic (In).

Table 5.10 Features of scintillators

Material	Type	Density (g/cm^3)	Refractive Index (n)	Wavelength of maximum emission (µm)	Light output (% Anthracene)	Decay time (µs)
Anthracene	Org	1.25	1.50	0.440	100	0.025
NaI (Tl)	In	3.67	1.775	0.413	220	0.25
CsI (Tl)	In	4.51	1.788	0.570	90	1.10
CsI (Na)	In	4.51	1.737	0.420	150	0.65
ZnS (Ag)	In	4.10	2.356	0.450	300	0.20
Stilbene	Org	1.15	1.49	0.410	75	0.037

The elements in the parentheses in the first column indicates the impurity added for enhancing the scintillation property. Thus, NaI(Tl) denotes thallium activated sodium iodide or sodium iodide with thallium additive. The presence of sodium and lithium induces hygroscopicity as in NaI, CsI(Na), LiI(Eu), and so on. NaI(pure) and CsI(pure) can be used with better light output and less decay time tabulated ones but only at temperatures below 80 K. Higher refractive indices are not very useful as a major part of the light emitted is entrapped by the crystals having high refractice index.

Inorganic scintillators have high atomic weight and radiation such as γ- and x-rays interact with these in three different mechanisms:

(i) photoelectric effect—where a light pulse is produced normally in proportion to the primary energy,

(ii) Compton effect—where a large portion of energy is scattered, amount of which is dependent on angle of incidence and scatter, and

(iii) pair production—in which the kinetic energy received is less than the incident energy. This difference in energy is used for producing the pair. In organic scintillators, the interaction occurs through Compton effect and the spectrum is a Compton distribution.

Of the organic scintillators, napthalene crystal was first used for γ-ray detection. Anthracene and stilbene have good scintillation properties. Stilbene specifically, is good for neutron detection. Organic scintillators in mixed mode such as solution of anthracene in napthalene, liquid solutions and plastic solutions, are easy to prepare and can be available in large quantities. Often, boron or godolinium are used in small quantities in organic scintillators to improve the detection efficiency though the light output shows small decrease. Other 'loading' materials are Sn and Pb specifically with liquid and plastic scintillators for γ-ray detection.

Plastic scintillators are polystyrene and polyvinyl-toluene 'loaded' with substances such as p-terphenyl. One other good plastic material is methyl methacrylate. Plastic scintillators can be given arbitrary shapes and they are inert towards water, air, and many other chemical substances.

Many alternatives are available in inorganic scintillators. $CaWO_4$, $Bi_4Ge_3O_{12}$, and $CdWO_4$ possess high density, are nonhygroscopic in nature, and have low decay times though less light output. But, they can be produced in smaller sizes than NaI(Tl).

5.4.5 Solid State Detectors

Semiconductor materials have also been used for radiation detection. To date, Ge and Si have been found to be the best-suited for the purpose. Other materials under investigation are cadmium telluride (CdTe), mercuric iodide (HgI_2), mercuric sulphide (HgS), gallium arsenide (GaAs), silicon carbide (SiC) and more. All of these are not suitable for detecting all kinds of radiations. For example, CdTe and HgI_2 are used for low energy X-radiations only.

Comparison of vacuum tube and semiconductor amplifiers is valid in this case in the sense that in gas, ionization requires energy about 10 times larger than that required in semiconductor solid materials; to be specific, it is 3 eV in Ge and 3.7 eV in Si. The conversion process in solid state detectors has already been described earlier. The radiation energy dislodges the electrons from the valence bonds and imparts sufficient energy to drive them to the conduction band. As a consequence, a field across the semiconductor forces a current to flow in the external circuit. The maximum fraction of energy that is transferred to the electron is given by

$$\frac{E_m}{E} = 4m \tag{5.77}$$

where

E_m is the maximum energy transferred,

E is the energy of the ionizing particle, and

m is electron to particle mass ratio M_e/M_p.

It is, however, not advisable to apply the field across the semiconductor detector through ohmic contacts or metal plate electrodes as is usually done. Application of the field across a slice of such material provides a current in the external circuit which is often huge enough to override any small change in it due to that produced by ionizing particles or radiation. Semiconductor junctions are made which act as electrodes as shown in Fig. 5.52. The diffused n^+ layer produced by phosphorus doping in Si and p layer produced by boron doping, impose a reverse bias with the supply. This reverse bias draws all holes to the n^+ layer and all electrons to the p layer from the actual detector region which then becomes the depleted region as shown. In absence of the ionizing particles, the bias current is negligible and with ionization, the electron–hole pairs produced are collected by the field-induced 'electrodes' and a signal is available. The thermal generation is, however, not eliminated unless operated at very low temperature. Doping is used either to increase the electrons or holes since the sum total of holes and electrons in a semiconductor at a fixed temperature is constant. This is known as *mass–action law*. For silicon, this value is $\approx 10^{20}$ at 25°C.

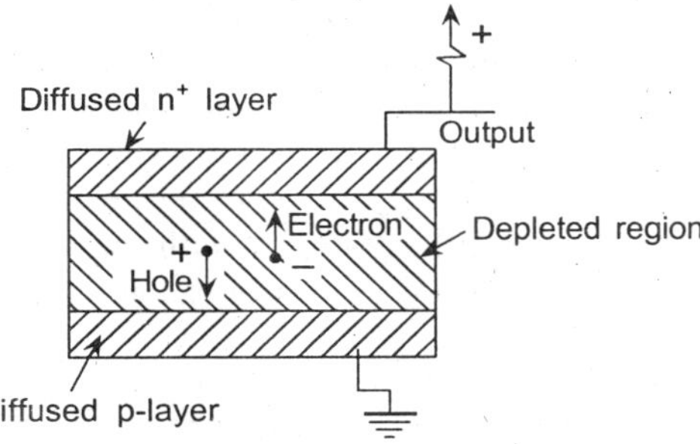

Fig. 5.52 Solid state ionizing radiation detector with output deriving network.

For high energy detectors, the acceptors or donors should be completely compensated in p-type or n-type silicon or germanium. Often, lithium is used as dopant for p-type Si/Ge. Large sized coaxial detectors have been made with such materials having high resolution.

Semiconductor materials or devices without forming the depletion region may be used as counters. With

d = distance between the electrodes,
x = average distance traversed by an electron produced in the detector,
n_e = number of electrons produced by ionizing radiation,
C = the detector capacitance, and
e = electron charge,

the pulse height is given by

$$V_{ph} = \left(\frac{n_e e}{C}\right)\left(\frac{x}{d}\right)\left[1 - \left(\frac{x}{d}\right)\left(1 - e^{-d/x}\right)\right] \tag{5.78}$$

The operation of the detector is affected to a certain extent by trapping and recombination and even polarization. Transit time of charge carriers is an important factor. For a supply voltage V_s and charge mobilities $\mu_{\pm}$, transit time $\tau_{\pm}$ is given by

$$\tau_{\pm} = \frac{d^2}{V_s \mu_{\pm}} \tag{5.79}$$

In this type of operation, high leakage current is a problem, as has been mentioned earlier.

This leakage current may be made low by increasing internal resistance and that is why, junction semiconductor reverse biasing principle has been proposed.

5.4.6 Plastic Film and Luminescent Detectors

High energy ionizing particles are known to cause damage to the molecules of polycarbonates. So, polycarbonate films irradiated with such particles leave tracks of damages which may be narrow or broad and the developed films may be studied with microscopes for the strength of the ionizing radiation—often comparing with 'file' or library standards.

Thermoluminescent property of certain materials such as fluorites and ceramics is sometimes used for radiation detection. Major portion of the ionizing radiation being absorbed by such materials is lost as heat, a small portion breaks the chemical bonds or is stored in metastable states which is subsequently available as visible light (since photons are emitted when the material is heated). This phenomenon is called *thermoluminescence*. The materials used for such thermoluminescent detectors are LiF, LiF activated with Mn, and CaF_2 activated with Mn. They can be used to measure small doses of x-rays, γ-rays, β-rays, elecrons, and even neutrons. The dose that can be detected lies in the range $1\,mR-10^5\,R$ with energy input ranging from $30\,KeV$ to $2\,MeV$. CaF_2 is used for detecting lower quantities of radiation as compared with LiF.

5.4.7 Factors Affecting Radiation Measurement

The factors that affect measurement of radiation by various sensors may be listed as scattering and backscattering; absorption and self-absorption, and detector/detector chamber geometry.

(a) *Scattering and backscattering:* Nuclear particles and photons are known to be scattered by materials with which they interact and, thus, apparently a loss of particle/photon energy occurs since the particles are deflected from the direct path between the source and the detector. The amount of scattering depends on the type of particles/photon, its mass, the energy content, and the type of material primarily its mass and density through which it passes. The phenomenon of scattering is, however, not that simple as stated. Some particles are absorbed by the material around the source and others are deflected away from the source–detector path, a part of which may be re-scattered back into the prescribed path and then into the detector. Some of these re-scattered particles may strike the detector at an angle of 90° to the beam path. These are said to be 'backscattered'.

Backscattering increases with increasing atomic number (Z) of the source and decreases source particle energy. Backscattering as well as scattering factors can be determined for different cases.

(b) *Absorption and self-absorption:* The intervening medium between the source and detector may absorb the particles/rays. To avoid this situation, the measuring medium may be an evacuated. The problem becomes more delicate for low-energy particles.

Sources also show self-absorption phenomenon, specifically when amounts of solid are present in them that are visible and 'foreign' to the source. This means, particles trying to be inside lower level of the source cannot always leave the surface to be emitted. Self-absorption factor is calculable from source thickness, range and the maximum fraction of range spent by the particles in the source without being completely self-absorbed. If these quantities are S_t, S_R, and f_R respectively then the self-absorption factor F_{sa} is given by

$$F_{sa} = \frac{S_R}{2S_t} \qquad \text{for} \quad S_t > f_R S_R \tag{5.80}$$

$$= 1 - \frac{S_t}{2f_R S_R} \qquad \text{for} \quad S_t < f_R S_R \tag{5.81}$$

(c) *Detector/detector chamber geometry:*   Source emits radiation in all directions but only a part of this radiation is collected by the detector. When window is used for transmitting the radiation from source to the detector, a geometry factor F_g can be obtained showing how much of the radiation is actually received by the detector. Figure 5.53 shows the scheme. Actually F_g is a fraction of the total solid angle 4π subtended by the source. From Fig. 5.53, for a window of radius r_w and angle θ subtended by the window at the source

$$F_g = 0.5(1 - \cos\theta) \tag{5.82}$$

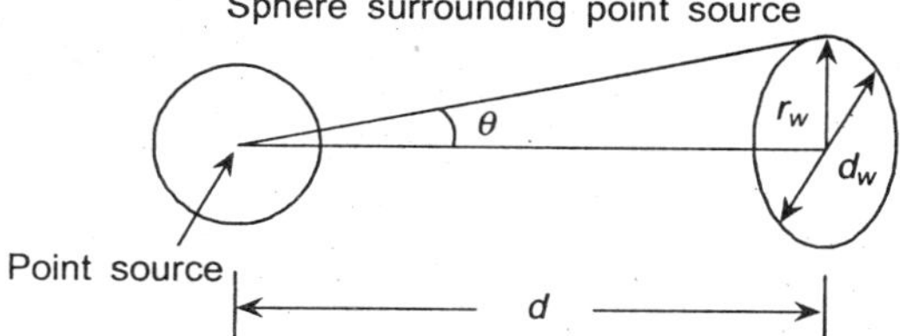

Fig. 5.53   Explaining geometry factor.

5.5   FIBRE OPTIC SENSORS

Fibre optic sensors could be classified as a separate group of sensors, as, although such sensors are in their prime, these are considered for sensing different types of variables such as temperature, liquid level, fluid flow, magnetic field, acoustic parameters, and so on. However, optical radiation happens to be the energy source in these applications with the fibre acting as medium as well as a sensor.

Optical fibres are basically considered as communication channels but it has been noticed that the optical transmission is affected by external parameters/stimuli such as temperature, acoustic vibration, magnetic field and many more. Study of these 'afflictions' or 'interferences' to the extent of the fibre being utilized as the sensor of such parameters has now been made and as a sensing device, fibre has been divided into two groups:

1. *Active*—the fibre is exposed to the energy source that affects the measurand and a consequent change in the optical propagation in the fibre is detected and related to the measurand. These are discussed in detail in subsequent subsections.

2. *Passive*—light transmitted through a fibre, called input fibre, is first modulated by a conventional optical sensor and this intensity-modulated light is propagated through a second fibre called the output fibre and then detected and corrected with the measurand.

5.5.1 Temperature Sensors

When two identical optical fibres are used to propagate radiation from a source, say, a laser source, and if one of these fibres is in a medium with temperature different than that of the other, the optical outputs from the two fibres would have a phase difference which is a function of the difference of temperature as mentioned. This phase difference is due to optical path length variations in the two paths occurring due to temperature difference and is so small that it can only be measured by producing interference patterns.

Two schemes are given in Figs. 5.54(a) and (b) that use He–Ne laser as source and the first one uses Mach–Zender interferometer as the detector while the second uses a Michelson interferometer. The beam-splitter (BS) and mirrors (M_i) in the first case have been dispensed with using fibre couplers in the second.

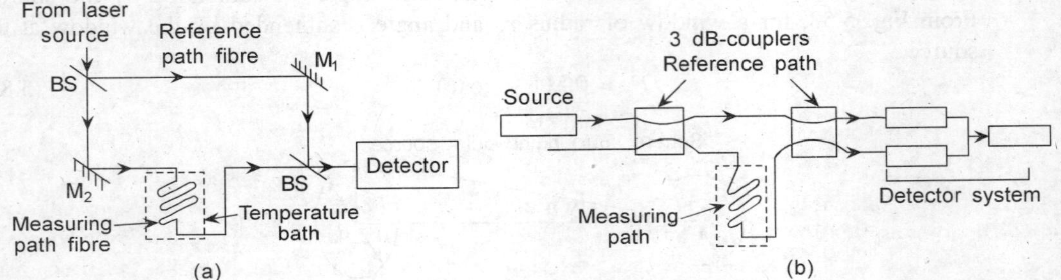

Fig. 5.54 Temperature measurement using optical fibres (a) phase difference method, (b) technique avoiding beam splitter and mirror.

Another optical fibre temperature sensor is used on the principle that a black body cavity changes radiance with varying temperature. Thus, at the end of a fibre a black body cavity is formed. The fibre is a high temperature fibre, usually a sapphire fibre, of diameter 0.25–1.25 mm. A thin film of iridium is sputtered onto the end-surface and a protective cover of Aluminium oxide(Al_2O_3) is then provided. This measuring fibre has a length usually within 0.3 m and not less than 5 cm. This propagates the radiation from the formed cavity which is being heated by the heat of the process. At the propagation end, another fibre, a low temperature fibre made of glass of about 0.6 mm diameter is coupled that has a length usually within 10 m. The detector system consists of one lens and two narrow band filters of close range middle wavelengths, two photomultiplier tubes in two measuring channels fed by a beam-splitter and a mirror. In fact, the filters have wavelengths of 600 and 700 nm respectively with a spread at the centre of 0.1 μm. The two channels are used to measure temperature by comparison over a range 500–2000°C. With an input power of 0.1 μW, for 1°C change there occurs 2% optical flux change and the system has a resolution of 1 in 10^8. This system is now being used as a temperature standard between 630.74 and 1769°C which are aluminium and platinum points respectively. Figure 5.55 shows one such temperature sensor.

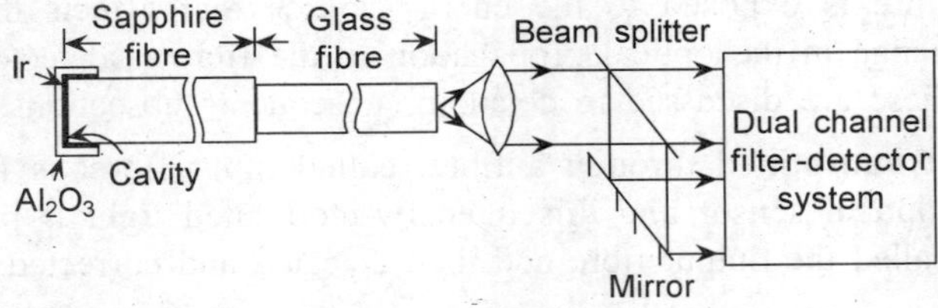

Fig. 5.55 Temperature sensor fibre black body cavity.

Optical fibre can be used for distributed temperature sensing. Optical pulse from a pulsed laser source is sent along a fibre over a distance covering a few kilometres. Any localized change in temperature somewhere along the fibre changes its backscattered intensity ratio (Stokes/anti-Stokes Raman). This backscattered light is filtered and Raman components are detected by photodetectors from which the temperature can be known. From the pulse delay time, the location can also be identified. Resolution of $1°K$ and 2–3 metres can be obtained in this system. A schematic representation of the system is shown in Fig. 5.56.

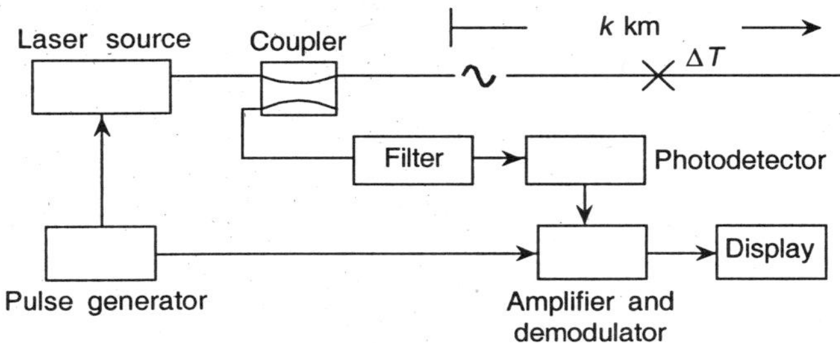

Fig. 5.56   Temperature sensing using backscatter in optical fibre.

5.5.2 Liquid Level Sensing

Usually, light propagates through a fibre by total internal reflection with appropriate cladding or even without that, if the light incidence angle is properly chosen. This is because the refractive index of air is such, with respect to that of the fibre, that no refraction can take place. If, however, the fibre is placed in a liquid medium of a different refractive index, it is possible that light refracts into the liquid and total internal reflection inside the fibre stops, stopping light propagation in it. This principle is utilized in measuring liquid level at specific values as shown in Fig. 5.57. The bottom end of the fibre is shaped like a prism so that with large difference in refractive indices of the fibre and the medium like air, there is internal reflection and the light travels to be detected as shown in Fig. 5.57(a). When liquid level rises to cover the bottom of the fibre, light refracts into the liquid and the detector fails to show any output, as shown by Fig. 5.57(b).

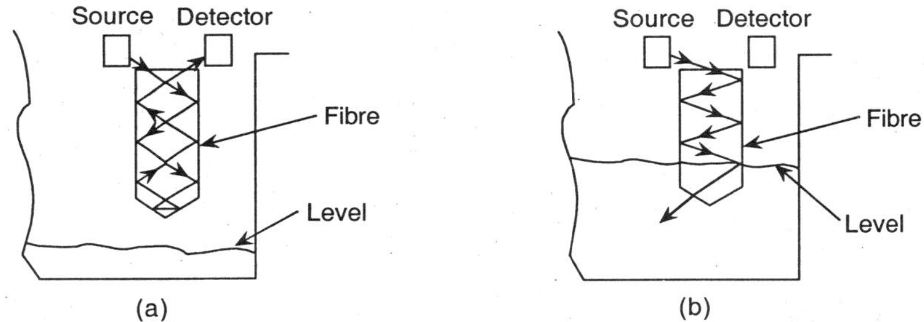

Fig. 5.57   Level detector using optical fibre: (a) level below sensor and (b) level covering sensor.

This single position level detection has been extended for discrete multistep detection covering the entire height of the tank. In this, a step-index multimode fibre is used and the fibre goes down carrying the light but in the return upward path, its cladding is exposed and the fibre is also given a zig-zag rise with small bend radius at regular intervals in length. When no liquid is there, cladding mode operation continues and a detector at the end of the return path of the fibre shows full intensity. But with liquid rising in the tank, refraction of light into liquid occurs at each bend and the intensity detected by the detector becomes less. Thus, for n bends there would be n-stepped intensity of signal, reducing in steps with rising liquid. Figure 5.58(a) shows the system and Fig. 5.58(b) depicts the intensity versus height plot.

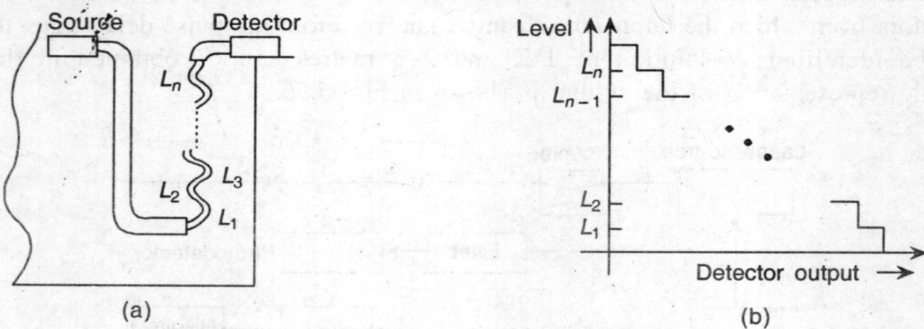

Fig. 5.58   Liquid level sensing in steps.

5.5.3   Fluid Flow Sensing

Fluid flow rate has been sensed by an optical fibre mounted transversely in a pipeline through which it flows. Because of the fibre, mounted across the flow, vortex shedding occurs in the channel and the fibre vibrates, which in turn, causes phase modulation of the optical carrier wave propagating through the fibre. The vibration frequency is proportional to the flow rate. Using multimode fibres of core diameter 0.2–0.3 mm and special detecting techniques, flow rates over a range of 0.2–3 m/s can be measured. Figure 5.59 shows the scheme to sense fluid flow. The fibre is kept under tension by a tension adjusting system and a fibre clamp. Flexible fillers are often used for small adjustment of tension.

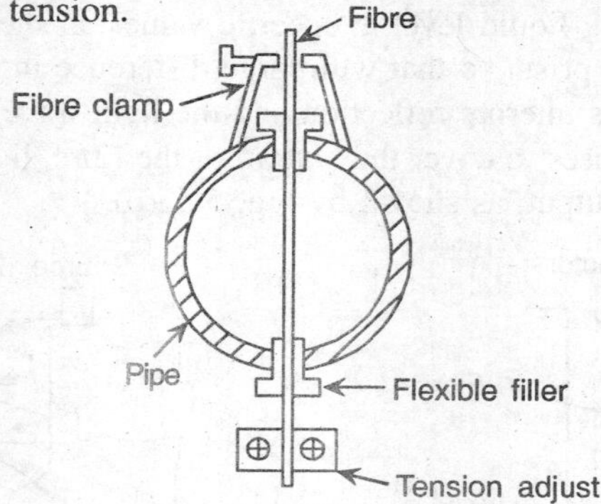

Fig. 5.59   Fluid flow sensing using fibre optics.

5.5.4   Microbend Sensors

Acoustic pressure sensing can be done by the microbending of a multimode fibre. Figures 5.60(a) and (b) show how light loss occurs in microbends of a fibre. The technique is utilized as shown in Fig. 5.61. Optical fibre is placed in two corrugated plates to form a transducer as shown. Applied force causes microbending in the fibre. Consequently, more light is lost and the receiver detector indicates less intensity. A calibration of force in terms of the intensity of detected light may also be made.

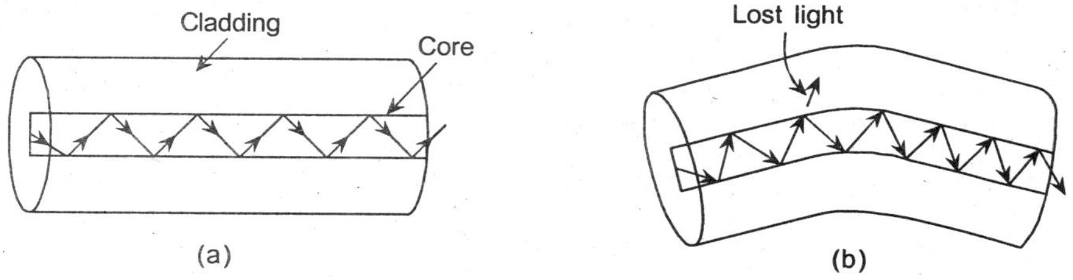

Fig. 5.60   Microbend sensors: (a) normal condition; no loss of light,
(b) bent condition; partial loss of light.

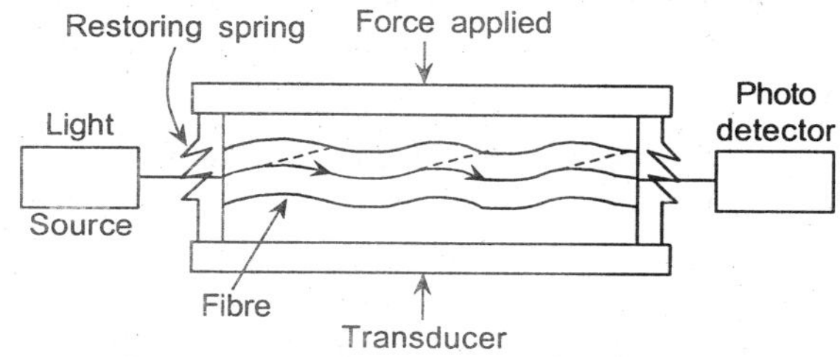

Fig. 5.61   Microbend force sensor using optical fibre.

REVIEW QUESTIONS

1. (a) What is photoelectric effect? How is it utilized in photosensitive devices for quantitative study of radiation? How do the intensity of radiation and its frequency range affect photosensors?

   (b) What are quantum efficiency and quantum yield? How are they related?

   If wavelength of a monochromatic radiation incident on a photoemissive surface of a sensor is 0.5461 μm, calculate the number of electrons that would be released for 1 watt of incident radiation.

   [*Hint:* 1 watt = 10^7 erg/s. Hence, electrons released = $((10^7\lambda)/12395)$ in erg. s/eV, or, $\{10^7 \times 10^8/(1.9857)\}(5461)$ per second. That is 2750×10^{15} electrons/s.]

2. How does a photomultiplier work? Sketch a photomultiplier tube having serially arranged dynodes and explain its operation. Compare the spectral response characteristics of cathode surface materials such as Na–Sb, Sb–Cs, and Cs–Cs_2O–Ag.

   In gas-filled photomultipliers, which gases are preferred and at what pressures, and, why?

3. What are the commonly used photoresistor materials? Discuss their spectral ranges and usable resistance ratio ranges.

   How are such cells characterized?

4. Draw the schemes of a simple and a planar semiconductor photoconductor structures. How do they respond to radiation of different wavelengths in terms of penetration depths?

Obtain an expression of photocurrent in such a sensor for a monochromatic radiation of frequency ν, power P, the sensor material radiation absorption coefficient α, sensor length l, and electric field impressed E. Assume, carrier life time as τ, transit time as t, and pair generated per unit volume as η.

5. (a) Why is noise so important in photoconductors? What are the different sources of noise and how are they quantified? Is *NEP* defined taking all such noise into consideration?

   What is detectivity? How is it related to *NEP*?

   (b) For an average of N incident photons, what is the probability that their number in a pulse stream be n? Calculate the probability for $N = 20$ and $n = 5$.
   [*Hint:* Assuming Poisson's distribution, $P(n) = N^n e^{-N}/n!$; and, hence $P(5) = 20^5 e^{-20}/120 = 0.53 \times 10^{-4}$]

6. (a) In what different modes can a p-n junction be used for radiation detection. Explain their operations with diagrams.

   (b) An unbiased p-n junction is modelled as a diode generator of voltage V with a series resistance inversely proportional to the radiation intensity (illumination) and the combination is in parallel with a resistance. Plot the load current versus illumination for different load resistors.

   (c) Can you assign two good reasons why selenium photovoltaic cells are still extensively used? What are the other materials used for making such cells?

7. Describe the operation of a junction photocell, also called a photodiode or a reverse biased p-n junction. Draw its equivalent circuit.

   Obtain an expression for the current density in the depletion region in terms of optical flux, incident radiation frequency, internal efficiency, air–semiconductor reflection coefficient, and absorption coefficient.

   Calculate the responsivity of a photodiode for an incident radiation wavelength of 0.8 µm, if its internal efficiency is 0.56. Assume that absorption coefficient is independent of the wavelength.
   [*Hint:* Responsivity $R_s = I_0/P_0 = \eta_e e\lambda/(hc)$; $\eta_e = 0.56$, $\lambda = 0.80 \times 10^{-6}$ m, $c = 3 \times 10^8$ m/s, $e = 1.6 \times 10^{-19}$ coul, and $h = 0.63 \times 10^{-34}$ js. Hence,

   $$R_s = 3.8 \text{ amp/watt}]$$

8. How are photocells used for position sensing purposes? Describe a three-layer type position sensing photocell and explain its operation. A Schottky barrier structure can also be used for position sensing.

   Describe such a cell and show how differential current output proportional to position is obtained from it.

9. What is an optocoupler? Where is it used? Describe different types of optocouplers with their operating principles and characteristics.

10. What are the commonly known ionizing radiations and what are the detectors used for their measurement?

    What is a Geiger–Muller counter? Discuss the variations in its design to suit different purposes.

11. Compare the performances of organic and inorganic scintillators. How does a scintillator act in radiation energy detection? On what factors does the output of a scintillator depend?

12. What materials are used for solid state radiation detectors? How do these detectors produce a measurable output with radiation incident upon them?

   A solid state radiation sensor gives a pulse height of 1.2 V, its thickness is 1 cm, and it has a capacitance 1.33 pF. Obtain the number of electrons produced by the ionizing radiation. Assume that the average distance traversed by an electron is 10^{-3} cm.

   [*Hint:* $1.2 = \{n_e \times 1.6 \times 10^{-19}/(1.33 \times 10^{-12})\} \times 10^{-3}[1 - 10^{-3}(1- e^{-1000})]$

   $$= 0.0012 \times 10^{-7} n_e$$

   Hence, $n_e = 10^{10}$]

13. Describe an optical fibre sensor for temperature measurement. Comment on its range, accuracy, and resolution.

14. How is optical fibre used for stress sensing? Describe a microbend sensor and discuss its operation.

Chapter 6

Electroanalytical Sensors

6.1 INTRODUCTION

A large part of analytical instrumentation systems uses sensors that can detect electrochemical actions in the analysis media. The detection is, however, neither a simple phenomenon nor a direct transduction. The basic approach is to use a pair of electrodes in the analysis medium that sense the actions and reactions in the medium in the form of current, voltage, or power. The electrodes themselves are specially developed devices, and in operation, they may have to produce electrical energy or consume it from an external source in the process of transduction. These are often termed as electrochemical cells—galvanic or electrolytic. Once the function of an electrochemical cell is understood, its variation in design can be taken up along with the 'sensing ranges'.

6.2 THE ELECTROCHEMICAL CELL

The electrochemical cell consists of two electrodes or may be just two conductors, immersed in suitable electrolyte solutions. The cell functions if these electrodes are connected externally by metal conductors and internally, the electrolyte solutions are in contact so that ion movement can take place between them. The electrodes are commonly known as anode and cathode where oxidation and reduction respectively, take place. Oxidation and reduction are 'interfacing mechanisms' where the ionic conduction of the solution is coupled to the electronic conduction of the electrode (metal) and thus, an electrical circuit is completed in a cell.

Each cell consists of two half-cells where each half-cell is said to be consisting of its electrode and electrolyte solution. If the electrolytes of the two half-cells are different in composition, they are not allowed to mix as this would decrease cell efficiency because of deposition, recombination, and so on. Instead, a liquid junction is created by special arrangement and as a consequence, a junction potential arises at this interface. The electrolytes in the anode and cathode compartments are often separated by a 'salt bridge' that is created when chemical reaction takes place in the cell. It has a variety of forms. A third solution (saturated) of a specific salt is often interposed for bridging purpose.

A typical representation of an electrochemical cell follows certain conventions adopted by the chemists. One such cell is shown in Fig. 6.1. The left part contains the anode and the information regarding the solution in its contact. Vertical line represents phase boundary. The cathode and its associated solution as well as its states are on the right. Any salt bridge is indicated by a discontinuous line.

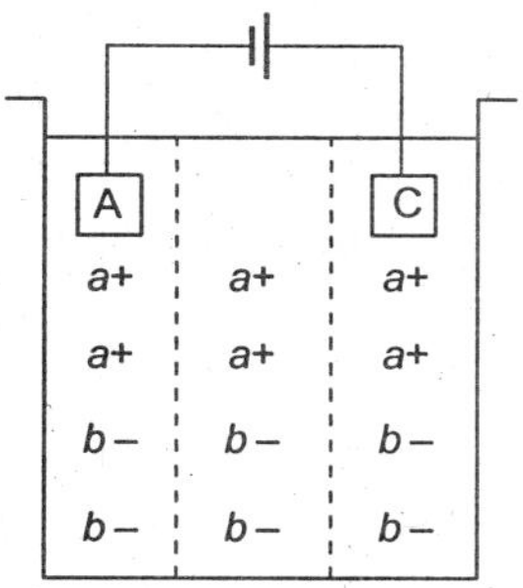

Fig. 6.1   An electrochemical cell.

The reactions in a cell occur in two phases: (i) the electronic ones at the metal electrode interfaces and (ii) ionic ones at the electrolyte level.

At the cathode, the reaction is

$$M_1^{k+} + 2e \rightleftharpoons M_1(s) \tag{6.1}$$

and at the anode

$$M_2(s) \rightleftharpoons M_2^{k+} + 2e \tag{6.2}$$

Therefore, the net cell reaction is

$$M_1(s) + M_2^{k+} \rightleftharpoons M_1^{k+} + M_2(s) \tag{6.3}$$

At the electrolytes, ionic movement takes place. The ions, however, have different mobility. The speed of an ion depends on its relative concentration as also on its inherent mobility. For example, in HCl, H^+ and Cl^- are produced but a proton (H^+) being five times more mobile than Cl^- ion current is largely contributed, in fact, 5/6th of it, by H^+ and only 1/6th by Cl^-. In Fig. 6.1, if a^+ is H^+ and b^- is Cl^-, after current flow, ion distribution will be different in the three compartments. The direct current flowing in a closed-circuited cell follows the Ohm's law except when polarization occurs. The electrolyte resistance is employed for the purpose.

For the dc current to flow with a dc potential applied between the electrodes, oxidation and reduction at the anode and cathode occur. Such a process is sometimes called *faradaic process*. If, an ac potential is applied, with cyclic change of positive and negative values at the same electrode, situation takes a new dimension. With sudden application of potential to the metallic electrode in an electrolyte, the electrode surface acquires an excess or deficiency of negative charge and the layer of electrolyte solution adjacent to the electrode acquires an opposing charge because of ionic mobility and a so called *electrical double layer* at the electrode solution interface develops. This double layer consists of an inner compact layer in which the potential decreases linearly with separation from the electrode and a relatively diffuse layer in which the potential decreases exponentially. When an ac potential is applied, the process is reversed every half cycle and then either positive or negative ions are attached to the electrode surface. Electrical energy is thus, consumed and converted to heat because of this ionic movement. Each electrode surface thus, behaves as a capacitor plate with the current increasing with surface area (capacitance) and frequency. This process is a nonfaradaic process.

Electrochemical cells can be reversible or nonreversible. A cell, a galvanic cell for example, formed with electrodes and an appropriate electrolyte solution develops certain amount of potential, say, E volts. If from an external source a potential greater than E volts is applied between the electrodes with the negative of the source connected to the anode, a reversal in electron flow direction is observed. This may cause a reversal of the electrochemical reaction and in such a case, the cell is said to be 'chemically reversible'. If the current reversal causes a different set of reactions, the cell is said to be 'chemically irreversible'.

6.3 THE CELL POTENTIAL

The potential of an electrochemical cell depends on the electrode potentials which are the characteristic of the half-cells with the concerned electrodes and the effect of concentration of reactants and their products in the solution. This effect of concentration is often termed as 'activity'.

From thermodynamical considerations, the maximum work obtainable from the cell at a constant temperature and pressure called the *free energy* or *Gibbs free energy* ΔG for a cell reaction, is given by

$$\Delta G = RT \ln B - RT \ln A \tag{6.4}$$

where

$R = 8.316 \, \text{Jmol}^{-1}\text{deg}^{-1}$, is the gas constant,
T is the temperature (in K), and
A is the equilibrium constant for the reaction given by

$$A = \frac{[a^+][b^-]}{[ab]} \tag{6.5}$$

where a stands for acid, b for base, and the $[ab]$ for activity.
The term B in Eq. (6.4) is similarly defined as

$$B = \frac{[a^+]_x[b^-]_x}{[ab]_x} \tag{6.6}$$

where the subscript x denotes instantaneous concentrations. In fact, Eq. (6.4) states that the amount of free energy is dependent on how far away is the system from equilibrium state. The cell potential E_c is related to the free energy as

$$\Delta G = -nFE_c \tag{6.7}$$

where n is the number of equivalents of electricity, that is, 'moles' of electrons associated with the oxidation–reduction process and F is Faraday which is 96487 coulomb/chemical equivalent.

Combining Eqs. (6.4), (6.5), (6.6), and (6.7), we derive

$$E_c = E_c^0 - \left(\frac{RT}{nF}\right) \ln \frac{[a^+]_x[b^-]_x}{[ab]_x} \tag{6.8}$$

where

$$E_c^0 = \left(\frac{RT}{nF}\right) \ln \frac{[a^+][b^-]}{[ab]} \tag{6.9}$$

which is a constant called the 'standard potential of the cell' that is, the potential of the cell when the reactants and products are at unit activity and pressure. In fact, the superscript 0 is used to denote standard in thermodynamical studies. Equation (6.9) is the Nernst equation and is widely used in analysis.

The activity α_m of a species is related to its molar concentration $[M]$ by the equation

$$\alpha_m = f_m[M] \tag{6.10}$$

where f_m is called the 'activity coefficient' which, in effect, implies the activity of $[M]$ and it varies with the ionic strength of the solution.

Electrode potentials basically go in to form the cell potential as the cell itself is made up of two half-cells. Thus, one can write

$$E_c = E_{\text{cath}} - E_{\text{an}} \tag{6.11}$$

where E_{cath} and E_{an} are electrode potentials for the corresponding half-cell reactions.

Nernst equation Eq. (6.8), is a very common equation for many transducers that use electrodes and a reaction medium which may be liquid or gaseous in states and the equation then is

$$E = E^0 - \frac{0.0591}{n} \ln \frac{[a_1][a_2]\cdots}{[b_1][b_2]\cdots} \tag{6.12}$$

where $RT/(nF)$ for half-cell at 298 K is given as $0.02568/n$ volts, and $[a_i]$ and $[b_j]$ represent partial pressures in atmospheres of reacting species when a_i, b_j are gases, and represent concentration in moles per litre when they are solutes with activity α. The half reaction for Eq. (6.12) is given as

$$\alpha_1 a_1 + \alpha_2 a_2 + \cdots + ne = \beta_1 b_1 + \beta_2 b_2 + \cdots \tag{6.13}$$

Here, a_i and b_j represent the 'chemical formulae' for reactants while α_i and β_j represent the number of moles or partial pressures and e, as usual, the electron.

6.4 STANDARD HYDROGEN ELECTRODE (SHE)

Since the absolute potential of a half-cell is not measurable and it is the potential difference that is measured, a second half-cell has to be formed. If this second half-cell can be made as a 'common reference electrode', this measured value would be available as a relative one but with respect to a standard reference. Such a standard reference is the *standard hydrogen electrode* (SHE) whose cell potential has been standardized by assignment. However, SHE is not easily reproducible and instead conveniently produced secondary standard electrodes or reference electrodes are used and the measured potentials using these references are easily converted to hydrogen reference standards.

As SHE is considered as the 'primary' standard, it is described here in some details. Its standard half reaction potential at 25°C, E^0, has been assigned a value exactly 0 (zero) volts by international agreement, assuming that the activity of H^+ and fugacity (partial pressure) of $H_2(g)$

are 1, and the relation is

$$2H^+ + 2e^- \rightleftharpoons H_2(g) \tag{6.14}$$

A typical hydrogen electrode is presented in Fig. 6.2. The platinum electrode uses a platinum foil coated with platinum black to provide a large surface area and a reversible reaction, Eq. (6.14) rapidly proceeds, with the solution near the electrode kept saturated with respect to the gas. The electrode can be either an anode or a cathode with hydrogen being oxidized to hydrogen ions, or vice versa.

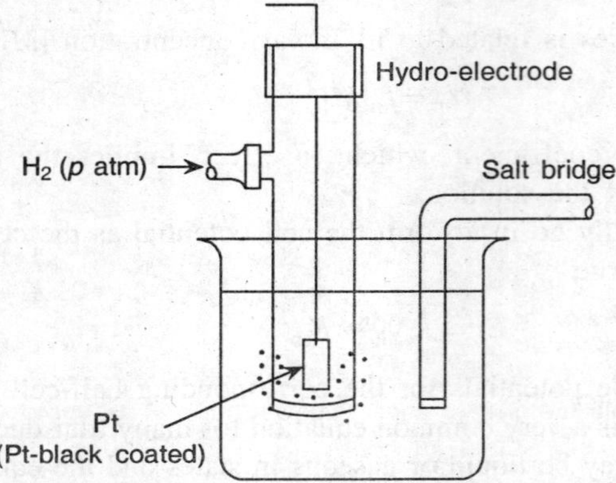

Fig. 6.2 Structure of a hydrogen electrode.

For measurement of potential using SHE or any second standard using Nernst equation, Eq. (6.12), two things are of importance, namely (i) the sign of the electrode potentials and (ii) the value of the standard electrode potential. The sign, in fact, is determined by the specific use of the electrode such as anode or cathode. If the electrode is used as an anode from which electrons flow through the external circuit to the SHE, this would be the negative terminal of the galvanic cell and its standard electrode potential (SEP) is negative; on the other hand, SEP is positive for the cathodes. International Union of Pure and Applied Chemistry (IUPAC) specifies that relative external potential is reserved for half-reactions as reductions. The magnitude of the SEP of an electrode is, as has already been mentioned, the value of the electrode potential of a half-cell reaction with respect to SHE when all reactants and products possess unit activity. Tables have been prepared by determining the SEP by actual measurements with SHE or other reference electrodes as the other half-cell.

6.5 LIQUID JUNCTIONS AND OTHER POTENTIALS

The liquid junction potential arises when two electrolyte solutions of different chemical compositions come in contact because of unequal distribution of positive (cations) ions and negative (anions) ions across the junction which, in turn, is due to the difference in their migratory speeds. The speeds are governed by the concentration difference between the electrolytes since migration occurs from higher to lower concentration of solutions. Besides, mobility is also a factor affecting migratory speeds of the ions. The two factors try to counteract and an equilibrium condition develops providing a specific potential value depending upon the above considerations, that is, half-cell specifications.

The junction potential may be quite large in value and in measurement of the cell potential, its contribution cannot be ignored. Neither can it be easily computed except in some simple situations. It can, however, be reduced by introducing a salt bridge, as has already been mentioned. Salt bridge is actually a concentrated electrolyte solution that joins the two half-cell electrolytes. Often, saturated potassium chloride is used as salt bridge because of its high concentration (4 M at room temperature) with respect to the half-cell electrolyte concentrations and its low ion mobility variation in many situations, specially with chlorides in half-cells.

Salt bridge solution should be selected in such a way that no interference occurs between this and the half-cell solutions. A typical cell structure with salt bridge is shown in Fig. 6.3.

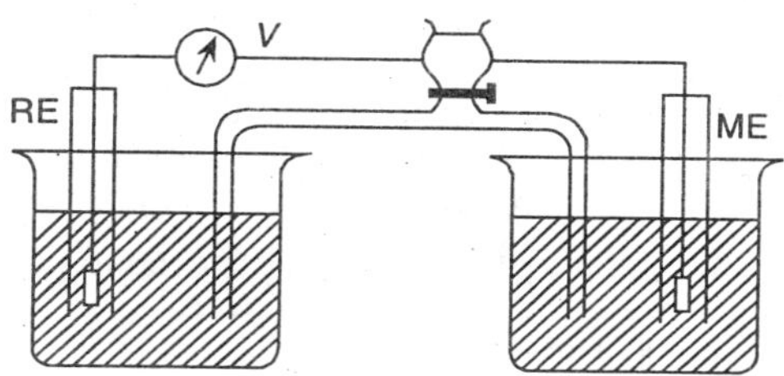

Fig. 6.3 Cell structure with salt bridge.

With a current flowing in the cell, there occurs an ohmic drop, following the Ohm's law, and the resistance of the cell contributes to the magnitude of this drop. In effect, the measured potential becomes less. The relations expressing the mechanism are as follows:

$$E_{thermodynamic} = E_{cathode} - E_{anode} \qquad (6.15a)$$

$$E_{cell} = E_{thermodynamic} - IR \qquad (6.15b)$$

or,

$$E_{cell} = E_{cathode} - E_{anode} - IR \qquad (6.16)$$

6.6 POLARIZATION

With electrode potentials being constant, as is usually the case, cell potential E_{cell} should be linearly related to cell current I as is seen from Eq. (6.16). But, sometimes this is not the situation. The nonlinearity that arises is due mainly to *polarization* which is manifested as reduction of current or corresponding overvoltage. There are four types of polarization, namely concentration polarization, reaction polarization, adsorption/desorption/crystallization polarization, and charge transfer polarization.

Concentration polarization

Oxidation–reduction at the electrode surfaces can occur normally when the movement of 'Ox' and/or 'Red' species across the bulk of the electrolyte by mass transfer is normal. If not, reaction rate decreases and correspondingly the current. This is due to what is known as concentration polarization.

Reaction polarization

If there is any intermediate chemical reaction in any half-cell for producing 'Ox' or 'Red' species that travel to electrodes and which participate in electron transfer, it is likely that formation of these species at the intermediate stage is not normal. This is known as reaction polarization.

Adsorption/desorption/crystallization polarization

Sometimes the current is limited by processes such as adsorption/desorption or crystallization of the reactants and hence, the name.

Charge transfer polarization

Charge transfer takes place from the electrode to oxidized species or from reduced species to the electrode. Often, the rate of charge transfer is reduced because of the existence of a charge surface film around the electrodes. This is called charge transfer polarization or electrical polarization. Figure 6.4 graphically shows the polarized and unpolarized conditions of a cell.

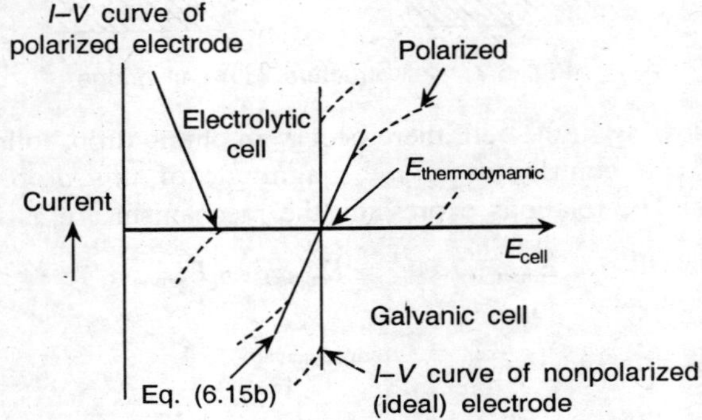

Fig. 6.4 Polarized and unpolarized conditions of a cell.

As has already been discussed, the concentration and/or type of an electrolyte solution can be measured by measuring the potential of a cell containing the solution, the cell being made up of two half-cells each with one electrode. Of these, one must be a reference electrode with, perhaps, zero electrode potential as in SHE. But, reproducible forms of reference electrodes are different than SHE. The other electrode is the measuring or indicating, or actual sensor electrode.

6.7 REFERENCE ELECTRODES

The reference electrodes are required to have a known and constant half-cell potential unaffected by solution composition. Their properties can be listed as they must

1. have known constant potential and follow Nernst equation,
2. have reversible cell-reaction,
3. have little or no hysteresis with temperature and small current cycling,
4. be non-polarized.

Some electrodes have been designed and used in practice which more or less follow these properties. The commercially used ones are:

- Saturated calomel electrode (SCE),
- Silver/silver chloride electrode, and
- Thallium/thallouschloride electrode.

Figure 6.5 shows the elements of a commercial type SCE. The chemists' nomenclature for the electrode function is given as

$$Hg(l) \mid Hg_2Cl_2(satd), \quad KCl(aq) \parallel \tag{6.17}$$

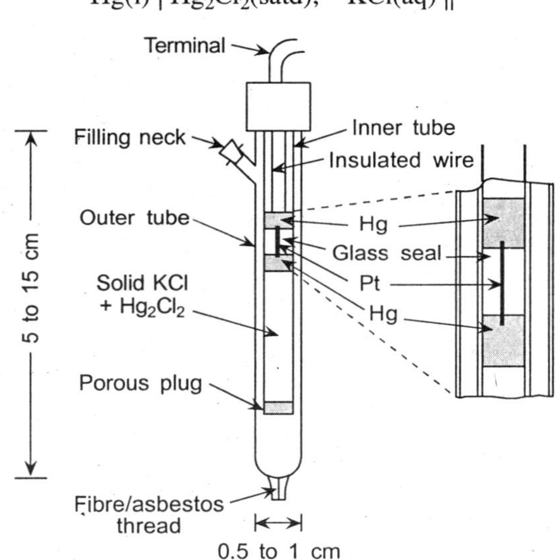

Fig. 6.5 Schematic view of a saturated calomel electrode.

The corresponding electrode reaction is

$$Hg_2Cl_2(s) + 2e \rightleftharpoons 2Hg(l) + 2Cl^- \tag{6.18}$$

The electrode consists of an inner glass/plastic tube containing Hg, Hg_2Cl_2, and KCl. This tube dips into a saturated solution of KCl and Hg_2Cl_2(calomel) contained in an outer tube made of the same glass/plastic material. The junction to the other half-cell for electrical continuity is made through a fibre of quartz or asbestos, a crack, a ceramic plug, or a glass sleeve. Platinum wire is used for external connection. Provision for refilling with KCl/aqua is also there. A normal calomel electrode containing 1 M KCl and decinormal type with 0.1 M KCl can also be made; but the SCE, the saturated electrode is most commonly used in practice. The glass-sleeved or ceramic plug junction type design has lower impedance than the fibre type where output impedance may be as high as 3 kΩ. The electrode potential at 25°C for different concentrations of KCl is tabulated in Table 6.1.

Table 6.1 KCl concentration versus electrode potential

KCl	*Saturated*	4.0 M	3.5 M	1 M	0.1 M
$E_{electro}$ (V)	0.244	0.246	0.250	0.280	0.336

Note: These potentials have been accurately measured with reference to SHE.

The Ag–AgCl electrode can be used as a reference electrode which responds to the concentration of an anion with which its ion either forms a precipitate or a stable complex ion. It is made by dipping a silver wire into a solution of specified KCl and into which AgCl is added till saturation. KCl strength may be 3.5 M or 1 M with electrode potentials 0.199 V and 0.222 V respectively at 25°C. Functional state and reaction relations can be given as

$$Ag \mid AgCl \text{ (satd), } KCl(kM) \parallel \tag{6.19}$$

and

$$AgCl(s) + e \rightleftharpoons Ag(s) + Cl^- \tag{6.20}$$

Thallium electrode is not as common as the other two but has the advantage that it attains equilibrium potential much more rapidly than the others after a change in temperature. Its functional state relation is

$$Tl(vo\%) \mid TlCl(satd), KCl \text{ (satd)} \parallel \tag{6.21}$$

The major source of error in the reference electrode is the contamination through the junction plugs and for this, the cell may behave erratically. Fortunately, the amount of contamination is so very negligible that it is of no concern for the electrode performance. But this leads to lower capacity cell as far as current rating is concerned.

6.8 SENSOR ELECTRODES

Sensor electrodes, or indicator electrodes as they are commonly known, are of two kinds: (a) metal, (b) membrane.

6.8.1 Metal Electrodes

Metal electrodes are similar to reference electrodes which can be subclassified to be of (i) the first kind, (ii) the second kind, (iii) the third kind, and (iv) the redox type.

(i) *The first kind electrode:* Such an electrode is in direct equilibrium with the cation derived from electrode metal. For such a case, denoting metal with μ,

$$\mu^{k+} + ke \rightleftharpoons \mu(s) \tag{6.22}$$

and the electrode potential is given by the standard equation

$$E = E^0_\mu - \frac{0.0591}{k} \log \frac{1}{\mu^{k+}}$$

$$= E^0_\mu - \frac{0.0591}{k} p\mu \tag{6.23}$$

where $p\mu$ denotes the negative logarithm of the μ-ion concentration. The metals in this category are Cu, Zn, Ag, Hg, Cd, and Pb. They show reversible oxidation–reduction behaviour.

(ii) *The second kind electrode:* As has already been mentioned in Sec. 6.7, this electrode is responsive to the concentration of an anion and its own ion makes precipitate with it or forms a complex ion. Ag, for example, can be used as a second kind electrode with respect to halide ions and for measurement purposes, the surface layer of the analyte has to be saturated with AgCl. Then one writes

$$AgCl(s) + e \rightleftharpoons Ag(s) + Cl^- \tag{6.24}$$

and

$$E = 0.222 - 0.0591 \log[Cl^-] \tag{6.25}$$

(iii) *The electrode of the third kind:* It responds to a different cation under certain circumstances. For example, Hg-electrode can be used to determine the Ca ion concentration in solutions containing calcium. This is possible, however, by introducing a complex of calcium and maintaining a certain concentration of the complex for standardization. The chain of processes makes the electrode a third kind.

(iv) *Redox electrode:* An electrode, usually made of inert metal, is immersed in a solution containing a substance in the reduced or oxidized state; the electrode would acquire a potential depending on the tendency of the ions in the solution to pass from a higher to a lower state of oxidation or from a lower to a higher state of oxidation. If the solution has a reducing property, the ions will tend to be oxidized losing electrons to the electrode and it becomes negatively charged with respect to the solution. The case reverses if the solution has oxidizing property. The potential value of the electrode will be a measure of the oxidizing or reducing power of the solution whereas the sign of the potential gives the actual characteristic. With α_+ as the activity of the oxidized ion and α_- that of the reduced ion, electrode potential is obtained as

$$E = E^0 - \frac{0.0591}{n} \log\left(\frac{\alpha_+}{\alpha_-}\right) \tag{6.26}$$

Platinum or gold is often chosen as the electrode metal but the choice of metal depends on the solution concerned. The metal should not be attacked by the solution, nor should it catalyse side reaction. The electron transfer process must not also be too slow for a reproducible behaviour. Two typical redox electrodes are shown in Figs. 6.6(a) and (b).

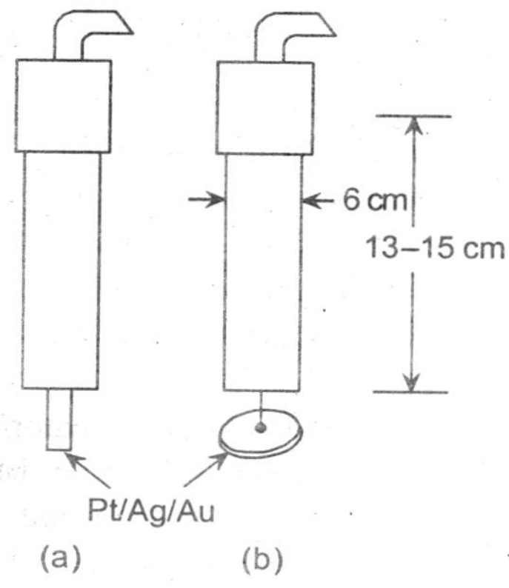

Fig. 6.6 Redox electrodes (a) rod type, and (b) plate type.

6.8.2 Membrane Electrodes

Membrane electrodes are of two different types: (a) ion-selective and (b) the molecular selective type.

The ion-selective type membrane electrode can be subdivided into two major categories: the non-crystalline membrane and the crystalline membrane. The molecular selective type has two major subgroups: the gas sensing type and the enzyme substrate type.

The ion-selective membrane electrodes

Ion-selective membranes must have some electrical conductivity which occurs, in general, due to ion transfer or exchange. The conductivity value may be quite small in some cases. The membrane should be so constituted that it must be capable of binding the analyte ion selectivity. The bindings are of three types, namely the ion-exchange, crystallization, and complexation. Complexation is not very common. The membrane should not dissolve to any extent in the analyte solution, which, in general, is aqueous in nature. This connotes that the membranes should be made of silica glasses or polymer resins or such other molecular aggregates. Some ionic inorganic compounds of low solubility can also be used mostly as crystalline membrane, single or poly- or mixed.

Conduction in a membrane cell, as has been mentioned already, takes place by ion transfer, unlike in aqueous solution where it is by migration of anions and cations or in metal electrode–liquid interface where it occurs through oxidation/reduction process. The membrane itself is ionic in nature. Some membranes are ion exchangers having numerous ionic sites which are capable of interacting with charged 'bodies' in a solution, such as silicate glass. It consists of a three dimensional 'lattice' form of oxygen atoms bonded to silicon atoms keeping open regions in the structure which can be occupied by cations and which can neutralize the negative charge of the O_2–Si network. If these cations have multiple charges as with Ca^{++} and Al^{+++} in Ca-glass and Al-glass, these positive ions are immobile but if they have single charge as in Na^+, K^+, Li^+, and the like, they exhibit mobility in the structure and this mobility allows the charge to be transferred through the glass.

There are two glass–solution interfaces and single mobile charges, like protons, are transferred once from glass to a solution at one side and from a solution to the glass at the other, making a current to flow. In fact the cation sites in the glass are mostly occupied by protons, H^+. If the current is absent, on the two sides of the membrane equilibria are reached which are due to relative hydrogen ion concentrations in the solutions on the two sides. When the positions of these equilibria differ, the surface of the membrane at which greater dissociation has occured would be negative with respect to the other surface resulting in a potential whose magnitude obviously depends on the difference in hydrogen ion concentration on the two sides of the ion-selective membrane. This can be called the *boundary potential* and can be measured in a cell for pH measurement.

Each cell consists of a membrane electrode as the indicator/sensor electrodes and a reference electrode, as has been described earlier, like calomel electrode. But the membrane is not amenable to be connected to the external circuit and hence, an internal standard solution with a second reference electrode in association with the membrane is used. The scheme is shown in Fig. 6.7. The second reference electrode has usually a different potential and its 'preparation' will largely be governed by the internal standard solution. As indicated in the figure, second reference electrode has a potential E_{r2}, between this and the membrane for the activity of the internal standard solution the potential E_{m2} is developed, on the other side of the membrane potential is E_{m1} for the activity of the analyte; first reference electrode junction has a potential E_s, and the corresponding reference electrode potential is E_{r1}. The overall potential would thus, be the algebraic sum of these potentials. The second reference electrode, the internal standard solution, and the membrane can be made a single unit and named the *measuring* or *analyte sensing electrode,* as will be discussed later.

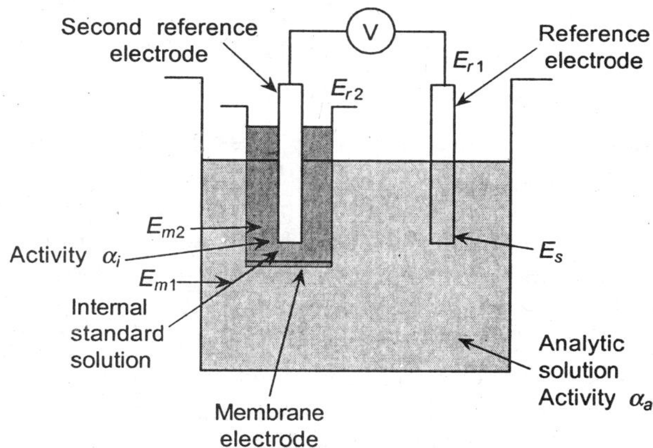

Fig. 6.7 Cell using a membrane electrode.

The potential of the entire cell is given by

$$E_c = E_{r1} + E_s - E_M^0 + \left(\frac{0.0591}{n}\right) \log\left(\frac{1}{\alpha_a}\right) \tag{6.27}$$

where E_M^0 is the value of E_M and is given as in Eq. (6.28) for $\alpha_a = 1$.

Now,

$$E_M = E_{m1} - (E_{m2} - E_{r2})$$

$$= E_b + E_{r2} \tag{6.28}$$

E_b being the boundary potential.

An important aspect of the ion-selective membrane electrodes is their selectivity. Selectivity coefficient has been defined for a membrane with respect to a specific analytic solution. Thus, if the activities of the cations of the membrane on the external side and the singly charged species of the analyte solution in it are α_m and α_s and on the surface of the membrane are α'_m and α'_s, the selectivity coefficient is given as

$$K_s = \frac{\alpha_m \alpha'_s}{\alpha'_m \alpha_s} \tag{6.29}$$

The boundary potential is then given by

$$E_b = E_b^0 + 0.0591 \log\left[\alpha_m + K_s\left(\frac{\mu_s}{\mu_m}\right)\alpha_s\right] \tag{6.30}$$

where μ_m and μ_s stand for mobilities of the membrane cations and charge species of the solution in the membrane, and

$$E_b^0 = -0.0591 \log\left(\alpha_m + \alpha_s\right) \tag{6.31}$$

Selectivity coefficient is actually a measure of the interference that the charged species of the solution cause to the cations of the membrane in potentiometric developments and, thus, can vary from zero to any arbitrarily large value. If there are several cations in the solution, multiplying charged species with activities α_{sj} and, with charges carried by the ions are q_{sj}, the generalized form of E_b is given by

$$E_b = E_b^0 + \left(\frac{0.0591}{q_m}\right) \log\left[\alpha_m + \sum_{j=1}^{n} K_{sj}\left(\frac{\mu_{sj}}{\mu_m}\right)(\alpha_{sj})^{\frac{q_m}{q_{sj}}}\right] \tag{6.32}$$

where q_m is the charge carried by the ion of activity α_m.

The pH sensor: The non-crystalline membranes are glasses and liquids. The discussion in the previous subsection although, is very general, pertains largely to glass electrodes. It is, in fact, the first membrane electrode to be developed. One of the purposes of a special type glass 'membrane' electrode is to sense pH, the negative logarithm of hydrogen ion concentration to indicate the amount of acidity or basicity. One such typical electrode sensor is shown in Fig. 6.8. A thin specially made pH-sensitive glass tip is sealed to the heavy-walled glass tubing as shown. The bulb so formed is filled with a buffer solution or a solution of 0.1 M HCl saturated with AgCl. A silver wire immersed in the solution forms reference electrode 2 of Fig. 6.7. Reference electrode for the complete cell is a calomel electrode. The overall measurement system, thus, requires two reference electrodes and a measuring glass electrode as a membrane.

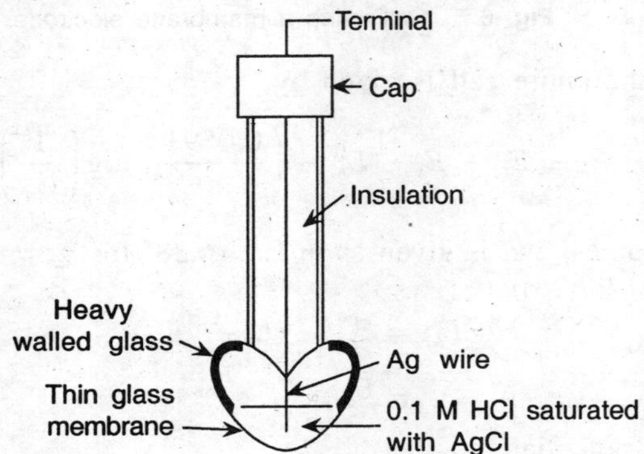

Fig. 6.8 The pH electrode.

The glass that is used for this membrane is, in general, the soda lime glass with $Na_2O(22)$, $CaO(6)$, and $SiO_2(72)$. It is quite selective for pH upto a value 9. Above this, lithium all purpose glass is more suitable which can be made when Na is replaced by Li and Ca by Ba in various proportions. This change increases the lifetime of the electrode as well. For pH sensing, the electrode needs to be hydrated. Hydration helps the ion-exchange reaction when singly charged cations of the glass are exchanged for protons of the solution. The reaction is given as

$$H^+ + Na^+G^- \rightleftharpoons Na^+ + H^+G^- \tag{6.33}$$

where G^- indicates one cation bonding site in the glass surface.

If the glass membrane has identical solutions on the two sides, it will have $E_{m1} - E_{m2} = 0$. But this is not always true and a small potential is seen to exist in some membranes which is termed as asymmetry potential. It is believed to develop because of strains within the two surfaces during manufacturing stage and contamination of the outer surface during continuous use. Frequent calibration with standard buffers eliminates this potential.

In an alkaline solution, the cation is the same kind as the Na^+ in the soda lime glass membrane; this condition reduces the selectivity and a negative error develops in the measured pH value which is called the *alkaline error*. Basically, the glass responds not only to H^+ but also to alkaline metal ions. It can be reduced to a certain extent using lithium all purpose glass. The error for different glasses is shown in Fig. 6.9(a). The alkaline error has been confirmed by taking test analytes of 1 M Na and 0.1 M Na. The error is seen to reduce by 60% from the former to the latter. As seen, there is an acid error also which has been obtained by testing with acid solutions but the cause is yet to be explained properly.

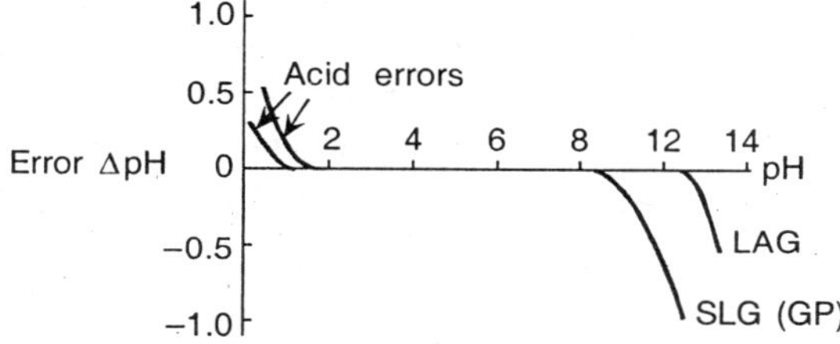

Fig. 6.9(a)   Error curves for different pH-sensitive glass membranes.

The pH electrode is now available in a combination form with the reference calomel electrode integrally designed in a single casing as shown in Fig. 6.9(b).

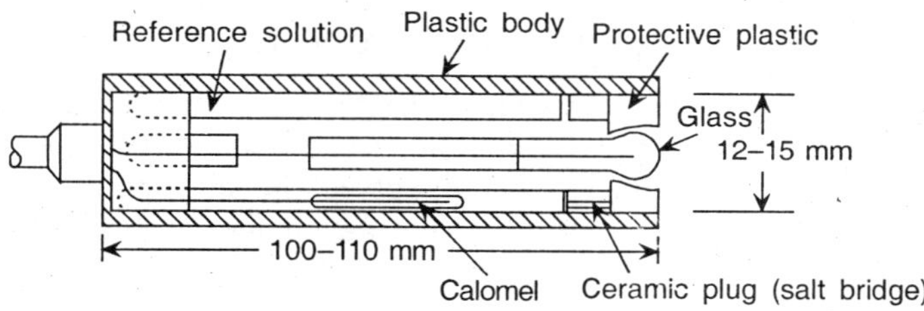

Fig. 6.9(b)   The integrated PH-sensor.

Liquid membrane electrodes: Also called liquid ion exchanger membranes, liquid membrane electrodes are formed from immiscible liquids that are selective in bonding and give rise to potentials with activities of some polyvalent cations and some other singly charged cations and anions. The schematic diagrams of the electrode is given in Fig. 6.10.

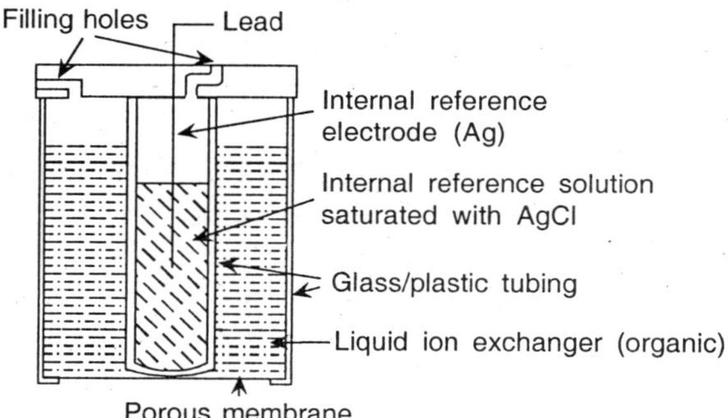

Fig. 6.10   Liquid ion exchanger membrane electrode.

As shown, the inner tube contains the internal reference electrode and solution saturated with AgCl, it also contains the aqueous standard solution of, say, MCl_x where M^{x+} is the cation whose activity is to be determined. The outer tube contains an organic ion exchanger liquid. The ion exchanger liquid and the inner liquid are held in place by a porous membrane usually made from some form of cellulose with pore diameter of around 100 nm. The membrane is made hydrophobic by chemically treating it. The ion exchanger liquid comes in contact with and permeates the membrane—actually wick action causes the pores to be filled with this liquid. With water repelled, only the cations M^{x+} from the inner tube liquid can be exchanged with the organic liquid producing a characteristic potential. However, proper choice of the exchanger liquid is essential. Such liquids consist of polar ionic sites in relatively large non-polar organic molecules. These sites are negative in a cation exchanger and positive in anion exchanger. For Ca^{++} and Mg^{++}, the exchanger is $(RO)_2 PO_2^-$, structure is shown in Fig. 6.11, and for Cu^{++} and Pb^{++}, this is $RSCH_2COO^-$; where R is an aliphatic group. Some of these are p-(1, 1, 3, 3-tetramethylbutyl)-phenyl, p-(n-octyl)-phenyl, and so on. With $MgCl_2$ in the analyte solution in internal tube, the ion exchange liquid separates the solution from the reference $MgCl_2$ solution establishing at each interface, the equilibrium which is represented as

$$[(RO)_2 PO_2]_2Mg \rightleftharpoons 2(RO)_2 PO_2^- + Mg^{++} \tag{6.34}$$

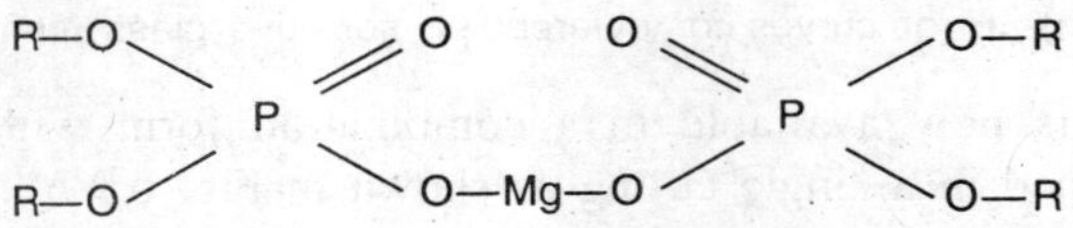

Fig. 6.11   The $(RO)_2PO_2^-$ structure.

In recent times, the liquid ion exchanger outer tube has been dispensed with by forming a membrane which has the ion exchanger immobilized in a polyvinyl chloride membrane. This membrane and the exchanger liquid are dissolved in an appropriate solvent such as tetrahydrofuran. When it is evaporated, a flexible membrane is obtained which behaves as a membrane with the ion exchanger liquid held in the pores. This new type of ion exchanger is actually being used extensively and is termed as immobilized liquid in rigid polymer.

The liquid membranes may have three types of active substances such as (i) cation exchangers, (ii) anion exchangers, and (iii) neutral macrocyclic compounds which are complex carriers and are used for cations being very selective. Selectivity for a particular ion depends on the ability of the electrode to extract the ion into the membrane. Basically, it is the ability of the ion to form a complex with the neutral carrier. After complexation and extraction, the species in such a membrane has the same charge as the extracted ion and this is the reason why it is called neutral carrier. Some typical neutral carriers are vanilonycin (for K^+), O-nitrophenyl-n-octylether (for Ca^{++}), and so on.

Crystalline membrane electrodes: Sometimes referred to as solid membrane electrodes, crystalline membrane electrodes may be single crystals or polycrystalline. A typical arrangement is shown in Fig. 6.12 with internal reference electrode. Crystalline membranes may be homogeneous or heterogeneous. Homogeneous crystalline membranes can be made cutting a single crystal or can be made as discs or pellets from the finely grounded crystalline solid by high pressure or by casting from a melt. The membrane is fitted by sealing to an inert plastic tube like teflon or polyvinyl chloride as shown in Fig. 6.12.

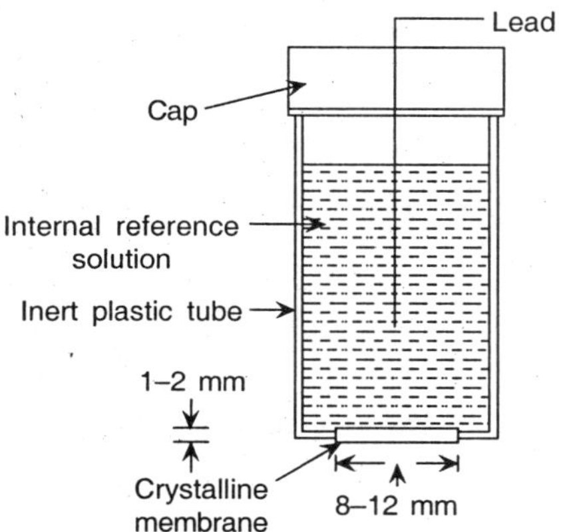

Fig. 6.12   Crystalline membrane electrode.

The heterogeneous variety is made from sparingly soluble salts. Finely grounded crystalline solids are dispersed/embedded in an inert matrix such as silicone rubber, paraffin, polyvinyl chloride and so on. Such membranes are selective to one of the ions present in the dispersed solid. If barium sulphate is dispersed in the matrix, it becomes selective to Ba ions as well as sulphate ions.

Most ionic crystals have very poor electrical conductivity at ordinary temperature. Some of which are conductive, have ions that are mobile in the solid phase such as Ag ions in silver halides and sulphides, fluoride ions in some rare-earth fluorides, and copper ions in cuprous sulphide. In such materials, the conducting ions jump to holes present as defect in the crystal lattice leaving oppositely charged holes. Only single kind of ions can participate in such a process because of mobility restriction in a solid crystal and hence, crystal membranes are so very selective.

As mentioned, single crystal membranes made from Ag halides become selective to silver and halide ions. If silver halides are mixed with crystalline silver sulphides in 1:1 molar ratio, the homogeneous membranes so prepared show good conductivity because of better mobility of silver ions in sulphide matrix. Silver sulphide in combination with Cd, Cu, and Pb sulphides also are used for selectivity of Cd^{++}, Cu^{++}, and Pb^{++}.

While Br^-, Cl^-, and I^- ions are determined by AgS–AgBr, AgS–AgCl, and AgS–AgI membranes; for detecting fluoride ions, LaF_3 has been found to be most suitable. LaF_3, or fluorides of neodymium and praseodymium are good conductors but their conductivity is further improved by doping with EuF_2. The single crystal formed is then cut according to the required size. The solid crystal membrane behaves as a glass electrode for pH measurement with charges created on the two surfaces by ionization with guiding equation

$$LaF_3 \rightleftharpoons LaF_2^+ + F^- \tag{6.35}$$

The potential across the membrane has its positive end where lower fluoride ion concentration occurs. Such membrane can be used for a concentration as low as 10^{-6} M in a temperature range of 0–80°C.

Metal salts with high electrical conductivity can be used directly in electrode form. This is called solid state electrode and has been reconsidered in recent years. Even pH electrodes have been developed on this basis. Figure 6.13 depicts such an electrode where an alumina (Al_2O_3) rod is coated with a layer of iridium oxide by sputtering. This electrode can be used in solutions where conventional pH electrodes tend to dissolve—such as a solution of hydrofluoric acid.

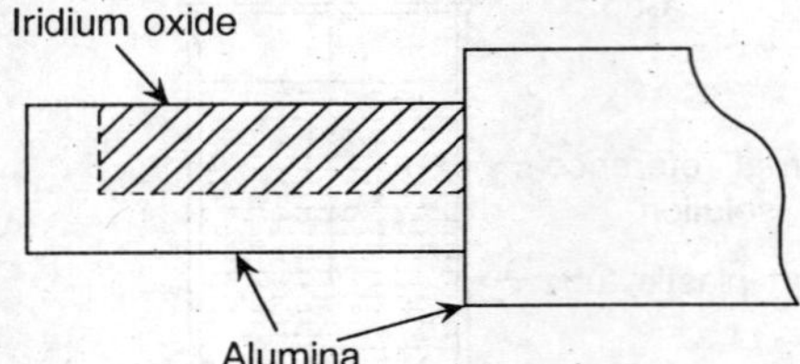

Fig. 6.13 The solid state electrode.

Molecular selective electrodes

Molecular selective electrodes have been developed for sensing concentration of certain molecular constituents in samples such as HCN, CO_2, glucose, urea, and so on. The two types of molecular selective electrodes are now briefly described.

Gas sensing electrodes: Gas sensing electrode is basically a measuring cell in which there is a reference electrode immersed in an internal solution which is held by a thin gas-permeable membrane and also there is a standard (Ag/AgCl usually) reference electrode as shown in Fig. 6.14.

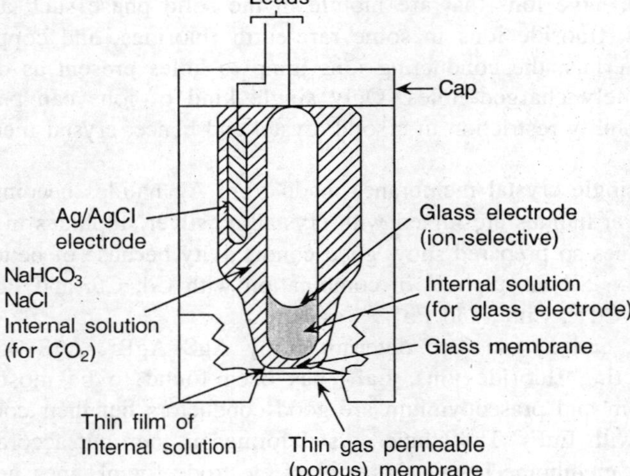

Fig. 6.14 Gas-sensing electrode.

A thin gas-permeable membrane separates the analyte solution from an internal solution; for CO_2 sensing, this is $NaHCO_3$ and NaCl. A pH-sensitive glass electrode is arranged so that a thin film of the internal solution is held between the gas-permeable membrane and the glass membrane. The reference electrode of Ag/AgCl is also placed in the internal solution. The pH of the film of liquid adjacent to glass electrode can be used to measure the gas content of the analyte. The gas passes on to the thin film of the internal solution and acts reversibly with it to form an ion to which the ion-selective electrode (the glass electrode here) responds. The activity of the ion in the thin film is proportional to the gas dissolved in the analyte and hence, the electrode response is directly related to the activity of the gas in the sample.

The gases, that are detected usually, are CO_2, SO_2, NH_3, and are detected by gas sensing electrodes based on the pH adjustment of the internal solution for equilibrium. The relevant equations are

$$NH_3 + H_2O = NH_4^+ + OH^- \tag{6.36a}$$

$$SO_2 + H_2O = HSO_3^- + H^+ \tag{6.36b}$$

$$CO_2 + H_2O = HCO_3^- + H^+ \tag{6.36c}$$

The pH indicated becomes a measure of the gas concentration.

The internal solution for SO_2 would be sodium hydrogen sulphite; for NH_3, it would be ammonium chloride solution. The stage-wise transfer, before the reactions given by Eqs. (6.36), is represented as

$$\underset{\text{analyte}}{CO_2(aq)} \rightleftharpoons \underset{\text{membrane pores}}{CO_2(g)}$$

$$\underset{\text{membrane pores}}{CO_2(g)} \rightleftharpoons \underset{\text{internal solution}}{CO_2(aq)} \tag{6.37}$$

The equilibrium constant K_e for the reaction Eq. (6.36c) is given by

$$K_e = \frac{[H^+][HCO_3^-]}{[CO_2(aq)]_{\text{analyte}}} \tag{6.38}$$

When concentration of HCO_3^- in the internal solution is high enough to be affected by the CO_2 from the sample, then one can write

$$K_n = \frac{K_e}{[HCO_3^-]} \frac{[H^+]}{[CO_2(aq)]_{\text{analyte}}} \tag{6.39a}$$

giving the hydrogen ion concentration in the internal solution due to CO_2 permeance in it as

$$[H^+] = K_n[CO_2(aq)]_{\text{analyte}} \tag{6.39b}$$

Hence, the voltage measured by the cell is

$$E = E^0 - 0.0591 \, \log\{K_n[CO_2(aq)]_{\text{analyte}}\}$$

$$= E_0' - 0.0591 \, \log\{[CO_2(aq)]_{\text{analyte}}\} \tag{6.40}$$

There are two types of membrane materials—microporous and homogeneous. The former is made from hydrophobic polymers such as polypropylene or polytetrafluorethylene with 70% porosity, each pore having a diameter of about 10^{-6} m and thickness 0.1 mm. Gases pass through these membranes by the process of effusion. In homogeneous membranes, the analyte gas dissolves, then diffuses through and finally desolvates in the internal solution. A common membrane material of this category is silicone rubber. Its thickness is required to be less for the whole process of gas transfer to be faster. A thickness of 0.01–0.03 mm is quite common.

Different sensing electrodes, other than pH are used for different gases. Their selectivity would depend on the internal ion-sensing electrode (ISE), the internal solution, and the type of membrane. The other gases that can be assayed are H_2S, HCN, HF, NO_2, Cl and so on with ISE's made of Ag_2S, Ag_2S–LaF_3, immobilized ion exchange type, Ag_2S–AgCl respectively.

Biomembrane or enzyme electrodes: Enzymes are highly selective biochemical substances often used as catalysts. Enzyme is selective in the sense that it catalyzes only a small number of reactions. An electrode that has a membrane coated with an enzyme containing acrylamide gel can act as an ion-selective electrode. The enzyme catalyzed reaction of the analyte with this electrode is monitored by another internal ion-selective electrode. Thus, a couple of ion-selective electrodes, as stated, can make the electrode sensing system highly selective and free from interferences.

The gel and enzyme are held on the membrane surface by an inert physical support, even cellophane or nylon-gauge may be used for the purpose. Immobilization can also be done by other means such as physical absorption in a porous inorganic support like alumina, covalent bonding of the enzyme to glass beads or polymers or copolymerization of the enzyme with a monomer. These electrodes are often termed as *enzyme reactor electrodes.*

A case of enzyme electrode can be explained with urea-selective electrode. The membrane, that is, an ammonia-sensitive glass electrode is coated with acrylamide gel and urease (enzyme). The electrode when dipped in a solution containing urea, produces a reaction

$$\underset{\text{urea}}{CO(NH_2)_2} + H_2O \xrightarrow{\text{urease}} 2NH_4^+ + CO_2 \qquad (6.41)$$

NH_4^+-selective electrode measures NH_4^+. Prior calibration of the electrode system is, however, necessary.

Biological electrode is a sort of enzyme electrode that uses living bacteria which excrete the enzymes in use. They can be replenished by proper nutritional treatment so that the bacteria may live on.

The ion-selective principle can be adopted in field-effect transistors to make ion-selective FET, ISFET which forms a class of ion-selective electrode. It is made by coating or depositing the ion-selective membrane on the gate of the FET. The membrane may be AgBr or a neutral carrier in PVC matrix. Prior to coating, a polyimide mesh is placed over the gate for retention of the matrix.

The analyte in contact with the membrane partially determines the potential of the membrane because of the activity of the analyte. This potential, as usual, determines the drain current in FET's. An ion-selective membrane adapted in an FET, that is, an ISFET is depicted in Fig. 6.15.

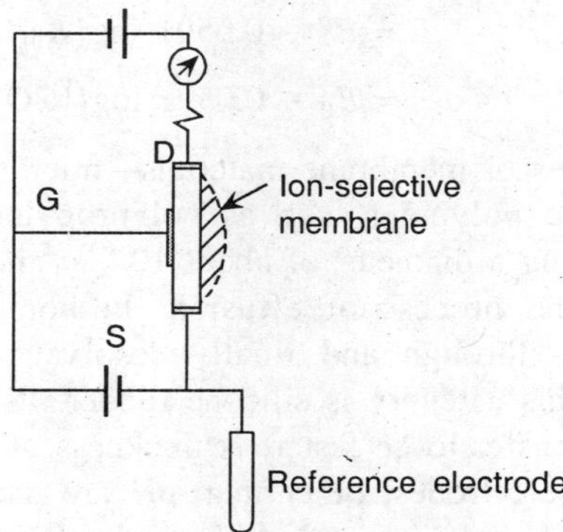

Fig. 6.15   The ion-selective membrane adapted in an FET.

6.8.3 Temperature Sensing Electrodes

The cell potential or electrode potential is highly dependent on temperature. Nernst equation includes the T and the value 0.0591 as in Eq. (6.12) is obtained for a temperature 25°C. This only prompts one to think that temperature could be measured using Eq. (6.8) when $E_c \propto T$.

In fact, in recent times a technique has been suggested for measurement of temperature using an internal reference electrode made of commercial variety of iron that consists of Mn(0.68), C(0.5), Si(0.19), Cr(0.18), Cu(0.13), Ni(0.07), P(0.024), and S(0.003). The internal reference solution is a solution of 0.5 M HPO_3, 0.1 M H_2O_2. The electrode is rod-shaped, the tip being kept open, the rest covered with a teflon sleeve. The scheme of sensing is shown in Fig. 6.16. The output voltage V_o shows a self-oscillating phenomenon with magnitude and period varying with temperature as shown in Fig. 6.17. This existence of oscillation is interpreted as a potential change associated with growth and dissolution of the passive film formed during anodic and cathodic reactions given respectively by

$$Fe \rightarrow Fe^{++} + 2e \tag{6.42a}$$

and

$$H_2O_2 + 2H^+ + 2e \rightarrow 2H_2O \tag{6.42b}$$

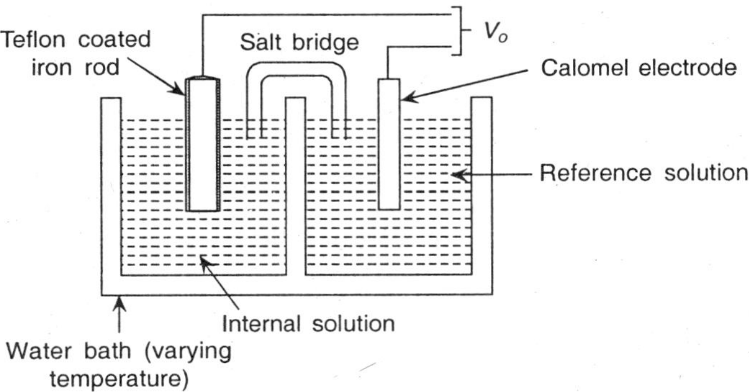

Fig. 6.16 Scheme of a cell for temperature sensing.

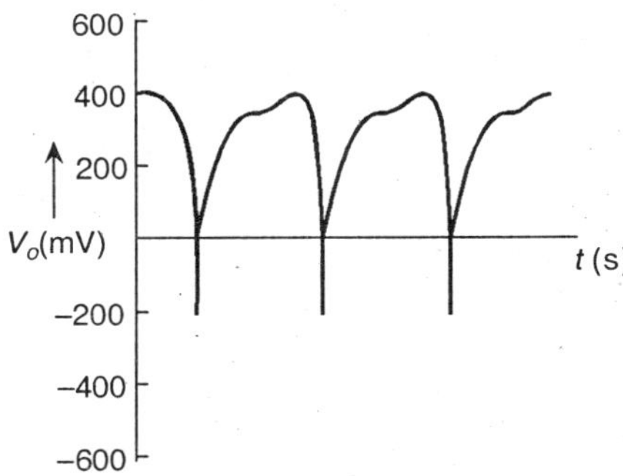

Fig. 6.17 Output waveform of the cell of Fig. 6.16.

Example 6.1

An electrode selective to ions (type I) and reference electrode is immersed in a solution containing type I ions of activity 1.5×10^{-3} M when the potential of the selective electrode is 0.21 V. If the same electrode is dipped in a solution that has activity of type I ion 1.6×10^{-3} M, and its selectivity coefficient with respect to another type, type II (with activity 1.1×10^{-3} M) in the solution is 2.5×10^{-1}, what would be the electrode potential? Assume solution temperature as 25°C and unit ion charges for both types.

Solution

Using Eq. (6.23),

$$-0.21 = E_1 = E^0 \frac{0.0591}{1} \log(1.5 \times 10^{-3})$$

so that

$$E^0 = -0.2693 \text{ V};$$

Now, using Eq. (6.32), with $(K_{s2}\, \mu_{s2})/\mu_m = 0.25$, $\alpha_{s2} = 1.1 \times 10^{-3}$, and $\alpha_m = 1.6 \times 10^{-3}$,

$$E = E^0 + \left(\frac{0.0591}{1}\right) \log\left\{1.6 \times 10^{-3} + (0.25 \times 1.1 \times 10^{-3})\right\}$$

$$= -0.2693 + 0.05935$$

$$= -0.20995\,\text{V}$$

Example 6.2

Calculate the half-cell potential of an Ag electrode dipped in a solution that has 1.5×10^{-2} M Ag^- concentration.

Solution

$$Ag^+ + e \rightleftharpoons Ag(s),$$

$$E^0 \text{ (from table)} = +0.799 \text{ V}$$

Hence,

$$E = 0.799 + \frac{0.0591}{1} \log(1.5 \times 10^{-2})$$

$$= 0.8583 \text{ V}$$

6.9 ELECTROCERAMICS IN GAS MEDIA

A number of ceramic oxides have been developed over the past two decades and some are still being developed which function to show some electrical properties such as develop an emf, change conductivity or surface ionic conductivity and so on, when placed in a gas media. A list of these properties with respect to the materials is given in Table 6.2 which is not comprehensive by any means.

Table 6.2 Functional properties and applications of electroceramics

S.No.	Functional properties and their representation	Material	Applications
1.	Ionic conductivity; uses Nernst Eq. $$Emf = \frac{RT}{4F}\ \ln\left\{\frac{P(O_2)_g}{P(O_2)_r}\right\}$$ $(P$ = partial pressure)	ZrO_2	Solid analyte, mainly as O_2 sensors
		Nasicon	Solid electrolyte, mainly as gas sensors
		$\beta,\ \beta''$-alumina	Solid electrolyte in potential generation
2.	Semiconductivity, conductivity depends on temperature and partial pressure of O_2, $$\sigma = \sigma_0\,e^{-E/kT}.\ \frac{1}{(P(O_2))^{1/m}}$$ $(E$ = activation energy, m = constant $4 < m < 6$, k = Boltzmann constant)	SnO_2, ZnO	Gas sensors, mainly hydrocarbon gases
		TiO_2	O_2 sensors
		Oxides such as $SrTiO_3$, $BaTiO_3$, $SrSnO_3$ (called pervoskites)	O_2 sensors
3.	Surface property (a) Varistor $$\frac{J_1}{J_2} = \left\{\frac{(\text{Electric field})_1}{(\text{Electric field})_2}\right\}^{\alpha}$$ α = varistor constant, $1 < \alpha < \infty$; typically 25–50	$BaTiO_3$	Electronic sensors
	(b) Surface ionic conductivity	SiO_2, $ZnCr_2O_4$	Humidity sensors

The transfer characteristic of a metal oxide sensor is approximated by the relation

$$\frac{R_j}{R_0} = \left\{a_j\,(\chi_j) + 1\right\}^{n_j} \tag{6.43}$$

where R_j is its resistance with gas j,

       χ_j is the concentration,

       R_0 is the resistance with air as the gas,

       a_j is a coefficient for the gas, and

       n_j is an index.

6.9.1 Ionic Conductors

Zirconia

ZrO_2 exists in the crystalline modifications in different temperature ranges.

In the range 2680°C (melting point) to 2372°C, zirconia crystallizes in face-centred cubic structure; between 2372°C and 1200°C, it appears as tetragonal; and from 1200°C to about room temperature, it has the monoclinic form (all these are at atmospheric pressure). At a high pressure, a stable orthorhombic structure is also known.

When phase transformation occurs from tetragonal to monoclinic structure, there is a volume expansion to the extent of 3–5% which leads to cracking in the material. Hence, zirconia is made only in small pieces by taking care of volume stability. Addition of oxides of yttrium, calcium, or magnesium by certain amounts leads to cooling of the cubic structure without transformation. Depending on the mole percentage of Y_2O_3 the structure at room temperature is determined. Thus with 1.8%, a monoclinic phase exists; upto about 2.5%, tetragonal phase with a little cubic phase exists; and beyond about 16%, only cubic phase can be available in stable form. With CaO and MgO, similar phase diagrams (Fig. 6.18) are known but these are not very well established. For application, the phase structure is required to be known, however, and the type and amount of coexisting phases, their distribution, transformation kinetics, added cation and so on, do determine the properties of zirconia.

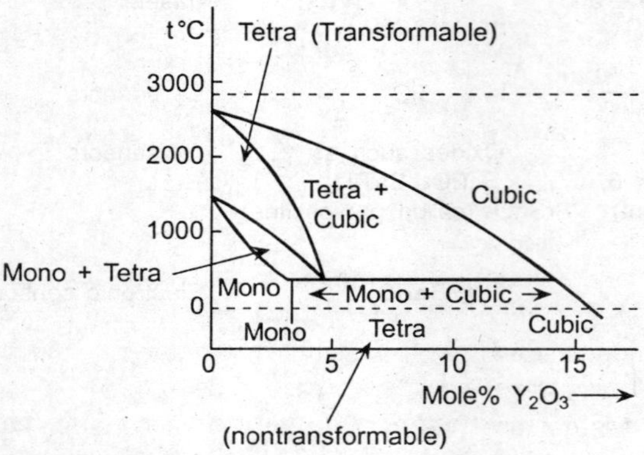

Fig. 6.18 Phase diagram of ionic conductors.

As zirconia based oxygen sensors are widely used in industries for combustion control, engine control and so forth, the ionic properties of zirconia with different cations varying with temperature are shown in Fig. 6.19.

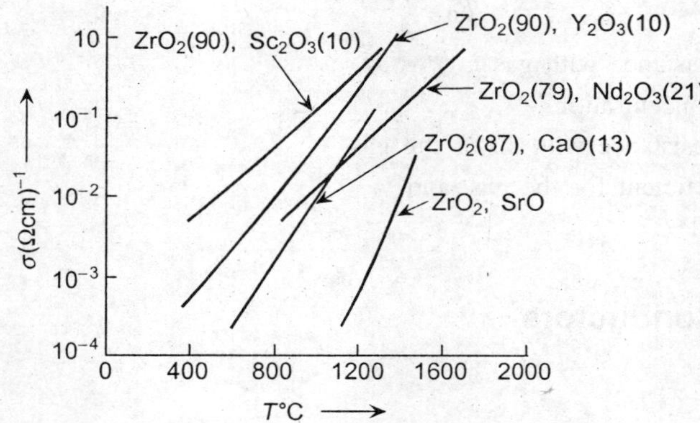

Fig. 6.19 Conductivity–temperature plots of some ionic conductors.

Conduction phenomenon in zirconia has only been recently realized. When Zr^{4+} cations are replaced by lower charge cations such as Y^{3+}, Ca^{2+} and so on, vacant spaces are formed in anionic oxygen lattice, but the distribution of these vacant sites and the occurrence of low charge cations are only statistical in nature. As these vacancies can move through the lattice, the charge transport thus occuring, produces conductivity. However, the size of cation that replaces Zr^{4+} determines the energy required for such charge transportation. Now as Ca^{2+} has a higher radius of 0.99 µm and Y^{3+} has a radius 0.92 µm whereas Zr^{4+} has that of 0.79 µm, replacement with Y^{3+} gives higher conductivity than with Ca^{2+}. The higher energy requirement is said to be due to the fact that oxygen moves only through a tetrahedral face formed by three cations. Further, the maximum ionic conductivity occurs for different mole per cent of either Y_2O_3 or CaO. In the former case, it is about 8–9% while for the latter it is about 12–13%. Even these maximum σ-values are also not same.

NASICON

It is a sodium solid electrolyte series with the chemical composition $Na_{1+x} Zr_2 Si_x P_{3-x} O_{12}$ with x varying from 0–3, the maximum being obtained at $x = 1.9$–2.2. Because of presence of sodium, its conduction property is very good. With $x = 2$, the structure of NASICON is rhombohedral. It consists of octahedral ZrO_6 groups separated by tetrahedral PO_4 or SiO_4 and a three-dimensional framework is made that appear as infinite ribbons which are quite loose and provide conduction channels for Na^+.

β-alumina

It belongs to the set of sodia-alumina (Na_2O–Al_2O) compounds that have been obtained from sodium aluminate ($NaAl_{11}O_{17}$). They are solid electrolytes and used at high temperature. They permit fast diffusion of Na^+, and hence, have high ionic conductivity but low electronic conductivity. They are quite inert chemically and possess good thermal and mechanical strength. A variation in the form is Na_2O–$5Al_2O_3$ which is called β''-alumina. These are the two most common types though there are others as well. In fact, β''-alumina is thermodynamically unstable and is made stable by 'ternary processing' by addition of cations such as Mg^{2+} and Li^+.

β-alumina is hexagonal and layered while Li^+-stabilized β''-alumina is rhombohedral. The latter has higher ionic conductivity than the β-alumina.

Figure 6.20 shows the comparative study of the variation of ionic conductivities of the different ceramics as discussed in the preceding paragraphs.

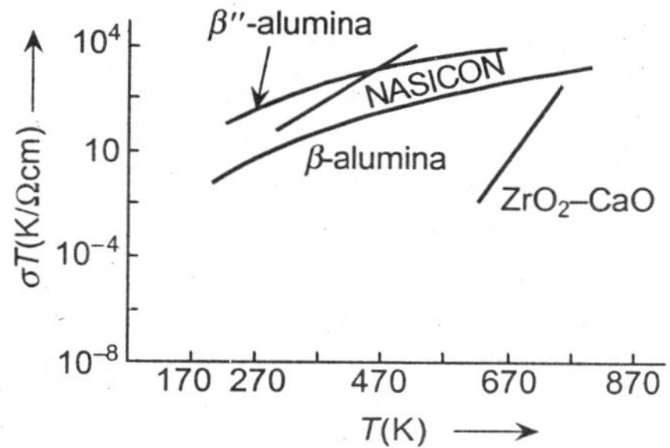

Fig. 6.20 Ionic conductivities of some ceramics.

Tin oxide

It is an oxygen deficient n-type semiconductor and crystallizes in a rutile structure. It is commercially used as gas sensors—particularly for reducing gases in industrial and domestic conditions, for alarm generation and known as *Tagushi Gas Sensors* (TGS). Various dopants and catalysts usually made of noble metals such as Pd, Pt, and so forth are used to increase sensitivity of such sensors. They also require temperature of 300–400°C for their operation and are affected by moisture content. These have low selectivity which is considered a disadvantage.

Zinc oxide

Zinc oxide crystal has hexagonal wurtzite form. It is made into n-type ceramic by doping with In (indium) and is then used to detect hydrocarbon gases. When platinum surrounds the n-type ZnO ceramic, it becomes specially sensitive to fuel gases such as isobutane and propane whereas using Pd as a catalyst, it may be made sensitive to H_2 and CO.

Materials of the kind Co_3O_4, WO_3, and Fe_2O_3 can also be similarly used for gas detection purposes.

When oxygen in air is chemically absorbed or 'chemiabsorbed' in metal oxides such as SnO_2 or ZnO, electrons become locally aligned at the surface that produces a negative surface potential and there a charge carrier depletion occurs. This means that the surface conductivity is low. This chemiabsorbed material when exposed to a gas, cause the gas to react with oxygen, decreasing the surface-bound oxygen as also the resistivity. The phenomenon is enhanced at an elevated temperature of a few hundred degrees when the reaction is rapid and also reversible. Figure 6.21 shows a typical arrangement of the sensor with heater.

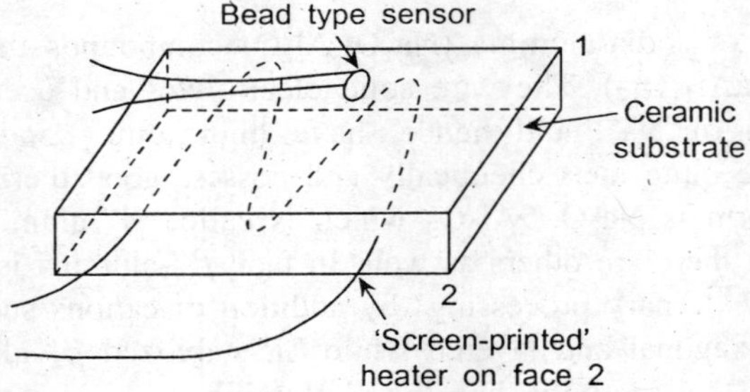

Fig. 6.21 A sensor using chemical absorption.

As mentioned already, Pd or Pt, or oxides and salts of noble metals increases the reactive selectivity as well as reactivity of different gases with oxygen. Selectivity and sensitivity are functions of temperature, peaking at specific values. This temperature dependence makes the sensors suitable for combustion monitoring purposes. They can be developed for a specific stoichiometric air-to-fuel ratio (λ) when a change of state is observed.

Titania (TiO$_2$)

Titania also forms a tetragonal rutile structure and is very useful as a flue gas monitoring sensor. Under reducing conditions, heating causes it to lose oxygen thus, generating oxygen vacancies which is then balanced by reducing Ti^{4+} to Ti^{3+}; and an electron donor situation arises. With increasing loss in oxygen with higher temperature, increasing number of electrons contribute to this conduction process and in the temperature range 300–1000°C, TiO_2 behaves as an n-type semiconductor. Other than temperature, partial pressure of oxygen also accounts for this conduction. Again, for TiO_2 with a Pt catalyst surface, oxygen mobility changes. If TiO_2 is porous, Pt coats a wall and sensitivity as well as response are high.

Pervoskites

Pervoskite structure is one structure in which materials such as $BaTiO_3$ can be found. It is cubic above Curie temperature (120°C). When temperature is brought below 120°C, spontaneous polarization takes place with all its Ti^{4+} ions shifting to O^{2-} ions and a permanent dipole moment is induced. The symmetry of the structure is now tetragonal making the polarized region ferroelectric. Below 5°C, the structure becomes orthorhombic which changes further below −80°C. All these changes in structure are reversible. If Ba^{2+} is replaced by Pb^{2+} or Sr^{2+}, the Curie temperature can be changed.

A special type of pervoskite is the $PbZrO_3$–$PbTiO_3$ called PZT family which is further modified by adding lanthanum, and is then called PLZT. PLZT is a solid solution but has a decreased ferroelectric stage, that is, lowered Curie temperature.

Varistors

These are ZnO based semiconductor devices and are used as electric field sensors. Above a certain value of applied field, which is breakdown value, they show large change in current flow for small voltage changes. The performance of the varistors are, to an extent, dependent on manufacturing processes. However, the conduction is through grain–grain interfaces. The details of operational mechanism are not very simple and are beyond the scope of present text.

6.10   CHEMFET

Chemically sensitive FET, in short ChemFET, is a specially designed FET which is adapted in IC technology with special encapsulation and packaging.

In a chemFET, the induced field is established by series combination of applied 'gate' potential and the interface potential due to chemical sensitivity of the 'source' or the substrate. The ordinary design is shown in Fig. 6.22. The stability of the system is ensured by coating it with silicon nitride. However, by depositing different electrochemically active materials on top of this silicon nitride layer, sensors for different ions can be obtained. Thus, aluminosilicate glass is used for sodium ions, valinomycin doped PVC may be used for potassium ions, and so forth.

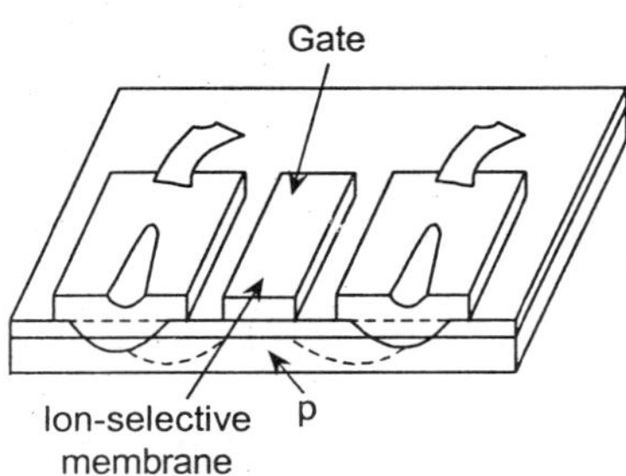

Fig. 6.22   A chemFET.

It must be remembered that for proper functioning, electrical isolation from the surrounding of the chemFET is essential. The active regions of the FET are isolated from the substrate by p-n junctions. A chemFET chip with polyimide mesh suspended over the gate region is sketched in Fig. 6.23.

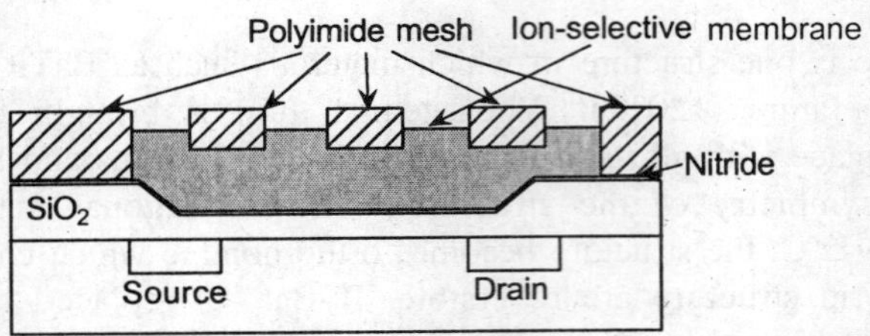

Fig. 6.23   The chemFET chip with polyimide mesh over the gate.

The main problem is the adhesion of the membrane to the FET. Well-shape was adopted for better adhesion, but the lifetime of this adhesion is still very short. A polyimide mesh was patterned over the active gate area and the mesh is 'asked' to entwine the solvent cast membrane and keep it anchored to the gate. For polymer gel, the problem is solved to a certain extent, but not for others, by selective deposition. The micromachined version shown in Fig. 6.24 solves this problem to a great extent.

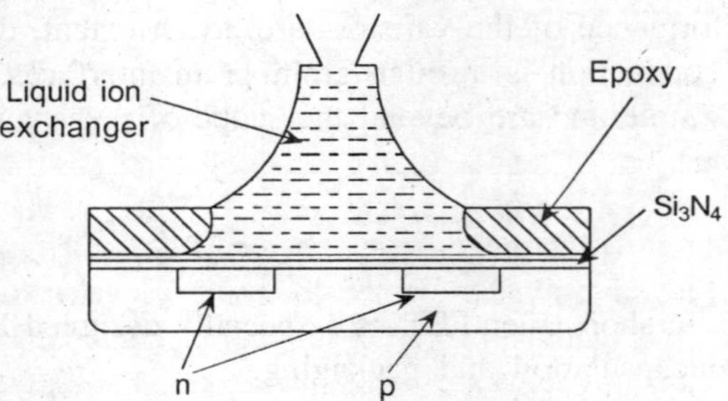

Fig. 6.24   Another version of the chemFET chip for polymer gel.

REVIEW QUESTIONS

1. Describe the basic construction and operation of an electrochemical cell. Define the terms—electrode potential, cell potential, and half-cell potential.

    What is Nernst equation? How does it account for the cell potential of an electrochemical cell?

    Calculate the potential for a half-cell consisting of Zn electrode immersed in a solution that is 0.030 M Zn^{2+}.

    [*Hint:* The reaction indicates $Zn^{2+} + 2e \rightleftharpoons Zn(s)$ and from standard table, $E^0 = -0.763$ V, so that

    $$E = E^0 - (0.0591/n) \log\{1/[Zn^{2+}]\}$$

    $$= -0.763 - (0.0591/2) \log(1/0.03)$$

    $$= -0.866 \text{ V}]$$

2. What is a standard hydrogen electrode? What is its utility in instrumental analysis?

    What is a junction potential? On what factors, does it depend? Why often a salt bridge is necessary in sample analysis through electrochemical cells? What is its function?

3. How is cell potential affected by polarization? What are the different types of polarization? What are practical means of countering it?

4. Why is a reference electrode needed in a sample analysis? What are the commonly used reference electrodes? Write the functional state relations of a silver reference electrode and explain the meaning of these relations.

5. What different types of sensor electrodes are known to be used commercially? To which type the first, second, third kinds, and, redox types belong? How are they different—construction-wise and operation-wise?

6. Distinguish between the operations of ion-selective and molecular selective membranes.
   What is selectivity coefficient of a membrane electrode?
   How is boundary potential defined in a membrane electrode set up? Obtain an expression for the same.

7. Explain the action of a liquid ion exchanger membrane for generating a cell potential. What are the specific types of materials used for such membranes?
   What is ion exchanger immobilized–in–polyvinylchloride membrane?

8. Show the constructional features of a gas-sensing electrode system using a gas-permeable membrane. What are the gases commonly detectable by such types of electrodes?
   Indicate what internal solutions are required for (i) SO_2, (ii) NH_3, and (iii) CO_2. Write the functional state equations for SO_2 analysis.
   What is an equilibrium constant? What is its function in solving for the output cell voltage?

9. Describe the special molecular selective electrode called biomembranes with due reference to materials, construction, functional state equations, and preservation.

10. Explain the characteristics of electroceramics such as ZrO_2, TiO_2, and (SiO_2, $ZrCr_2O_4$) and show how do they use their ionic conductivity, semiconductivity, and surface ionic conductivity respectively for measuring oxygen content and humidity.
    Discuss on the recent developments of such electroceramics as gas sensors.

Chapter 7

Smart Sensors

7.1 INTRODUCTION

A sensor producing an electrical output when combined with interface electronic circuits is said to be an intelligent sensor if the interfacing circuits can perform (a) ranging, (b) calibration, and (c) decision making for communication and utilization of data.

Both sensors and actuators are used as intelligent components of instrumentation systems. In fact they are used as field devices. The block diagram of one such intelligent equipment is shown in Fig. 7.1(a). Figure 7.1(b) shows the simplified version with facilities of processing that can be incorporated.

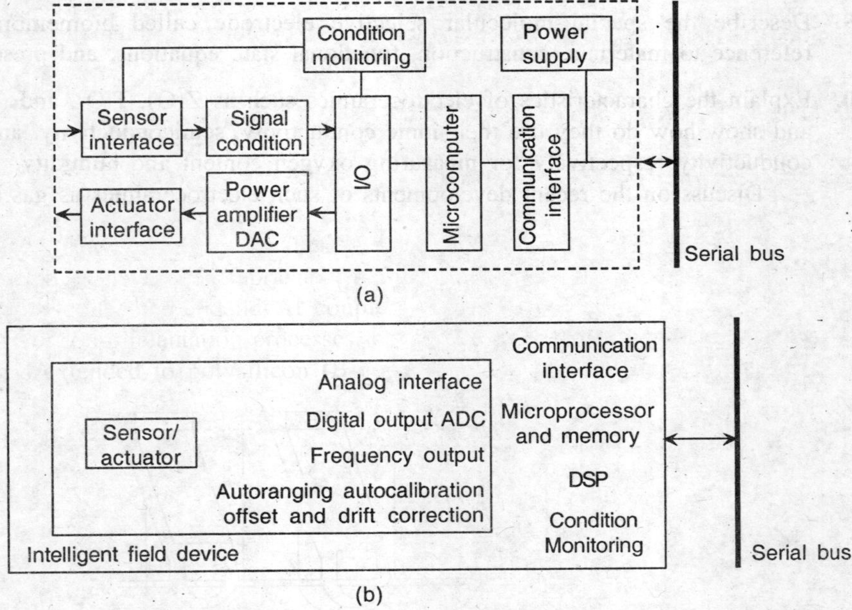

Fig. 7.1 (a) Typical intelligent sensor and actuator and (b) simplified version of (a).

An intelligent field device possesses the following properties:

1. automatic ranging and calibration through a built-in digital system,
2. auto-acquisition and storage of calibration constants in local memory of the field device,
3. autoconfiguration and verification of hardware for correct operation following internal checks,
4. autocorrection of offsets, time, and temperature drifts,
5. autolinearization of nonlinear transfer characteristics,
6. self-tuning control algorithms, fuzzy logic control is being increasingly used now,
7. control programme may be locally stored or downloaded from a host system and dynamic reconfiguration performed,
8. control is implementable through signal bus and a host system,
9. condition monitoring is also used for fault diagnosis which, in turn, may involve additional sensors, digital signal processing, and data analysis software, and
10. communication through a serial bus.

Intelligent sensors are also called smart sensors which is a more acceptable term now. The initial motivation behind the development of smart sensors include (i) compensation for the non-ideal behaviour of the sensors and (ii) provision for communication of the process data with the host system. Traditional sensors that are being used, have varying requirements of compensation and signal processing objectives and the number of measurands in industrial establishments is growing each day. The variety of variables, both physical and chemical, is also increasing and newer sensing mechanisms are being exploited increasing the load on signal processing.

Thus, for each type of variable a different kind of processing is required and with increasing number of types of variables in industries, centralized computers have been overloaded with processing load. The smart sensor is intended to sense as well as do the sensing-related processing within itself. Further, it communicates the response to the host system so that the efficiency and accuracy of information distribution are enhanced with cost reduction.

Advanced processing technologies have now replaced earlier ones used for developments of smart sensors. Sensor elements are open to process although they are now being built in the smart system itself. Certain sensors require supply, constant voltage or constant current along with comparison capabilities; the feature is included in sensor subsystem. Amplification is necessary which usually analog, may also be controlled digitally. Earlier analog filters were employed which have now been replaced by digital counterparts. These three systems, namely the supply, amplification, and filters, comprise the analog signal processing unit (ASPU). Smart sensor also requires a data conversion module either from analog to digital (A/D) or from frequency to digital (F/D) which interfaces with the microprocessors for information processing and bus interfacing for communication. Figure 7.2 shows a stack-block simplified version of the scheme.

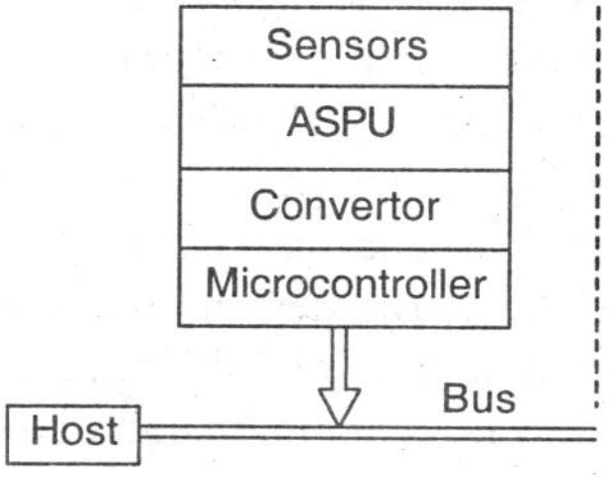

Fig. 7.2 A sensor interfaced with a host system.

The smart sensor devices integrate complementary trends such as

   (a) new sensing methods,
   (b) improved computing capability, and
   (c) digital communication.

New sensing methods are realized through synthesis of those from individual sensors with combined technologies and integration techniques. Digital correction in such new techniques improves performance by

   (i)   compensating for sensor non-linearity,
   (ii) permitting a larger proportion of sensors to meet specifications,
   (iii) incorporating programmable gain,
   (iv) changing sampling rate,
   (v) changing interaliasing filter frequency, and so on.

Digital communication, on the other hand, plays an important role in reducing or overcoming noise and quantification errors to send error-free data.

7.2 PRIMARY SENSORS

Existing sensors of all kinds with a cascaded block for providing electrical output in the form of voltage or current can be adapted to an integrated processing system but the system can then be hardly called a smart sensor.

External stimuli such as strain/stress, thermal/optical agitation, and electric/magnetic field change the behaviour of materials at atomic/molecular level or in crystalline state. This concept is utilized in designing a primary sensing element for particular stimulus or a specific physical variable so that, in response to this, the considered material yields a maximized output and its response to other stimuli is minimized. This is not an easy task as a particular material block has to be developed as a controlled system responding maximally only to a single set of variables yielding electrical output which is amenable to be processed by integrated information schemes.

One way to understand response maximization in electrical forms to one or a set of target variables while ignoring others on the part of a sensor element is to state that it should show negligible reaction to interferences and parasitic effects. For reliable operation of a sensor, environmental conditions have to be maintained where parasitic effects do exist though limited. In some cases, these effects are eliminated by correcting in the processing units. In fact, a sensor has its own characteristics which can be broadly be classified as (a) static, (b) dynamic, (c) reliability, and (d) response/sensitivity (to environmental effects).

For integrating processing and sensing units, attention has long been on the type of materials that could be so used. Since electrical/electronic circuits are now largely silicon-based, silicon has been an element of interest for primary sensing elements. Also, it has been well established that electrical behaviour of silicon changes with change in temperature, electrical and magnetic fields, stress/strain, radiation, and even doping. Silicon-based designs of some such sensors have already been discussed in earlier chapters. New technologies and techniques have evolved for realization of such integrated sensing elements. Micromachining of silicon, for example, has been used to produce vibrating systems of the kind of cantilevers, diaphragms and so on, which are small yet robust and serve as high frequency devices.

Silicon thermosensors and chemical sensors have also been produced. A single chip realization of primary sensors and processing elements has, therefore, been advanced to the extent of developing smart sensors and further extension of the same to smart transmitters where communication between these and the control gears receives equal emphasis.

Silicon-based microsensor technology has been of great use in the in vivo adaptation of various types of sensors. Pressure sensor is an example at hand. It consists of a thin, deformable silicon diaphragm with piezoresistors arranged along the edges of the diaphragm. These piezoresistors are then connected in the form of a bridge circuit. A single chip pressure sensor with signal conditioning unit may look like the one shown in Fig. 7.3.

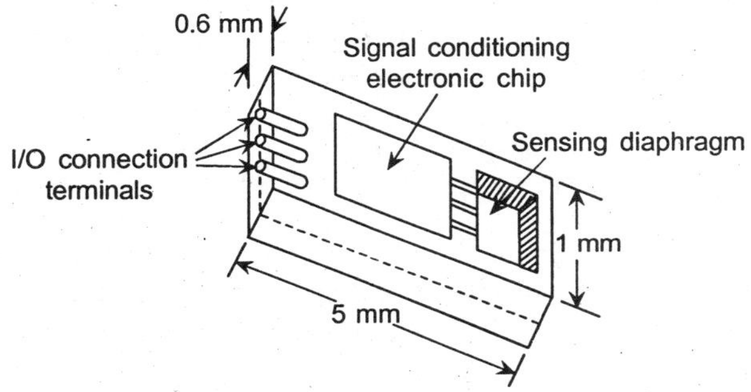

Fig. 7.3 Single chip pressure sensor with signal conditioning units.

A single transistor temperature sensor is well-known now, although for desired proper output in specific relation to the variable, a single smart sensor is also commercially available but it is only a dedicated type device. This has been described in the chapter on thermal sensors.

Thermal sensors based on thermoemf or Seeback effect in the form of thermopiles have also been made in ICs. They are now being batch-fabricated with addition of on-chip signal conditioning electronics.

Two semiconductors are coupled together with a difference of temperature ΔT between the junctions, the open ciruit emf ΔV is given by the relation (Fig. 7.4)

$$\Delta V = \alpha_s \Delta T \tag{7.1}$$

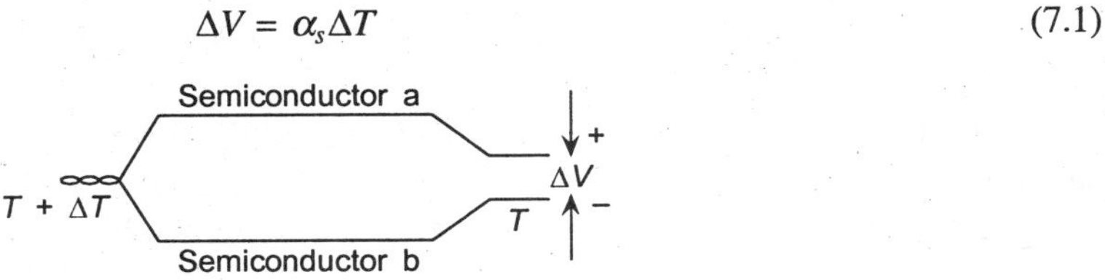

Fig. 7.4 Semiconductor thermoemf elements.

where α_s is the Seebeck coefficient.

If E_f is the Fermi energy so that with charge q, the electrochemical potential ϕ_f is given by

$$\phi_f = \frac{E_f}{q} \tag{7.2}$$

and, hence, for known E_f, α_s for silicon is obtained as

$$\alpha_s = \frac{1}{q}\left(\frac{dE_f}{dT}\right) \tag{7.3}$$

As temperature increases, silicon becomes more intrinsic; charge carriers attain a higher average velocity; and the temperature difference causes phonon flow from hot to cold 'space'.

For conduction band edge energy E_c, conduction band density of charge N_e, electron density (by doping level change) n, and Boltzmann constant k_B, increase in intrinsic behaviour of silicon with temperature rise causes

$$\frac{1}{q}\left(\frac{dE_f}{dT}\right)\Bigg|_{E_c - E_f} = -\frac{k_B}{q}\left\{\ln\left(\frac{N_e}{n}\right) + \frac{3}{2}\right\} \tag{7.4}$$

When the average velocity of charge carriers increases with increase in temperature, a charge builds up on the cold side of the silicon; also, scattering occurs. A parameter called the mean free time interval between two successive collisions of charge carriers, τ is important. If λ is an exponent to denote relation between τ and the charge carrier energy, then,

$$\frac{1}{q}\left(\frac{dE_f}{dT}\right) = -\frac{k_B}{q}(1 + \lambda) \tag{7.5}$$

With the net movement of phonons from hot to cold part, it is possible that a transfer of momentum from these to charge carries occurs if silicon is nondegenerate. This momentum drags the charge carriers towards the cold portion of the silicon and for this

$$\frac{1}{q}\left(\frac{dE_f}{dT}\right)\Bigg|_{\phi_n} = -\left(\frac{k_B}{q}\right)\phi_n \tag{7.6}$$

where ϕ_n denotes the relevant drag effect.

Thus, the coefficient α_s is given by

$$\alpha_{sn} = -\left(\frac{k_B}{q}\right)\left\{\ln\left(\frac{N_e}{n}\right) + \frac{5}{2} + \lambda_n + \phi_n\right\} \quad \text{for n type} \tag{7.7a}$$

and

$$\alpha_{sp} = \left(\frac{k_B}{q}\right)\left\{\ln\left(\frac{N_p}{p}\right) + \frac{5}{2} + \lambda_p + \phi_p\right\} \quad \text{for p type} \tag{7.7b}$$

In these relations, λ varies from -1 to 2 and ϕ ranges from 0 (for high doping) to 5 (for low doping) at around room temperature. For lower temperatures, the value of ϕ for low doping increases to 100 or even more. In any case, the approximation of α_s is done by a simplified relation

$$\alpha_s = \left(\frac{mk_B}{q}\right)\ln\left(\frac{\rho}{\rho_0}\right) \tag{7.8}$$

where

$m = 2.5\text{--}2.6$

resistivity $\rho_0 = 5 \times 10^{-6}$ Ωm.

Both m and ρ_0 have been obtained experimentally.

However, integrated thermopiles have been produced with strips of deposited aluminium forming junction with silicon. Figure 7.5 shows a schematic representation, where n-silicon has been used as epilayer.

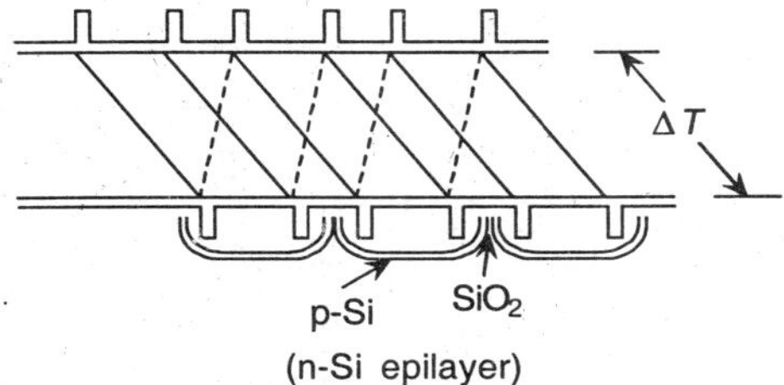

Fig. 7.5 An integrated thermopile system.

7.3 EXCITATION

Although *excitation* is a generalized term used for supply to the primary sensors, when necessary this also means the supply for the entire chip including the processing units. This supply may be required to provide different output to different stages of the system. In the thermocouple form of sensors, no excitation to the sensors is needed while for resistive bridge, an extremely stable supply is required. In stages of electronic processing units, ac supply or else pulsed form supply may be required for phase sensitive detection in the processor unit. In any case, as per requirement the facilities are to be made available for the entire chip to be self-sufficient.

7.4 AMPLIFICATION

Considering the output of the sensor to be generally small, amplification is essential in all smart sensors. If the gain requirement is very high, noise becomes a problem. However, stage-wise approach with adequate compensation realizes the requirement, the design and layout being critical as well.

7.5 FILTERS

Analog filters are often resorted to although filters are necessary at conversion stages, mainly because the digital type, consume large real time processing power.

7.6 CONVERTERS

Conversion is the stage of internal interfacing between the continuous and the discrete processing units. The conversion, in most of the situations, does not have one-to-one correspondence. Often, controlled conversion through software is provided with range selection and so on.

Data conversion from analog amplitude to frequency is often done for convenience of signal transmission, internally or externally, and/or for subsequent digital conversion. Voltage-controlled oscillators are used for these purposes. One such converter is a multivibrator shown in Fig. 7.6. Analysis shows that the time period of the generated square wave is given by

$$T = 2RC \ln\left(1 + 2\,\frac{R_2}{R_1}\right) \tag{7.9}$$

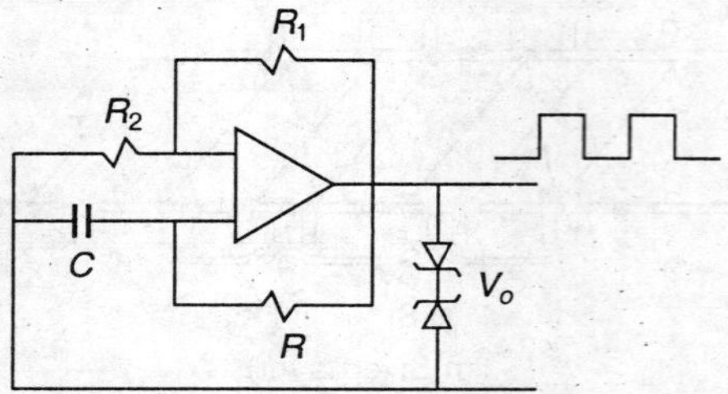

Fig. 7.6   A multivibrator.

The parameters R and C can be related to the input voltage. Fixing R_2/R_1 at 0.859, T is obtained as

$$T = 2RC \tag{7.10}$$

or, frequency f is given by

$$f = \frac{1}{2RC} \tag{7.11}$$

In fact, the capacitance or resistance may be the sensed instead of the input voltage or measurand/sensor output voltage. Ring oscillator realized with MOS technology is one popular V–f converter (or signal-to-frequency converter). A scheme of the V–f converter is shown in Fig. 7.7 which consists of an odd number of cascaded NOT, NOR, or NAND gates with its last gate-output fed back to the first stage to form the ring. With the gain of each stage being greater than one, the circuit is self-oscillatory with the frequency determined by the number of gates and their delays. Supply frequency and chip temperature need be controlled on which also depends the frequency.

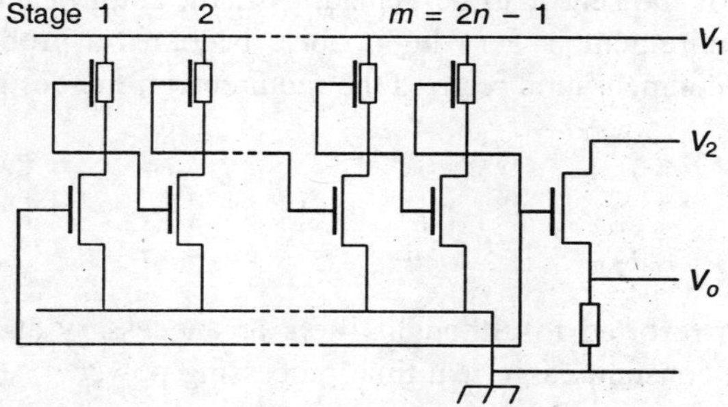

Fig. 7.7   An integrated ring oscillator.

If the MOS channel resistance is a piezoresistance whose value may be made dependent on the pressure exerted on it; this would change the gate delay and there is a frequency change. Supply frequency and temperature changes are usually compensated by using two ring oscillators and the ratio of two frequencies is taken as the output.

Next is the frequency to digital conversion. It must be remembered that when a voltage output is straightaway obtained from a sensor, other direct digital converters such as ADC's can be used. But, there are instances where the sensor is so designed that it inherently provides analog frequency output as in the case discussed in precedence, with the ring oscillators integrated on Si-diaphragm and pressure sensors utilizing the piezoresistive effect. Some other examples are (i) capacitive/inductive sensors controlling oscillator frequencies, (ii) photoresistances for illuminance sensing used in harmonic/relaxation oscillators, and (iii) quartz tuning fork as frequency standard.

In digital conversion, frequency from the 'sensor oscillator' is 'counted' by actually counting clock pulses in a pulse-width of the oscillator. There are various ways of doing it. One arrangement is shown in Fig. 7.8. Over the time period $T_x = 1/f_x$, f_{ref} would be counted; dividing f_x by a suitable factor n, this time interval is suitably increased to obtain a better resolution. In fact, the resolution R_n is given by

$$R_n = \frac{1}{n}\left(\frac{f_x}{f_{ref}}\right) \tag{7.12}$$

where $1/R_n$ is the actual count.

It must be remembered that there are variations of this circuit incorporating facilities required for different applications.

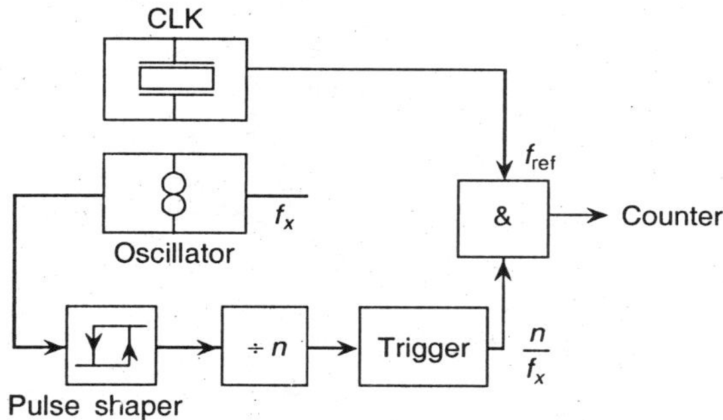

Fig. 7.8 A typical digital conversion method.

7.7 COMPENSATION

Compensation is an attempt to counter all sorts of nonideality in the primary sensor characteristics as well as environment of measurement. The commonly encountered sensor defects are:

(a) nonlinearity,
(b) noise,
(c) response time,
(d) drift,
(e) cross sensitivity, and
(f) interference.

Manufacturing tolerance may be combined under drift whereas temperature and/or other environmental effects are accommodated in noise.

7.7.1 Nonlinearity

Analog processing shows serious nonlinearity which at one time, was solved by piecewise linear segment approach modelled by linear electronic circuits. With digital processing methods in use now, more readily available general techniques are there to be used for the purpose. One very common technique is to refer to look-up tables while others are polygon interpolation, polynomial interpolation, and cubic splines interpolation techniques of curve fitting.

(a) *Look-up table method:* In this method, the sensor characteristic is described by a number of reference points very close to each other which are stored in ROM with linearized values. Response of the sensor for a measured value is referred to the ROM to look up for the corresponding linearized value which is then passed on for display or further processing. For good accuracy, this requires a large storage capacity or memory.

(b) *Polygon interpolation:* It is intended for soft nonlinearity where sectionalized linearization can be adopted. This method assumes that the nonlinear range is divided into a few linear sections and hence, a fewer reference points serve the purpose of linearization since between these stored reference points, the sensor is considered to behave linearly. For hard nonlinearity, the technique fails because the reference points are numerous. Figure 7.9 shows the technique of polygon interpolation.

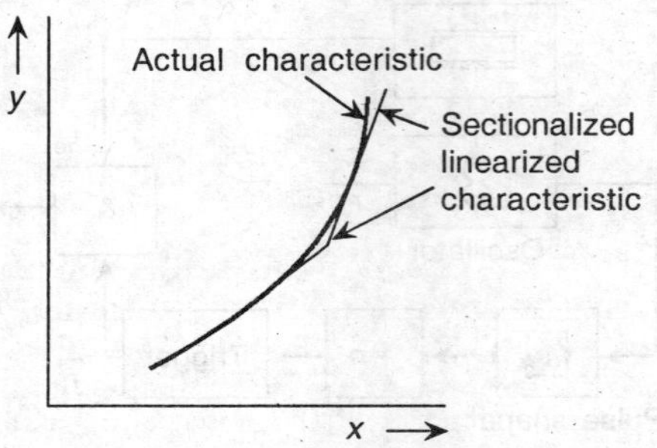

Fig. 7.9 Polygon interpolation.

(c) *Polynomial interpolation:*   This technique is again a standard technique which is based on the functional relationship between n selected measured points on the sensor characteristics and a polynomial of order $\leq (n - 1)$ over the range covering the characteristics. Lagrange's interpolation technique is a very common such technique. The curve is represented by the formula

$$y = \sum_{i=0}^{m} a_i x^i \tag{7.13}$$

The modification of this method for full-scale linearization is to generate a complementary curve for this characteristic as

$$y_c = \sum_{j=0}^{m} b_j x^j \tag{7.14}$$

and then, obtain the arithmetic, geometric, or root mean square mean as,

$$y_{\text{linear}} = \frac{1}{2}(y + y_c) \tag{7.15a}$$

$$y_{\text{linear}} = (yy_c)^{1/2} \tag{7.15b}$$

or,

$$y_{\text{linear}} = \left(\frac{y^2 + y_c^2}{2}\right)^{1/2} \tag{7.15c}$$

Figure 7.10 shows the linearization principle graphically. The polynomial interpolation method is usable under limitations of order. Increase in order often leads to oscillations.

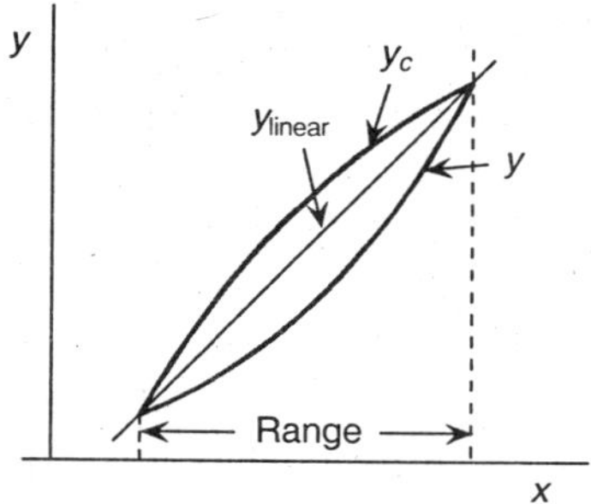

Fig. 7.10 Linearization using complementary function.

(d) *Cubic spline interpolation:* This method is so named as the sections of the characteristic curve of the sensor between a selected pair of reference (measured) points are represented by cubic spline functions as

$$S_i(x) = a_i + b_i(x - x_i) + c_i (x - x_i)^2 + d_i(x - x_i)^3 \qquad (7.16)$$

with $x \in [x_i, x_{i+1}]$ and $i = 0, 1, 2, \ldots, (n-1)$.

Each section on two sides, except the first and the last sections (Fig. 7.11) which have one end free, have junction points that are also represented by the adjacent spline functions. Both these functions must coincide with each other in function values, gradient, and curvature at these points, from which, conditions for the polynomials are derived. The end-points or the range binding points possess separate features—often it is considered that at these points, curvature is zero. With all these specifications, we obtain

$$S_i(x_i) = y_i, \qquad i = 0, 1, \ldots, n$$

$$S_i(x_i) = S_{i-1}(x_i), \quad i = 0, 1, \ldots, n, \text{ for function values;}$$

$$S_i'(x_i) = S_{i-1}'(x_i), \quad i = 0, 1, \ldots, (n-1), \text{ for gradients;} \qquad (7.17)$$

$$S_i''(x_i) = S_{i-1}''(x_i), \quad i = 0, 1, \ldots, n-1, \text{ for curvatures;}$$

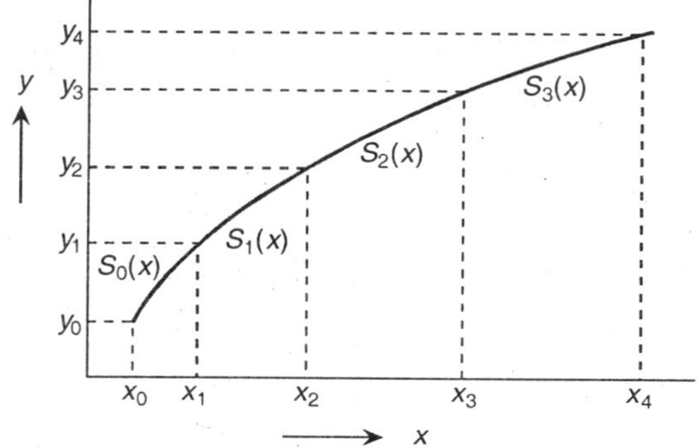

Fig. 7.11 Cubic spline interpolation.

for the points other than the end-points. It is advantageous to choose y-direction increments equal for the reference points and from the given conditions, the coefficients are evaluated. For success of the technique, at least five reference points, including the end-points, are to be taken.

Basically, interpolation is to fit a polynomial through the points around the point y where the function value is to be found. This polynomial is an approximation of the function and is used to find $f(r)$.

Assuming a second order polynomial of the form

$$f(x) = a_1(x - x_2)(x - x_3) + a_2(x - x_1)(x - x_3) + a_3(x - x_1)(x - x_2) \tag{7.18}$$

even by inspection, one easily gets,

$$\left.\begin{aligned}
a_1 &= \frac{f(x_1)}{(x_1 - x_2)(x_1 - x_3)} \\[2ex]
a_2 &= \frac{f(x_2)}{(x_2 - x_1)(x_2 - x_3)} \\[2ex]
a_3 &= \frac{f(x_3)}{(x_3 - x_1)(x_3 - x_2)}
\end{aligned}\right\} \tag{7.19}$$

so that the nth order polynomial can be expressed as

$$f(x) = \sum_{i=1}^{n+1} f(x_i) \prod_{\substack{j \neq i \\ j=1}}^{n+1} \frac{(x - x_j)}{(x_i - x_j)} \tag{7.20}$$

which is known as the Lagrange's polynomial.

Approximation and regression

Appropriate choice of the reference points for obtaining an efficient interpolation is very important. The coefficients obtained for the interpolated characteristic should be such that they have minimum deviation at 'each' point of the characteristic from the actual characteristic. The obtained function as it is an approximation of the actual function, is likely to deviate from the actual one and the errors between the 'approximate' values $\bar{y}$'s for the approximate function are given as

$$\bar{y} = f(a_1, a_2, \ldots, a_m, x) \tag{7.21}$$

and the reference points y_i, as measured, can be written as

$$d_i = y_i - \bar{y}_i = y_i - f(a_1, a_2, \ldots, a_m, x_i) \tag{7.22}$$

For deviation to be minimum, it is proposed that certain principles be adopted and the minimization should not be individual point to point process. Some of the proposed principles are

$$1. \quad \text{Min}\{d(a_1, a_2, \ldots, a_m)\} = \sum_{i=1}^{n} w_i \left| y_i - f(a_1, a_2, \ldots, a_m, x_i) \right|$$

$$= \left[\min \sum_{i=1}^{n} \left| (y_i - \bar{y}_i) \right| \right] \tag{7.23a}$$

2. $$\text{Min}\{d(a_1, a_2, \ldots, a_m)\} = \sum_{i=1}^{n} w_i \{y_i - f(a_1, a_2, \ldots, a_m, x_i)\}^2$$

$$= \left[\min \sum_{i=1}^{n} (y_i - \bar{y}_i)^2 \right] \tag{7.23b}$$

3. $$\text{Max}|\{d(a_1, a_2, \ldots, a_m)\}| \le \sum_{i=1}^{n} w_i |y_i - f(a_1, a_2, \ldots, a_m, x_i)|$$

or $$\le D \tag{7.23c}$$

The left hand side of Eq. (7.23a) is called R function and the approximation is known as L_1 where one or two wayout points in the 'fit' are ignored. Similarly, in Eq. (7.23b), it is called S function in L_2 approximation when minimization is in the least square sense, and in Eq. (7.23c), it is T function when maximum deviation is allowed but within specified limits. This is known as *Chebyshev approximation*. The term w_i is the weight factor for points (y_i, x_i).

Minimization in the least square sense is an approximation method and often called *regression*. This is very often used in the calibration of sensors and instrumentation systems. One specific kind of regression is linear regression.
Polynomial regression begins with

$$\min S = \min \sum_{i=1}^{n} (y_i - \bar{y}_i)^2 \tag{7.24}$$

where

$$\bar{y}_i = a_n x_i^n + a_{n-1} x_i^{n-1} + \ldots + a_1 x_i + a_0$$

$$= \sum_{j=0}^{n} a_j x_i^j \tag{7.25}$$

Here, S becomes a function of $(n + 1)$ unknown variables $a_0, a_1, \ldots, a_n$. Taking partial derivatives of S with respect to $a_0, a_1, \ldots$ and setting these to zero, a set of $(n + 1)$ equations is obtained as

$$\frac{\partial s}{\partial a_j} = \sum_{i=1}^{n} 2\left(y_i - \sum_{j=0}^{n} a_j x_i^j \right)(-x_i^j) \tag{7.26}$$

These $(n + 1)$ equations, called *normal equations* for polynomial regression, are solved for $(n + 1)$ coefficients by Gaussian elimination procedure. Higher the number of coefficients, more severe becomes the numerical difficulties and hence, simpler techniques such as transforming the nonlinear function into a linear function and then using the linear regression, are adopted. The following example makes the process clear. Let the function be exponential

$$y = \alpha \exp(-\beta x) \tag{7.27}$$

$$\Rightarrow \qquad \log y = -\beta x + \log \alpha$$

But, log α is a constant, (say a_0) and $-\beta$ is another constant, say a_1, so that using the regression analysis

$$a_0 = \frac{\Sigma(\log y_i)\,\Sigma x_i^2 - \Sigma x_i\,\Sigma(x_i \log y_i)}{n\,\Sigma x_i^2 - (\Sigma x_i)^2} \qquad (7.28a)$$

and

$$a_1 = \frac{n\,\Sigma(x_i \log y_i) - \Sigma x_i\,\Sigma(\log y_i)}{n\,\Sigma x_i^2 - (\Sigma x_i)^2} \qquad (7.28b)$$

Coefficients α and β are now given as

$$\alpha = \exp(a_0)$$

and

$$\beta = -a_1 \qquad (7.29)$$

7.7.2 Noise and Interference

Thermal noise is important in almost all sensors. Besides, there are other unwanted signals that may be picked up due to external magnetic fields (sort of an interference) when the structure is not adequately screened. Noise is also introduced at different stages of signal processing such as data conversion, analog to digital interfacing by stray effects, and so forth.

The methods of minimization of noise are appropriate signal conditioning techniques that include filtering, signal averaging, and correlation among others. If the signal is periodic as in the case of the output of the frequency converter, the correlation technique improves the signal-to-noise ratio by a large value. This is due to the superposition property of autocorrelation.

Again, if the input is corrupted at any stage by noise, specifically white noise, a cross correlation technique can be used to obtain the system response/function without this corruption. This is obtained in Fig. 7.12.

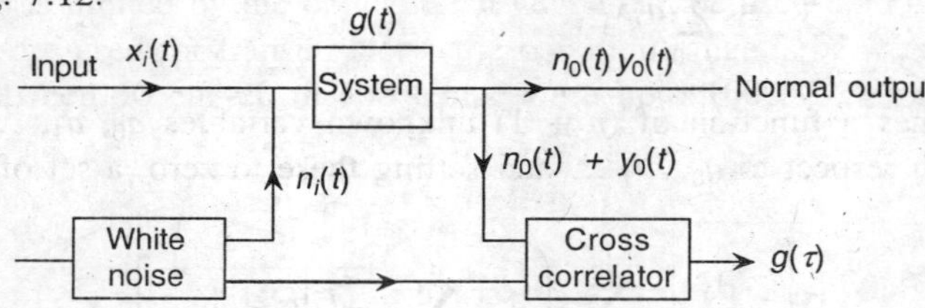

Fig. 7.12 The cross-correlation technique for noise reduction.

For a signal $f_1(t)$, the autocorrelation function is

$$\phi_{11}(\tau) = \lim_{T \to \infty}\left(\frac{1}{2T}\right)\int_{-T}^{T} f_1(t)\,f_1(t + \tau)\,dt \qquad (7.30a)$$

If the output for a signal $f_1(t)$ is $f_o(t)$, the cross correlation function is

$$\phi_{12}(\tau) = \lim_{T \to \infty}\left(\frac{1}{2T}\right)\int_{-T}^{T} f_1(t - \tau)\,f_o(t)\,dt \qquad (7.30b)$$

7.7.3 Response Time

Because of the presence of storage and dissipative elements, a sensor is likely to have quite inferior time response characteristics and the 'dynamic correction' of sensor becomes necessary. This is possible with the use of microprocessors/microcomputers with suitable algorithm if the dynamic parameters are known through solving the convolution integral. In fact, it is the facility of the inverse operation of deconvolution that is available in such processes and makes such a correction possible. If the sensor function is given by $f(s)$, the signal processing unit should have a function $1/f(s)$ as shown in Fig. 7.13, so that we obtain

$$x_i(t) = \int_0^t x_o(t - \tau)\, g(\tau)\, d\tau = x_o(t) * g(\tau) \tag{7.31}$$

where $g(\tau) = \mathscr{L}^{-1}\left\{\dfrac{1}{f(s)}\right\}$

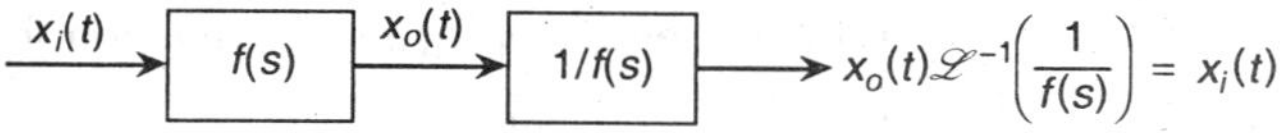

Fig. 7.13 Cascading complementary processing function.

Since $x_i(t)$ can be written in terms of the output $x_o(t)$, the correction can be easily made. In fact, for a second order system, the input $x_i(t)$ in terms of $x_o(t)$ is

$$x_i(t) = \frac{1}{K}\left\{x_o(t) + \left(\frac{2\zeta}{\omega_0}\right)\dot{x}_o(t) + \left(\frac{1}{\omega_0^2}\right)\ddot{x}_o(t)\right\} \tag{7.32}$$

where ζ is the damping factor and ω_0 is the natural frequency of oscillation. Thus, $x_i(t)$ is expressed in terms of the output and its derivatives. The same polynomial interpolation can be used with $x_o(t)$, $\dot{x}_o(t)$, and $\ddot{x}_o(t)$ as the reference points. The cubic spline polynomials are advantageous for second order systems.

Another method using the difference equation is also useful in digital systems for obtaining $\dot{x}_o(t)$, and $\ddot{x}_o(t)$ as

$$\dot{x}_o(t_j) = \frac{1}{T_s}[x_o(t_j) - x_o(t_{j-1})] \tag{7.33a}$$

and

$$\ddot{x}_o(t_j) = \frac{1}{T_s}[\dot{x}_o(t_j) - \dot{x}_o(t_{j-1})] \tag{7.33b}$$

T_s being the sampling interval.

7.7.4 Drift

Drift appears in a sensor because of slow changes in its physical parameters either due to ageing or deterioration in ways of oxidation, sulphation, and so on. Drift is a kind of noise and should be counteracted. As drift tends to change the sensor characteristics, the reference points for polynomial interpolation also tend to drift. These are required to be updated and hence, the coefficients are re-evaluated through an algorithm.

7.7.5 Cross-Sensitivity

A sensor, while responding to a specific variable, responds to others as well, may be, with much less sensitivity. It is therefore necessary to maximize the sensitivity for the desired measurand and minimize that for the others. A common undesired interfering variable is temperature for non-thermal sensors.

If the interfering variable is denoted as z, output as y and measurand as x, then the nominal or rated (constant) z, z_0 is taken as the base value of the interfering quantity while with varying z_0 from z, the characteristics are changed as shown in Fig. 7.14. The function $y(x, z)$ can be expressed as a series of the base characteristics $y_0(x, z_0)$ given by

$$y(x, z) = \alpha_0(z) + [1 + \alpha_1(z)]\, y_0(x, z_0) + \alpha_2(z)\, y_0^2(x, z_0) + \dots \tag{7.34}$$

For $z = z_0$, the function $\alpha_i(z)$, $i = 0, 1, \dots, n$ becomes zero, otherwise it describes the effect of interference by z. This function $\alpha_i(z)$ can be written as a polynomial function that can be written as

$$\alpha_i(z) = \sum_{j=1}^{m} \beta_{ij}(x - z_0)^j \tag{7.35}$$

By approximation methods and regression algorithm, β_{ij}'s are evaluated by sensor characteristics at different values of interference quantity z. The correction can be affected by measuring the sensor output and calculating its effective value with the base characteristics. Thus, one obtains a measured characteristic and also one evaluated at a value $z \neq z_0$. There is not only zero shift but may also be changes in gradient and curvature. The relative deviations for various x's are then obtained by calculating $100(\Delta y / y_{max})$ from the two curves shown in Fig. 7.14.

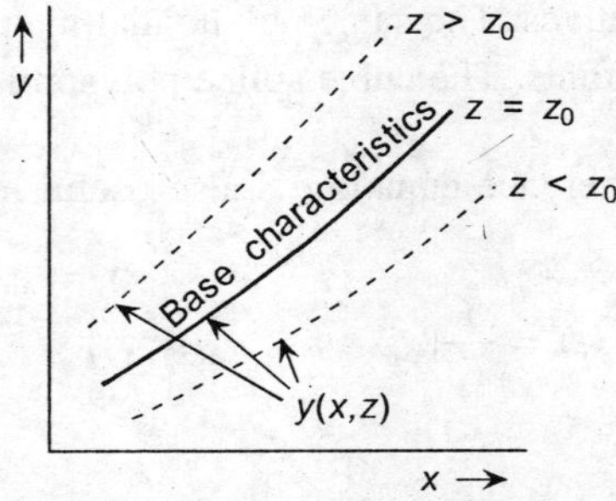

Fig. 7.14 Curves pertaining to analysis of cross-sensitivity.

Compensation takes account of many undesired interfering influences and is, therefore, critically examined. As has been discussed, the compensation made through devising algorithm by monitoring the change in response characteristics because of any interfering quantity, is quite common as it is possible to develop the algorithm from 'measured' data. Such a compensation is often termed as *monitored compensation* and is very common in structuring modern day sensors where a number of sensors are possible to be produced on a common substrate. Also, in Si-technology, an in-built monitoring sensor can be used within the main sensor.

Besides, such compensation methods, it is attempted to design the sensor to be least responsive towards interfering quantities. In fact, the idea is to provide 'structural compensation' by giving symmetry to the sensor so that the desired output is derived through differential mode while the interfering signals are derived through common mode and are rejected.

Even taking proper care in the design with symmetry of the sensor, manufacturing or production tolerance may lead to error which needs be compensated. This means that individual sensor needs to be compensated depending on its performance and response to inputs. Such individual compensation is called *'tailored compensation'* and a dedicated algorithm has to be developed for the purpose or a specialized analog module has to be incorporated.

When none of the discussed compensation methodologies can be adopted because of 'physical inaccessability' in some cases, model reference data sets are considered and compensation values are deducted and incorporated. It is, to a certain extent, inferential and subject to errors for error in the model itself. However, such compensation is known as *deductive compensation*.

7.8 INFORMATION CODING/PROCESSING

It has so long been assumed that signal from a sensor is processed providing correction, compensation, linearization, freedom from cross-sensitivity and drift, and so on. It is also true that such a processed signal is finally to be made available in digital form and, perhaps, in a serial form. It is good to remember that smart sensors are generally multi-sensor systems and a number of signals are available for either display or further processing subsequently to be connected to the 'communication bus'.

Information, the state of the process in the form of a processed signal through sensor and signal processing systems, is first received by the information coding system. Some of these signals are released, some stored, some destroyed, and some restructured.

For indication purposes only, the signals are coded and displayed over appropriate display modules as is done in digital meters, indicators, recorders, and so forth. A typical IC temperature sensor-based smart sensor is depicted in Fig. 7.15.

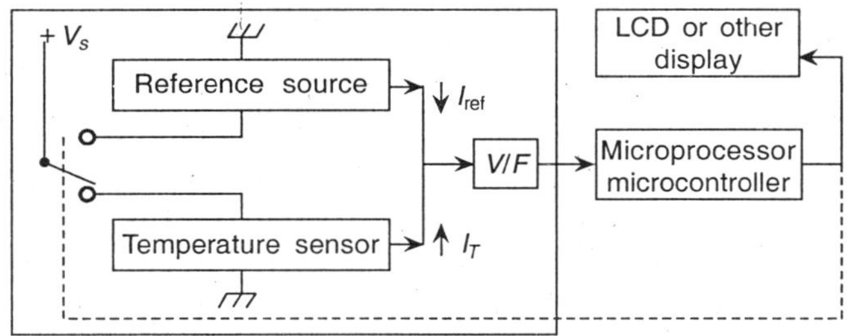

Fig. 7.15 A typical IC-temperature based smart sensor.

When these signals are required to be used for system control and surveillance as is usually the case, in addition to display, control system should be able to read the signals for their functioning. Information processing assembly in a smart sensor is basically an encoder, the encoded data from this are fed to the communication unit. As is usual, the conventional signal processing provides an output of 4–20 mA. One way is to get a corresponding voltage range which is then parallelly encoded into digital signal through a converter. When necessary, the 4–20 mA output is also drawn. Voltage-to-frequency converter is another kind which is quite extensively used (see Fig. 7.15), then using a reference frequency generator, frequency difference encoding is employed.

Mark-to-space ratio control of a square wave is another coding technique but not often resorted to. There are many other techniques and choice is largely based on the specific requirement and associated conditions.

7.9   DATA COMMUNICATION

Data communication is essential in smart transmitters where the sensor outputs are communicated with the host through bus-system. Coded data are processed for communication by a software processor and a suitable interface system communicates between the processor and the bus. The bus was, till lately, being standardized. Commercial versions available for quite sometime used their own protocol. Each smart sensor/transmitter has always been provided with a local operating system in a ROM, that consists of an application programme and library modules, for ADC and DAC hardwares, bus driving hardware, local interface hardware, and LCD/keyboard hardware.

Earlier manufacturers preferred to develop their own protocol. One such protocol is HART (Highway Addressable Remote Transducer) offered by Rosemount which superposes a digital transmission protocol on the standard 4–20 mA loop. A typical transmitter with HART protocol appears as the one shown in Fig. 7.16. Some other protocols that find use are High Level Data Link Control (HDLC), Synchronous Data Link Control (SDLC), Factory Instrumentation Protocol (FIP), and so on which are sufficiently advanced.

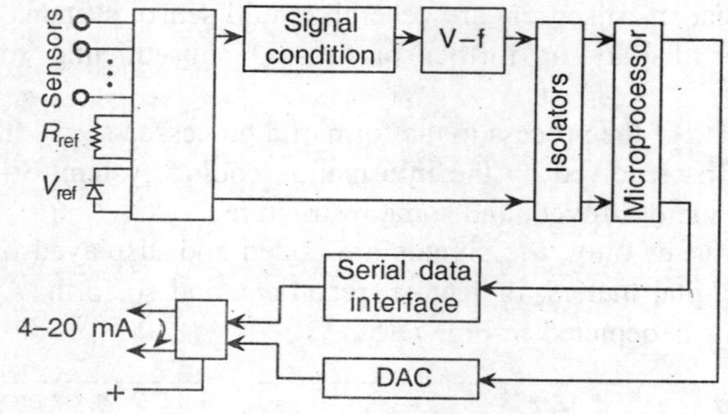

Fig. 7.16   A smart transmitter.

Recently, an international standard in protocols has been reached which permits any host to be in communication with the smart sensor/transmitter system. This ensures common field bus standard. This, however, is for the standardization of the communication unit. The actual smart sensor remains open to development for better operation with existing and emerging sensors and underlying technologies.

The HART protocol has been designed for direct use of 4–20 mA output device having facilities of digital communication with superimposed modulation between the field device and a host system. Such devices can be connected in parallel. The addressing procedure allows each unit to set its output for power supply at 4 mA and the device is forced to communicate only digitally. The parallel connection coverts the twisted pair into a multiloop bus but the number is limited to 15 as specified by this protocol. The power source, therefore, supplies a maximum of 60 mA. The basic multiloop connection method is presented in Fig. 7.17 while Fig. 7.18 shows the hardware requirements for microprocessor-based field devices.

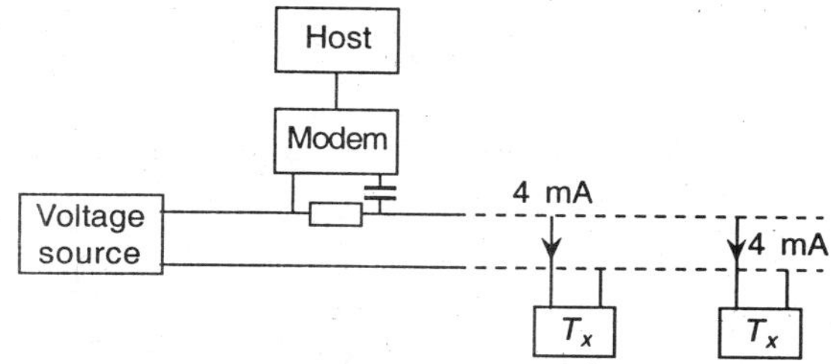

Fig. 7.17 The basic multiloop connection.

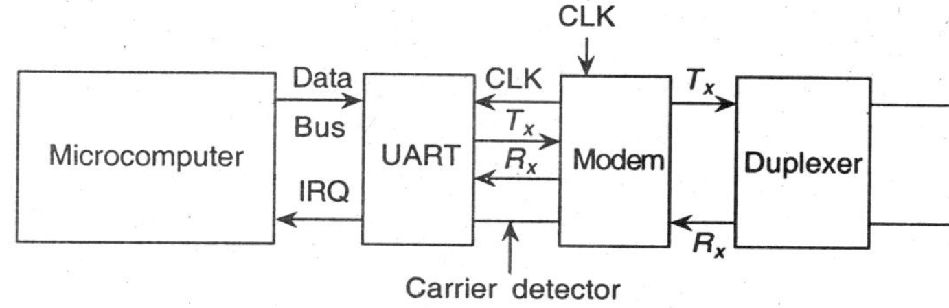

Fig. 7.18 Demonstration of hardware requirement of an intelligent field device.

Frequency shift keying (FSK) is used for coding digital information. Logic 1 is represented by 1200 Hz and 0 by 2200 Hz both with sine wave of amplitude 0.5 mA. Data rate is 1.2 Kb/s. The implementation of this digitally signalling technique can be done by using a modem of telephony standard.

In HART protocol, it is the master–slave proposition that works—the field device responds only when it receives instruction from the bus and in every reply message, the status of the field device is included to check its state.

Application specific integrated circuits (ASIC) are receiving attention more and more for the internal operation of the sensor and signal processing system of the smart sensor. ASIC and its supporting technology make available a host of ready items from which those required can be selected, incorporating variety in the smart sensor design and enhancing its capability.

7.9.1 Standards for Smart Sensor Interface

The ultimate goal of the standards is to provide the means for achieving transducer-to-network interchangeability and interoperability. The objectives are to define a set of common communication interfaces for connecting transducers to microprocessor-based systems, instruments, and field networks in a network-independent environment.

Figure 7.19 shows a scheme of communication using IEEE 1451. Here, NCAP (Network Capable Application Processor) information model is intended for defining a common object model for the components of the smart transducer working in networked mode and also to develop the software interface specifications for them. Such an object model provides two interfaces (i) to the transducer block with details of transducer hardware implementation and simple programming model—this resembles an I/O driver and (ii) to the NCAP block and ports with details of different network protocol implementation schemes, this is IEEE P 1451.1

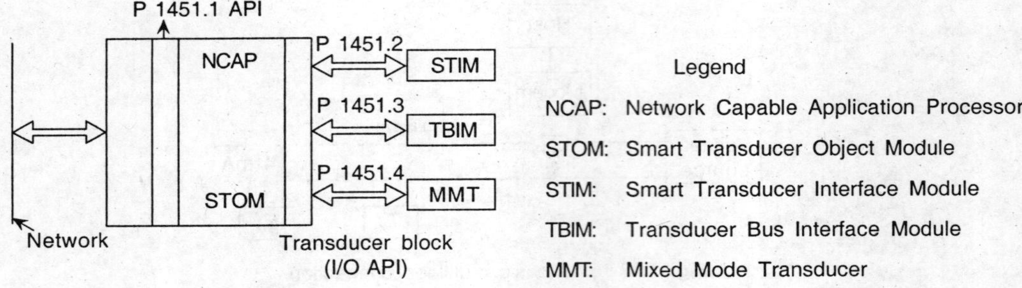

Fig. 7.19 A communication scheme through IEEE 1451.

The IEEE P 1451.2 provides the transducer-to-microprocessor communication protocols and transducer electronic data sheet (TEDS) formats. It also provides the digital interface and communication protocols between the transducers and microprocessors.

IEEE P 1451.3 provides digital communication and TEDS formats for distributed multiloop systems. This is basically intended to develop a standard digital interface for multiple physically isolated/separated transducers in multidrop configuration.

IEEE P 1451.4 provides mixed mode communication protocols and also the TEDS formats. This is intended to develop bidirectional communication of digital TEDS in addition to an interface for mixed mode transducers.

7.10   THE AUTOMATION

In modern control systems, signal communication standards have been of tremendous significance. The first signalling standard (IEC Technology Committee TC–65, 1971, namely IEC 381–1) established was 4–20 mA. In 1981, work on International standards for PLC; in 1985, for field bus; and in 1987, functional safety for programmable electronic systems started but proprietory standards still continue to exist.

Hierarchical structure of control of large complex processes has specific advantages. However, distributed control structure reduces the cost significantly by eliminating the need for long transmission lines between the controller, and the sensors and actuators. A typical scheme of such a structure is shown in Fig. 7.20. By connecting field-located devices with a serial bus and the field bus, cabling costs can be reduced further. An example is presented in Fig. 7.21.

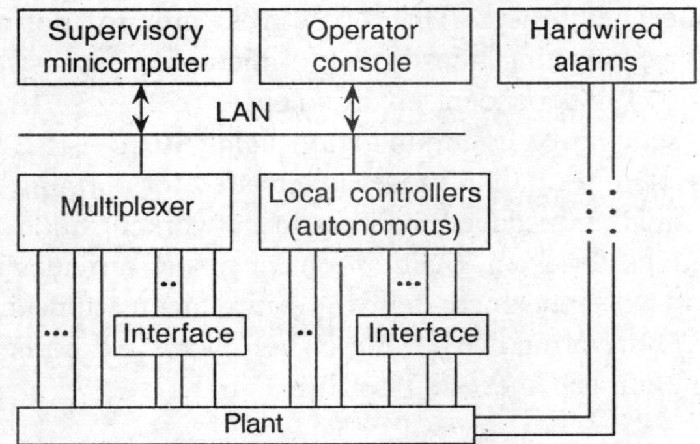

Fig. 7.20 Distributed control structure.

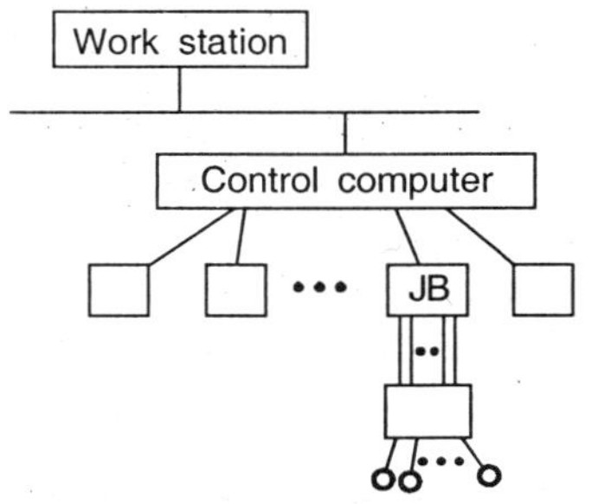

(a) Instruments and actuators star-connected to junction box (JB)

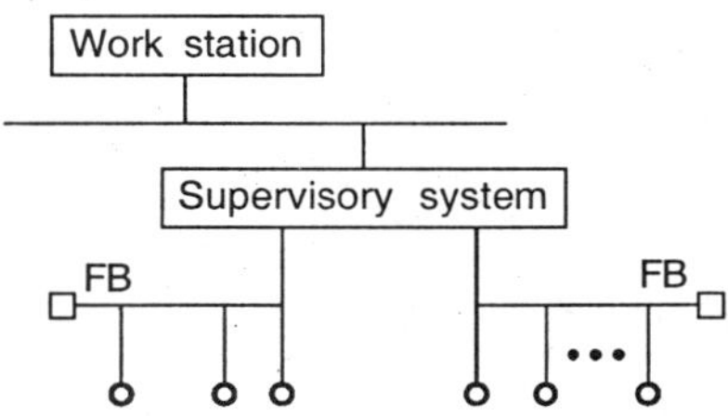

(b) Intelligent instrumentation and actuators linked by a field bus system

Fig. 7.21　Cost reduction in a field bus system.

Automation entered the area of flexible manufacturing satisfying quality specifications. This kind of manufacturing covers the aspects of disciplined (Just-in-time) production with enforced environmental legislation. Other than negative feedback, the strategy has now incorporated programmability and communication technology. The advancement of semiconductor technology has paved the way for all these to be integrated and applied at relatively low cost in the industrial processes. Thus, one can represent the system as

$$\text{Instrumentation} + \text{Programmability} + \text{Communication} \Rightarrow \text{Automation}$$
$$\text{(negative feedback)} \quad \text{Integrated low cost} \Leftarrow \text{Semiconductor technology} \quad \lrcorner$$

In recent years, process automation and factory (manufacturing) automation are using similar automation systems for closed loop control, man-machine interfaces, and for networking. Such convergence has been made possible by the use of IEC field bus standard designed for automation applications.

REVIEW QUESTIONS

1. What is basically the concept of 'smart sensors'? What are the essential elements in such an unit? Show with the help of a diagram, the arrangement of these elements.

2. Show, with the help of diagrams, how the primary sensors are being integrated with signal processing ensembles. What are the essential components in the signal processing unit?

   What is a ring oscillator? How has it been integrated? Where, in a smart sensor, is it likely to be used?

3. What are the different deviations that need be compensated in sensor systems? How is a nonlinearity taken care of in a present day smart sensor?

4. Explain the cubic spline interpolation method. What are its limitations?

5. Explain regression, cross-sensitivity and interference. How can specific nonlinearities be handled with linear regression analysis? Show this using the function $y = y_0 \exp(-ax)$.

6. Describe the principles of a 'smart transmitter'. What and where does it transmit? Discuss some aspects of its development in recent years.

8

Recent Trends in Sensor Technologies

8.1 INTRODUCTION

Although conventional sensors are commercially still very much in vogue, over the last three decades or so solid state sensors have been inching towards a state that threatens many of the older types. In this category, the semiconductor micro- and nano-sensors, ceramic and chemical sensors using newer materials and technologies such as IC technology, VLSI chips, and micromachining techniques are included.

For semiconductor microsensors, the IC technology comprising of photolithographic etching, deposition, metallization, and assembling is essential and this also is the basis for thick and thin film, chemical and electrochemical, and biological sensors. IC elements are now quite extensively used in the measurement of temperature, flow, and magnetic field. However, for generating three-dimensional features, precise micromachining techniques are required which offer better performance. Besides, reliability, sensitivity, uniformity, and stability are also better. The miniaturization associated with these new technologies is one special advantage in the smart sensing and intelligent instrumentation that has already become an unavoidable part in all walks of life.

This chapter is intended to provide the idea of development of the newer types of sensors such as thick/thin film sensors, standard methods of semiconductor sensor technology, and microelectromechanical systems (MEMS).

8.2 FILM SENSORS

Basically, such sensors are produced by film deposition of different thicknesses on appropriate substrates. The deposition techniques used are different for the thick and thin film sensors. Sensors produced through these techniques have varying electrical and mechanical properties while a variable is being sensed.

8.2.1 Thick Film Sensors

Thick film deposition is a mature technique and there has not been substantial improvement whilst thin films are being developed almost at the same pace as microelectronics incorporating latest technology. It is to be noted that thick film process had been in use for producing capacitor, resistor, and conductors—and has subsequently been adopted in sensor development. The processing of a sensor can be expressed schematically as

Step 1:   Selection and preparation of a substrate.

Step 2:   Preparation of the initial coating material in paste or paint form.

Step 3:   Pasting or painting the substrate by the coating material or screen printing it.

Step 4:   Firing the sample produced in step 3 in an oxidising atmosphere at a programmed temperature format.

The substrates used for developing thick film over them are alumina (96% or 99.5%) and beryllia (99.5%). These are fired at about 625°C. Others used are enamelled steel which is low carbon steel coated with low alkali content glass frit that are fired at around 850°C. Alumina or beryllia have dielectric constants around 9.5 and 7 respectively with dielectric strength around 5600 V/μm. Thermal expansion coefficients are 6.5×10^{-6} and 7.5×10^{-6} respectively with bulk resistivity being almost the same for both, at about 10^{14} Ωcm, thermal conductivities are 0.36 and 2.5 W/(cmK) respectively. Enamelled steel has better strength and machinability being almost double for those of alumina or beryllia which have values around 175 MPa. Though it has better machinability and improved thermal conductivity, enamelled steel is less costly.

For thin film deposition, alumina and beryllia can also be used. Besides, special glass, quartz, fused silica and sapphire are often used which have similar properties and sometimes even better.

It is to be understood that the compatibility between the substrate and the transducing element in film sensors is very important. For example, there should not be difference of thermal expansion coefficients which would induce stress between them and correspondingly result in zero offset, drift, and instability.

Sensors which are produced through thick film deposition ($\sim$20 μm) are used for sensing temperature, pressure, gas concentration, and humidity.

Temperature: Thick film sensors such as (i) thermopiles (usually of gold and gold-platinum alloy), (ii) thermistors (usually with oxides of manganese, ruthenium. and cobalt), and (iii) temperature dependent resistances based on gold, platinum, and nickel are used for temperature sensing.

Pressure: Sensing pressure is possible by making thick film diaphragms or capacitive devices made with alumina (Al_2O_3) and $Bi_2Ru_2O_7$, or piezoresistive devices made of same materials.

Concentration of gases: Gases such as methane (CH_4), CO, and C_2H_5OH can be checked for concentration using films of SnO_2 + Pd, SnO_2/ThO_2 + hydrophobic SiO_2. H_2, CO, C_2H_5OH, and isobutane are sensed by SnO_2 + Pd, Pt, Ba$^-$, Sr$^-$ and $CaTiO_3$ (Nasicon). Oxygen and hydrogen gases also are separately sensed by these types of films

Humidity: It is sensed by (i) resistive films made from RuO_2 (spinel type)/glass and (ii) capacitive films made from glass ceramic/Al_2O_3. On the other hand, dew point is sensed by films made from ($BaTiO_3$/RuO_2)-glass.

Starting from the same basic material, say $SnSO_4$, one can produce SnO_2-based sensors for H_2, CO, and NH_3, as mentioned in the preceding paragraphs. The host material (1% by weight), $PdCl_2$ mixed with SnO_2 as catalyst and $Mg(NO_3)_2$ (also 1% by weight) is mixed presumably for sensitivity range. The combination is fired at about 800°C for one hour. Selectivity is obtained by a second firing process at almost the same temperature by adding different ingredients for different gases. For H_2 detection, for example, it is mixed with Rh (6% by weight) and fired for 1 hour at 800°C. For CO, ThO_2 is added (5% by weight) and for NH_3, ZrO_2 is added (5% by weight) and processed in the same manner as explained.

For control of the porosity of the films which determine the overall sensor sensitivity, organic materials are added in a selective manner. For example, alcohol is added for H_2 and sometimes, inorganic materials work well with appropriate selection. Silica of different varieties is added for CO and NH_3. The materials so produced are now painted on the substrate and dried, then calcined at controlled temperature for varying times.

The other thick film variety is the ceramic-metal or *cemet* which consists of gold/silver/ ruthenium/palladium based complex oxides in an insulating medium, mainly glass (lead borosilicate). There are thick film resistors of the cemet which require precise control of heat treatment. The resistivity is controlled by the size, concentration, and distribution of the metallic (conductive) component, that is, their own resistive properties, and the insulating medium. Pure metal powders and resistor pigments differ in so far as changes in their resistive values are concerned and hence, their embedding in per cent weight changes the resistivity of the sensor developed. Figure 8.1 shows the difference for two typical cases.

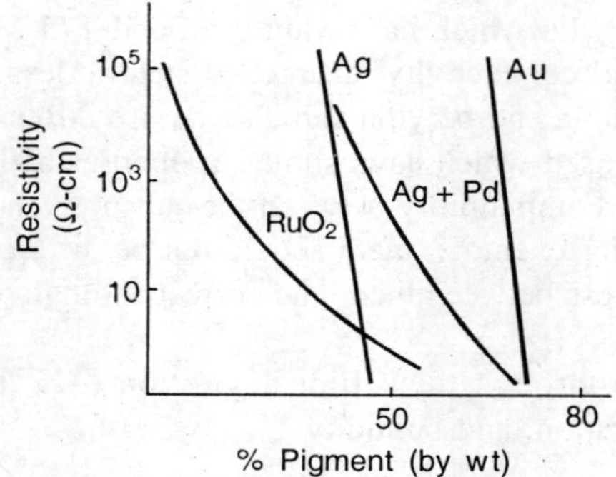

Fig. 8.1   Resistivity variation with change in pigment.

8.2.2 Thin Film Sensors

Thin film sensor processing differs from thick film technology mainly in the film deposition techniques. This technology is similar to that used in silicon micromechanics. A number of techniques are used for thin film deposition, such as:

(a) Thermal evaporation
   (i) resistive heating
   (ii) electron beam heating

  (b) Sputter deposition
     (i) DC with magnetron
    (ii) RF with magnetron

  (c) Chemical vapour deposition (CVD)

  (d) Plasma enhanced chemical vapour deposition (PECVD)

  (e) Metallo-organic deposition (MOD)

  (f) Langmuir–Blodgett technique of monolayer deposition.

Of these, the thermal evaporation and sputter deposition are decades old. However, in the sputter deposition technique, magnetron sputtering is an improved form where a magnetic field perpendicular to the applied electric field is applied. This increases the ionization probability of the electrons as the Lorentz force $\mathbf{E} \times \mathbf{B}$ restricts the primary electrons near the cathode. As a result, sputtering efficiency is also enhanced.

Plasma enhanced chemical vapour deposition (PECVD) has been found to be particularly suitable for sensor fabrication. This is a low temperature process in which plasma is introduced into the deposition chamber to enhance the pyrolytic process which in normal CVD process is performed by thermal decomposition that requires high temperature. In this process, the volatile compound of the material to be deposited is thus vaporized, decomposed, and made to react with gaseous species over the substrate to produce a nonvolatile amorphous product on the surface of the substrate. The deposition level is controlled by controlling the flow-rates of the vapours. A parallel plate, radial flow type PECVD processing chamber is shown in Fig. 8.2.

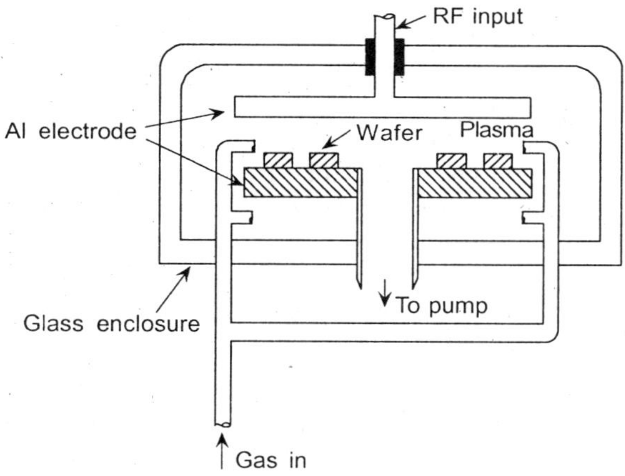

Fig. 8.2   A PECVD processing system.

Metallo-organic deposition (MOD) is another very versatile technique which can be used both for thick and thin film sensor fabrication. It consists of applying ink of metallo-organic compound to the silicon substrate consisting of silicon wafer coated with silica, then spinning the asembly at about 3000 rpm and finally heat treating the deposit. Metallo-organic compounds consist of a central metal ion bonded with a ligand through a heterobridge containing oxygen, sulphur, nitrogen, phosphorus, arsenic, and so on. It is prepared by dissolving the compound in organic solvent. Specially prepared thin films, by this technique, are barium titanates ($BaTiO_3$) and their derivatives that are mostly used in pyroelectric measurement, tin-oxides for gas sensors, superconducting oxides such as yttrium-barium-copper oxides ($YBa_xCu_yO_z$) for high temperature and ZrO_2, TiO_2 stabilized by yttrium for oxygen sensors.

Thin film sensors measure the same variables as done by thick film counterparts with variations in principles and materials. Table 8.1 shows the variable, sensing element, and principle of sensing for certain different variables.

Table 8.1 Working principles of the materials

Variable	Material	Principle
Flow	Au	Thermoanemometry
Humidity	Ta_2O_5	Capacitance change
Magnetic field	$Ni_{81}Fe_{19}$, NiCo, $Co_{72}Fe_8B_{20}$	Magnetoresistive effect
Oxygen	ZnO	Variation in electrical conductivity
Pressure	Polysilicon	Piezoresistive effect (Diaphragm)
Radiation	Au	Bolometry
Strain	CrNi	Piezoresistive effect
Temperature	Pt	Resistance variation

Langmuir–Blodgett film can be fabricated from materials which have a polar hydrophilic head and a hydrphobic tail, that is, an amphilphilic materials. Fatty acids such as palmitic (16), magaric (17), stearic (18), arachidic (20), and so on satisfy the requirement. They are generally represented by

$$H_3C—(CH_2)_{n-2}—COOH \qquad 16 \leq n \leq 20$$

It is a monolayer film. Fatty acid films are soft with low melting points (<70°) and are not robust for electronic sensors/devices. Their derivatives which can be polymerized are used instead. Examples are vinyl stearate and diacetylenic acid which can be polymerized by γ- or UV-radiation usually after deposition. However, polymerization before deposition can also be done.

Solid film generation or deposition on the substrate is done by a process known as *dipping* which is a special mechanism and is either vertical or horizontal. However, orientation of the film on the substrate can be different with respect to hydrophilic and hydrophobic ends. Considering the circle-end as hydrophilic and line-end as hydrophobic three different orientations are obtained which are shown in Figs. 8.3(a), (b), and (c). These are marked as types X, Y, and Z respectively.

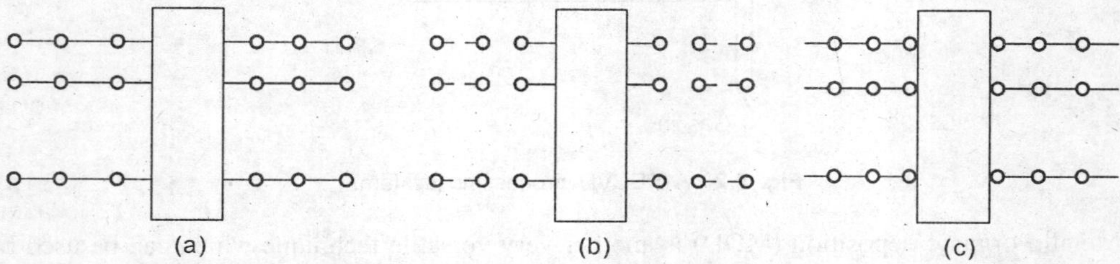

Fig. 8.3 Type X, Y, and Z orientations.

LB films are good as ion-selective membranes when deposited on the insulator face of the ion-sensitive FET. Majority of LB films of different materials have found convenient use as biosensors.

LB gas sensors are classified as chemiresistors and surface acoustic devices (SAW). In the former, the change in conductivity of the sensor is related to the concentration of the gas and the materials used for films are phthalocyanines and porphyrines.

Surface acoustic devices such as general SAW devices sense change in frequency due to absorption of gas by the film deposited on the piezoelectric substrate. Same materials can also be used for film deposition.

8.3 SEMICONDUCTOR IC TECHNOLOGY—STANDARD METHODS

Over the last three decades or so, solid state sensors have been developed using a new technology. These sensors are semiconductor micro- and nano-sensors, ceramic and chemical sensors using new materials, and optical fibre sensors. These sensors are developed through standard IC technology as used in VLSI design and micromachining techniques. In fact, the IC processes are to be complemented by a precise micromachining technique for developing three-dimensional structures of the sensors for better performance, reliability, sensitivity, manufacturing uniformity and stability, and cost reduction.

The processing of sensors involves certain necessary steps in semiconductor sensor fabrication using IC technology. These are shown in Fig. 8.4. Starting with a polished Si, Ge, or GaAs wafer (having appropriate flats for identification and orientation) on which film is deposited by

   (a) epitaxial growth, or
   (b) oxidation, or
   (c) polysilicon and dielectric deposition, or
   (d) metallization.

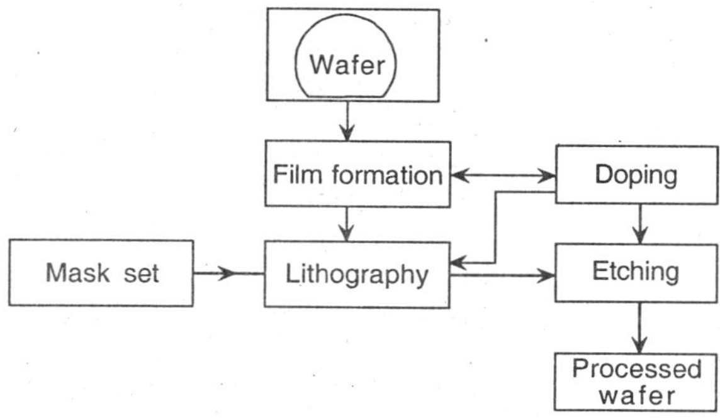

Fig. 8.4 Processing steps in semiconductor technology.

Doping (imparting impurity) is done usually by ion implantation or diffusion. At this stage, the mask patterns are transferred to the film surface by lithographic process. The unwanted film and substrate parts are then removed by etching. The process may be repeated for n number of times for transfer of n mask patterns. A finished wafer would contain thousands of identical chips (features) which are then separated by diamond sawing or laser cutting.

Single crystal silicon wafers or VLSI wafers have certain specifications to be satisfied and adhered to—both with regard to dimensions and materials properties. Single crystal and polycrystalline silicon have been grown on insulator surfaces such as sapphire (silicon-on-sapphire (SOS)) and SiO_2. GaAs can also be grown on silicon by epitaxy. The process is important as optical sensors can be developed in this way.

Oxidation of Si wafers can also be employed as it passivates the wafer surface and serves as diffusion and ion implantation masks. Besides, it acts as the dielectric for integrated MOS capacitors. Oxidation can be dry (in dry oxygen) or wet (in steam vapour). Oxidation layer thickness achieved is a function of time and temperature. Oxidation in high pressure, however, requires less time and temperature.

Lithography transfers the pattern desired to a layer of resist which transfers the pattern to the films or substrates through etching. Resist is the radiation sensitive material. Lithography can be classified as

(i)   photolithography (with optical radiation),
(ii) X-ray lithography (with X-radiation),
(iii) E-beam lithography (with electron beam), and
(iv) Ion-beam lithography (with ion-beam as radiation)

The radiation generates a pattern and transfers the same to the wafer. Figure 8.5 shows the steps followed in photolithography used in microsensor design laboratories. Computer-aided designing from pattern generation is common now. Besides, direct writing with E-beam to fabricate master masks is also gaining ground.

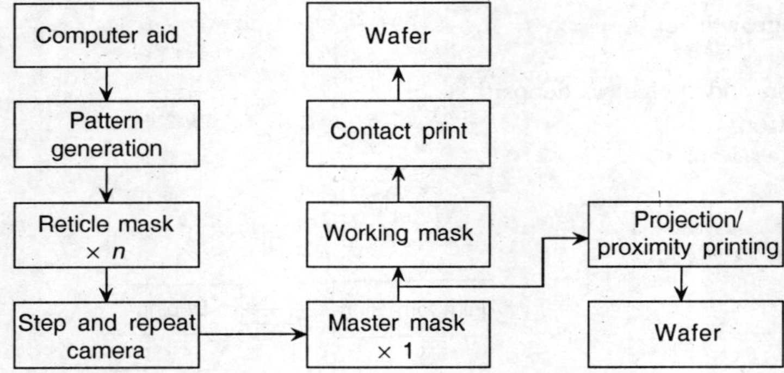

Fig. 8.5 Flowchart demonstrating the photolithographic technique.

As shown in Fig. 8.5, two methods of pattern transfer are (1) contact or shadow printing and (2) projection printing. The minimum line widths l_w that can be printed by the optical method can be given by

$$l_{w(\text{shad})} = \sqrt{\lambda g_w} \tag{8.1}$$

where

λ = wavelength of the light and

g_w is the gap between the mask and the wafer plane.

For projection printing, numerical aperture $n\sin\theta$ divides λ for the minimum line width, thus

$$l_{w(\text{proj})} = \frac{\lambda}{n\sin\theta} \tag{8.2}$$

Figure 8.6 shows the technique of pattern transfer. θ is half angle of the cone of light, focussing a point image on the wafer surface. As would be seen, the size of the chip (feature) varies as the wavelength changes and hence, for VLSI design, that is, reduction of feature size E-beam, ultraviolet or X-ray is used. For a wavelength of 450 nm and a gap of 14×10^3 nm, the feature size obtained by shadow printing is

$$l_w = (450 \times 14 \times 10^3)^{1/2} = 2.4 \times 10^3 \, \text{nm} \tag{8.3}$$

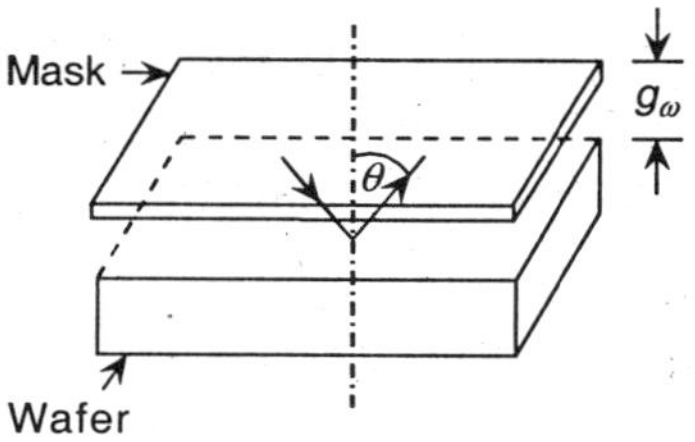

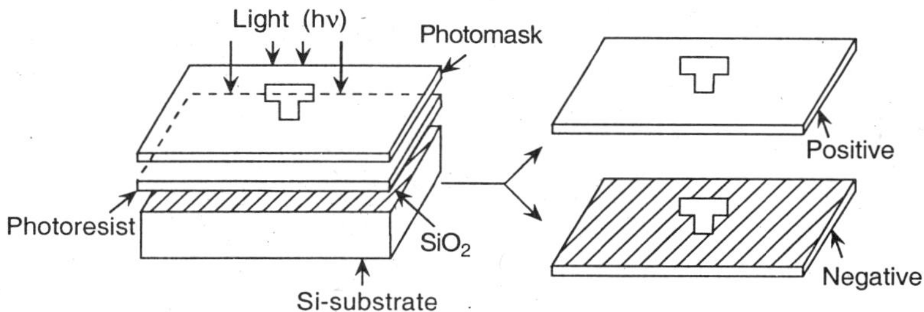

Fig. 8.6 The printing process.

Actually, pattern is transferred to the photoresist which is of two types: positive and negative. In the positive type, the clear zone of the mask is exposed and the resist, after development process, produces the same pattern on the mask. For negative type, the reverse process is followed. The distinction is shown in Fig. 8.7. The pattern initially comes on photoresist then the wafer is etched to transfer the pattern on the substrate, completing the lithographic process.

Fig. 8.7 Distinction between positive and negative photoresists.

Etching: It is essential for surface polishing, removing contamination, drawing pattern, and opening windows in the in-between insulator (SiO_2, say) and fabrication, specifically three-dimensional features by micromachining techniques. Substrates used for etching are Si, GaAs, metals, and insulators.

Etching is of two types: wet and dry. In wet etching, appropriate etchants or reactants are diffused into the reaction surface for chemical reaction to occur at the surface and the reaction products are diffused out. Etching is controlled by controlling parameters such as temperature and agitation. Different etchant materials are there for different substrates. For SiO_2, for example, a reactant containing 48% HF and NH_4F in 1:10 part ratio for an etch rate of 0.1 μm/min is used. For polysilicon, aqua regia (or HF, HNO_3 and H_2O in 3:50:20 ratio) for an etch rate of 0.8 μm/min is used. For platinum, aqua regia for etch rate 20 μm/min and for gold aqua regia at an etch rate 25–50 μm/min are common.

Dry etching can either be physical or physico-chemical. Physical dry etching includes sputtering and ion-milling while physico-chemical one includes plasma etching, reactive ion etching, and reactive ion-beam etching. Sputtering is a process of removal of surface atoms/molecules from a cathode by bombarding positive ions, that is, ions are derived from rare gas discharge. Some of the released atoms/molecules deposit on the substrate to form a thin layer of film. A few variations to this are DC sputtering, RF sputtering, and magnetron sputtering. These methods are also used in micromachining and thin film deposition.

In plasma etching, after the plasma is initiated by free electrons, the electron to gas molecule concentration at 1 Torr is roughly of the order $1:10^6$ imparting a reactive nature to the electrons, as their temperature is very high (≈ 10 kK) while the gas molecules are only at around 50–100°C. This reactive nature makes them suitable for dry etching. Figure 8.8 shows the steps involved in reactive plasma etching (RPE). Gas molecules which are ionized react physically with the wafer surface, playing the role of catalyst for chemical reaction to occur on larger gas surface.

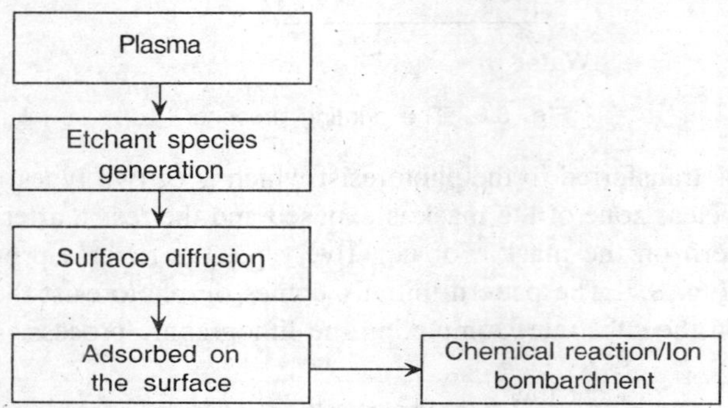

Fig. 8.8   Reactive plasma etching.

The variation in physico-chemical dry etching process lies mainly in the design of the reactor system and chamber configurations such as

(a)   the basic plasma reactive etcher uses two parallel plates as electrodes,

(b)   the reactive ion etcher has the wafer placed on the RF electrode, and

(c)   the reactive ion-beam etcher also known as the ion-milling etcher uses the RF to a gas mixture to release special etchant species while the wafers are kept downstream receiving them.

Selectivity is an important parameter in etching which is defined as the ratio of etch rate of the material to be etched to the etch rate of the material of the mask. As plasma is produced by energizing inert CF_4 gas by electrons to release fluorine that dissolves in insulating Si_3N_4 layers, fluorine becomes the active species which converts N_2 into SiF_4. For increasing the selectivity, H_2 gas is added to the CF_4 plasma which produces a subsidiary reaction to produce flux of HF which reduces the concentration of fluorine and this reduces the etch rate. This can be used for SiO_2-on-Si (SOS)—a technique quite often used in micromachining. The reaction is given by

$$2F + H_2 \rightarrow 2HF \tag{8.4}$$

Metallization is the process of metal film deposition usually employed for ohmic contacts, diode formation, and interconnections. The processing can be done by vacuum evaporation, chemical vapour deposition, or sputtering.

Diffusion sand ion implantation: These are the two very impotant processes by which dopant impurity atoms are introduced in controlled quantities into the selected regions of the wafer, to make the semiconductor substrate regions n or p-type. Selectivity is ensured by masking the top surface of the wafer impurities.

In the diffusion process the wafers are placed in a furnace with temperature ranging from 800–1200°C and an inert gas containing desired dopant is flushed through the furnace. Diffusion occurs in two steps. In the first step, the impurities are transferred from the vapour source onto the surface of the semiconductor. However, the concentration at surface is required to be regulated strictly as is done by the control of the vapour source or rather concentration of the dopant in the vapour source and the source itself. These deposited impurities then diffuse onto the wafer surface which is regulated by the solid solubility of the dopant in the wafer. In the second step, the wafer is heated up in the diffusion furnace at a temperature higher than that mentioned. Appropriate inert gas or oxidizing gas acting as the carrier is present in the furnace which helps to redistribute the diffused impurity atoms in the wafer to reach a desired depth. This process is often called the *drive-in* or *re-oxidation* process.

In the ion implantation process, the surface is bombarded with an ion-beam of a few hundred keV energy. This process causes the layer of impurities to be formed just below the surface layer. The process is quite suitable for silicon where, by this process (i) precise control of impurities can be done and (ii) impurity layer can be produced below the surface layer. Choice of the mask materials is wide as the process is a low temperature process. The mask materials may be oxides, nitrides, and photoresists. But the process is complex and costly and unless adequate care is taken, there are possibilities of crystal lattice being damaged. Figures 8.9(a) and (b) provide comparison of the two processes with regard to properties such as layer formation.

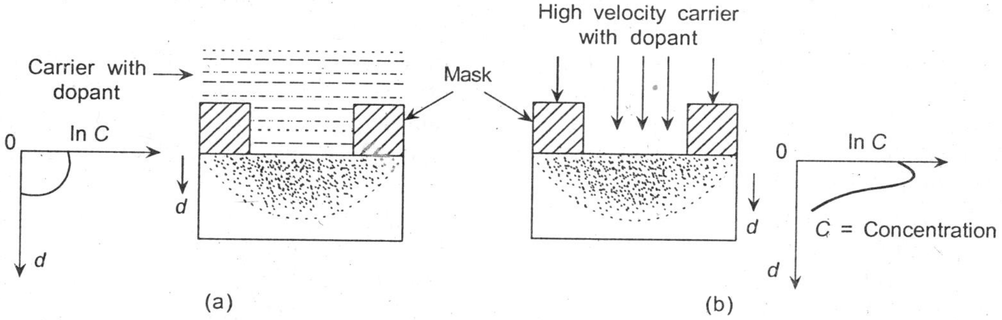

Fig. 8.9 Comparison between (a) diffusion and (b) ion implantation processes.

The theory of diffusion, as proposed by Fick and the Fick's law often fail to hold good in the two step process as the diffusion rate changes with depth as well as with number of times of operational steps.

However, in case of ion implantation, the distribution is approximately Gaussian. Thus, if the bombarding ions per unit area is i_b, and the layer to be formed below the surface is of width (called range), R, depth of the wafer is d, and $(d - R) = \pm\Delta R$. Then the number of distributed ions in the wafer is

$$n(d) = \frac{i_b}{2\Delta R} \exp\left(\frac{-(\Delta R)^2}{2\Delta R^2}\right) \tag{8.5}$$

Boron is a popular p-type dopant; arsenic and phosphorus are popular n-type dopants for diffusion process and all these are introduced as BN, As_2O_3, P_2O_5. A typical reaction is given by

$$2P_2O_5 + 5Si \rightarrow 4P + 5SiO_2 \tag{8.6}$$

Chemical vapour deposition (CVD):   It is the process of deposition of solid material onto a substrate in a state of raised temperature. Deposition occurs either by decomposition or chemical reaction of compounds. For sensor development, the materials deposited, though not many, are

single crystal silicon (epitaxy),
polycrystalline silicon (polysilicon),
SiO_2, and
Si_2N_3.

8.4   MICROELECTROMECHANICAL SYSTEMS (MEMS)

MEMS are basically miniature devices on a silicon chip which have found a major use in sensors. These find an application in modern times in the car air bag systems. In UK and the European continent, these are often referred as microsystem technology (MST). This discipline is termed as microengineering and the terms micromachining and micromechanics are associated with it. Microengineering uses materials such as semiconductors, metal, glass, polymer-based materials, and so on.

A *microsystem* is conventionally described as an assembly whose overall dimensions do not exceed 30×10^{-3} m and that includes structures with individual features having dimensions not more than 100 µm. The device is visible to the eye but not its individual components though a low-powered microscope can make the features visible.

Microsystems and miniaturisation are gaining ground because of the contention that (i) transistors work faster with shorter gate length and (ii) more devices can be accommodated in a given area. Obviously, the latter reason is more relevant here. In microsystems, there are not too many individual elements to be interconnected and choice for doing so is left open. For all such designs, initial process begins with the silicon substrate.

While other kinds of processing such as deposition follow the conventional techniques it is the micromachining technology that has given special significance to the new technology. It comprises machining conventional silicon, the use of silicon-on-insulator (SOI) as the initial requirement, and the use of deep profile lithographic processes. The technique is adaptable to developing three-dimensional structures and physical sensors are being produced by the process of micromachined silicon.

8.4.1   Micromachining

Micromachining can be done in many ways. More important ones include:

(a)   *Bulk micromachining*—It is now well-known that there are differences in etch rates between the crystallographic directions of silicon with particular etchants and using this property, features can be fabricated in particular crystal planes. Generally, (111) planes etch slowest of all and (100)-oriented substrates are preferred for features. The substrate is masked by SiO_2 when ethylene diamine pyrocatechol is used as etchant, or Si_3N_2 is used for KOH as the etchant.

(b)   *Surface micromachining*—Differences between the etch properties of polysilicon and SiO_2 are used for feature development. The process is based on CMOS technology. Polysilicon layer is deposited on top of SiO_2 (sacrificial layer) and then etched. The thickness of the deposited layer is limited to a few microns only. The name sacrificial layer is often used as it is ultimately not of any use in the design except its application in growing or properly etching other layers.

(c)   A process known as LIGA from the words **LI**thographie, **G**alvanoformung, **A**bformung, is an alternative to the process of surface micromachining. It uses the lithographic exposure of thick photoresist and then electroplating is carried out for building mechanical parts. This process fabricates thicker structures than that by surface micromachining. The exposure source in LIGA is the synchroton radiation of wavelength between 0.1 and 2 nm that can penetrate deep down to about 500 μm. Lasers and UV sources have also been used when the penetration depths are limited to 200 μm and 20 μm respectively.

(d)   A process, comparatively newer in development, is based on bonded silicon-on-insulator (BSOI) where silicon wafer is thermally bonded to an oxidized silicon (SiO_2) substrate. The bonded wafer is polished to the desired thickness, between 5 μm and 200 μm, and the etching is done by Deep Reactive Ion Etching (DRIE). The DRIE technique uses an inductively coupled plasma etcher with special etching techniques to achieve high etch rates. In this process, anisotropy in the material is used to form very deep features with the vertical sidewalls. Figures 8.10(a)–(d) show the machining processes (etch) schematically.

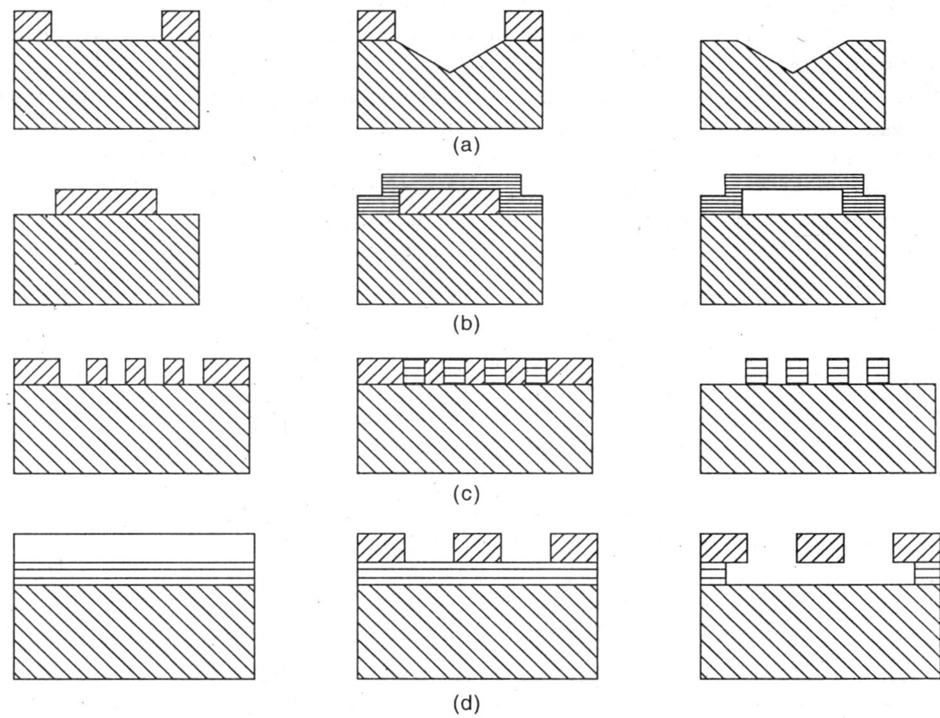

Fig. 8.10   (a) Bulk micromachining, (b) surface micromachining, (c) LIGA, and (d) DRIE of BSOI.

Processes (a) through (d) as discussed here, however, require surface patterning. As a result, the structures produced turn out to be quasi three-dimensional. Complete three-dimensional microstructures can be produced in both polysilicon (process (b)) and BSOI (process (d)) by rotating the surface micromachined parts out of the plane of the wafer and latching them into position required for the design.

It must be remembered that technologies are continuously changing and newer ones are evolving, making the design complex though realization is easier.

8.4.2 Some Application Examples

In application of the MEMS, electrical, mechanical, and chemical properties are exploited as known over a wider range of areas. Sensors, microelectronic and optoelectronic subsystems are the ones discussed in this subsection.

MEMS technology has been used to develop silicon based inertial sensors such as accelerometers and gyroscopes. A typical accelerometer has been produced as a device with a polysilicon proof mass suspended above a single crystal silicon substrate. This requires the removal of a sacrificial layer of SiO_2 making perforations in the mass. The mass remains attached to the substrate at two anchor points. This allows the mass to bend in the plane of the wafer on one axis. The position is detected by capacitive effect and the electrostatic force is fed back to operate the device as an accelerometer. For obtaining a large proof mass, it is laterally extended to several microns. This also provides an appreciable capacitance value. The processing is carried out as in the case of surface micromachining.

However, such accelerometer structure can be developed through the BSOI technique as well. A single crystal silicon is taken as the proof mass in this case which is supported by two silicon beams attached to the substrate. The sensitive axis is now perpendicular to the plane of the wafer. Instead of capacitive sensing, this design employs electron tunnelling method for sensing. Tunnelling takes place across a narrow oblique gap formed by focussed-ion-beam (FIB) which machines through a third beam. Figure 8.11 shows a simplified sketch of such a nonresonant sensor.

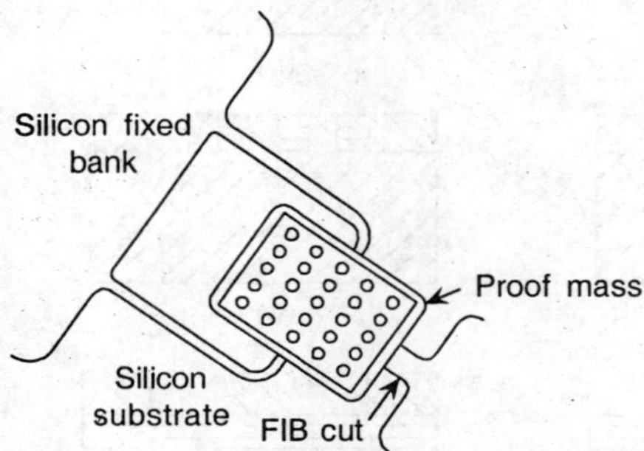

Fig. 8.11 Simplified sketch of a nonresonant sensor.

Resonant sensors: Sensors based on vibrating elements have also been produced by BSOI process to measure many physical and chemical parameters. This technique can also reduce the size of the resonant sensors by the process of miniaturization of the sensors producing them on silicon substrate and in the process, making them resistant to shock. Internal losses can also be reduced by using high quality materials. The design is possible in the single crystal or the polysilicon type in both the cases high Q-factor is obtainable.

Ring gyroscope structures have also been produced/fabricated using BSOI and DRIE technologies. In this fabrication, electrostatic actuation and detection is utilized. The signals derived with the application of turn are coupled between two naturally tuned resonant modes via Coriolis effect. A Q-factor of the order of 10^4 can be obtained by this technique.

Basically, the idea is to arrange the microresonators such that they affect the mass or stiffness of the vibrating system. In pressure sensors, for example, the resonant system is mounted/supported on the top of the diaphragm. As the external pressure bends the diaphragm, the microresonator is strained with a consequent change in its resonance frequency. The resonator thickness is less than 50 μm and is generated by fusion bonding of the layer on to an etched diaphragm.

The resonant mass can be adjusted by using the focussed-ion-beam (FIB) machining. The same technique is used to make narrow cuts at the roots of a flexure suspension by which suspension length can effectively be altered. Figure 8.12 shows the displacement versus drive frequency without and with trimming as suggested in precedence.

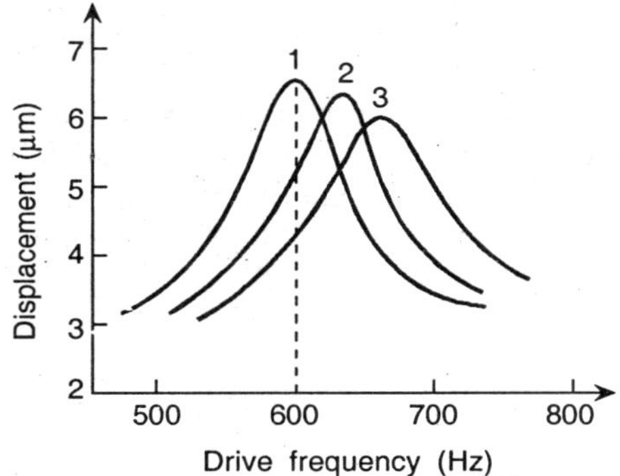

Fig. 8.12 Displacement versus drive frequency: (1) machined twice (2) trimmed once, (3) without trimming.

Micrototal analysis systems: For consumer gas sensing, microfabricated chemical sensors with total analysis facility, referred to as micrototal analysis system (μTAS), are now available. In this system of sensing, a silicon-based metal oxide chemical sensor chip is used. The chip is fabricated from the oxide, is in the thin film form, and has an integrally produced polysilicon microheater buried into the system. The gas sensitive layer converts gas concentration into electrical conductivity when exposed to air. The heater and the sensing element are built on a precision generated diaphragm micromachined from a silicon oxynitride ceramic. This material has thermal properties which minimize both response time and sensor power.

Operation of the integrated sensor can be understood as follows. Ambient oxygen molecules chemisorb onto the film surface of the sensor only to dissociate into oxygen ions. This requires the heater to be at a temperature of around 450°C. Formation of these oxygen ions removes free electrons from the tin oxide grain surfaces. This process increases the sensing film resistance. Figure 8.13 shows such a system. If now, carbon monoxide is present in the gas, it being a reducing agent, reacts with oxygen to produce CO_2 and releases electrons such that the conductance of the sensor is restored. As CO concentration increases, the film resistance decreases further. This change is called the 'detecting principle'. If instead, an oxidizing agent like NO_2 is present in the gas, the effect reverses and the resistance of the film sensor increases. The response of the sensor can be made to be dependent on the constituents of the gas by appropriate control of the heater temperature. This phenomenon has been utilized to optimize the sensor response to a specific gas against a background of interfering gases. Thus, a single sensor can be used to speciate a mixture of gases such as hydrogen, CO, and methane. It must be remembered, however, that the gas mixture cannot be arbitrary.

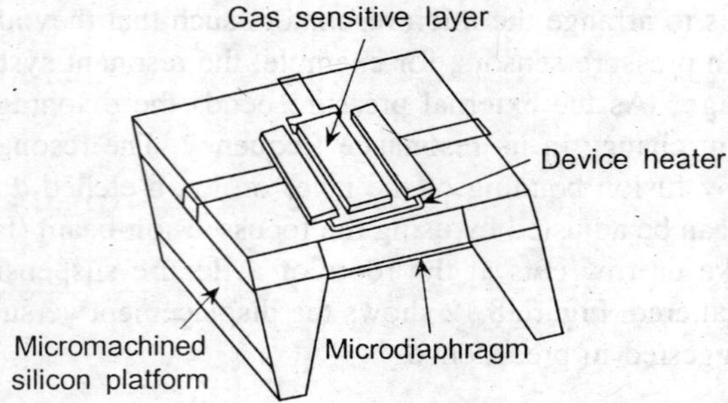

Fig. 8.13 A gas sensor.

The important thing is that the sensitivity of the metal oxide film depends on the grain-size as well as the structure. Smaller grain-size provides greater surface area and increased film porosity, the factors that improve sensitivity and long-term stability of the film.

Recently nanostructured metal oxide films with 20 nm grain-size have shown four-fold improvement in sensitivity. It is expected that selectivity, gain, response, and stability would also improve gradually.

8.5 NANO-SENSORS

Microelectronics naturally leads to nanoelectrons for realizing nano-devices which are expected to create an impact in the enhancement of energy conversion, control of pollution, production of food, and improvement in the conditions of human health and longevity. At present, however, the best examples of nano-devices, as far as applications are concerned, are associated with information technology industry.

The development of nanostructures has been accelerated over the past decade because of unexpected breakthroughs in the synthesis and assembly of nanometer scale structures. Besides, capabilities of manipulating matter from the top down have been there during this period. Some areas that have seen such unexpected developments are:

1. Discovery and controlled production of carbon nanotubes and the use of proximal probe and lithographic schemes,
2. Success in placing engineered individual molecules onto appropriate electrical contacts and in measuring transport through these molecules,
3. Availability and use of proximal probe techniques for fabrication of nanometer-scale structures,
4. Development of chemical synthetic methods to realize nanocrystals and also implantation technique for putting these crystals into a variety of larger organized structures and making assembly,
5. Implantation/incorporation of biological motors into nonbiological environments,
6. Integration of nanoparticles into gas sensors,
7. Fabrication of photonic level band gap structures,
8. Development of quantum effect, single electron memory, and logic elements that can operate at room temperature.

Many other offshoots have also been developed that have enhanced the research and actual commercialization of nano-devices in general and nano-sensors, in particular. One important area is the development of the nanometer scale objects which can manipulate and develop other nanometer scale objects economically and efficiently. Scanning tunnelling microscopy (STM) or atomic force microscopy (AFM) are the techniques used for the purpose. Work is in progress to integrate nanometer scale control electronics and micromachines for developing the ideas postulated.

While progressing towards the development of fast and miniaturized memory structures, giant magnetoresistance structures have been produced using Thomson effect. These giant magnetoresistance (GMR) structures consist of layers of magnetic and nonmagnetic metal films wherein the critical layers have thickness of the order of nanometers. They are used as extremely sensitive magnetic field sensors. The spin-polarized electrons are transported between the magnetic layers on the nanometer length scale and this process, during operation, allows the structure to sense magnetic field. These are already in use in sensing magnetic bits stored on computer discs.

Organic nanostructures have been developed combining chemical self-assembly as mentioned earlier in the chapter, with a mechanical device. The organic sample is reduced to a size that consists of a single molecule and this is connected by two gold leads. This structure has been successfully used to measure the electrical conductivity of a single molecule. Figure 8.14(a) shows the microstructure, while Fig. 8.14(b) shows the operation mechanism of a GMR.

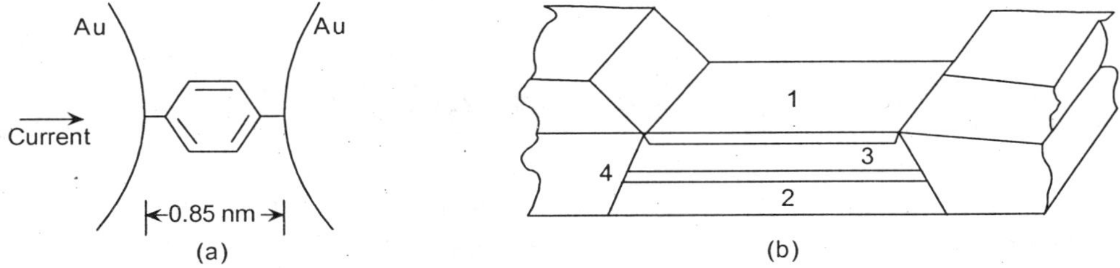

Fig. 8.14 (a) The microstructure (b) operation scheme of a GMR (1) Antiferromagnetic exchange film, (2) Ni–Fe GMR free film, (3) Co-GMR pinned film, (4) Cu-spacer.

As mentioned already, molecular electronics is receiving a boost in the area of sensor development. Molecules which are quantum electronic devices and through batch processes of chemistry such devices are designed and synthesized and assembled in useful circuits. The final stages, however, require processes of self-organization and self-alignment. Molecular switches have already been produced by these processing machanisms. The circuits use carbon nanotubes as connecting wires. These devices can be used as power-efficient memories with extraordinary high speed, perhaps a billion times faster than the CMOS devices. The operation of the switches depends on the electronic states of the molecules. In the closed state, current flows by resonant tunnelling through the states. The switches are opened by applying an oxidizing voltage across the device. Figure 8.15 shows the voltage–current characteristics of the switches in the on–off states.

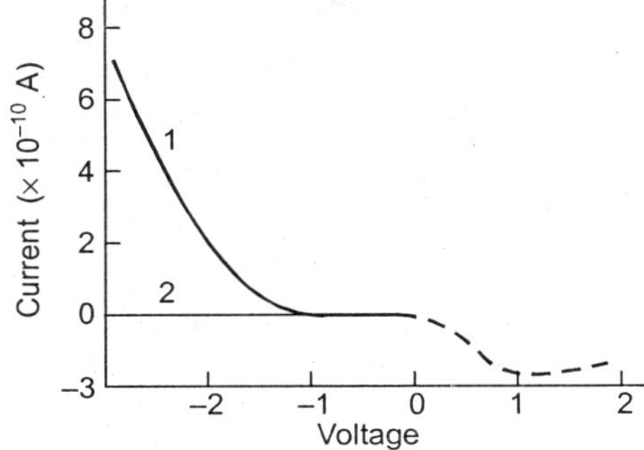

Fig. 8.15 Voltage–current characteristics in 'ON and OFF' stages of the switch.

Research groups have succeeded in fabricating electronic switches such as FET from single-walled carbon nanotubes. Figure 8.16 shows a single-walled carbon nanotube (dia $\approx$ 1.6 nm) placed on metal contacts using AFM. It then behaves like a FET turning 'on' or 'off' depending on the applied gate voltage. Silicon substrate is itself used as gate.

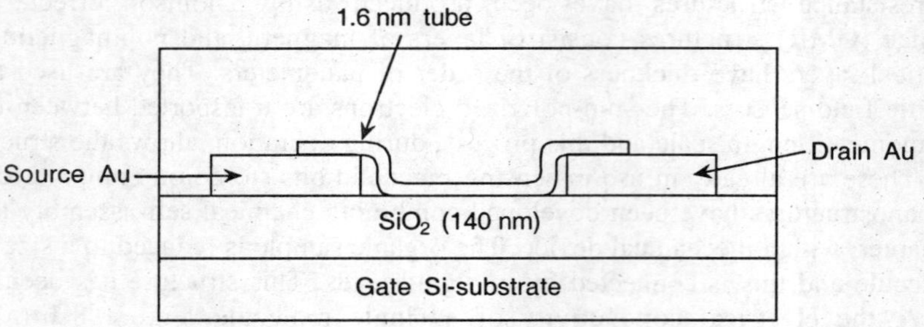

Fig. 8.16   Single-walled carbon nanotube FET.

REVIEW QUESTIONS

1. Describe the steps of processing in the production of thick film sensors. What are the substrates and initial base materials used in the process? In what atmosphere is the firing done at the final stage?

2. What are the characteristic properties of the three commonly used substrates in thick film technology? Discuss it with special reference to dielectric constant, dielectric strength, coefficient of thermal expansion, bulk resistivity, and machinability.

3. Discuss how selective gas sensors can be produced from the same initial base material. Explain taking examples of gases such as H_2, CO, and NH_3. Does selectivity change by changing the firing temperature?

4. List six commonly used techniques for producing thin film sensors. How are these different from the ones for thick films?

   Describe with appropriate diagram how does plasma enhanced chemical vapour deposition (PECVD) technique improve with respect to simple chemical vapour deposition technique?

5. What principles are used in sensing flow-rate, humidity, radiation, and strain by thin films of gold, Ta_2O_5, and CrNi respectively?

   What are hydrophillic and hydrophobic ends and how do they change the orientation of the thin film on the substrate? In what type of processing is it considered?

6. What are the major processing steps in developing the standard semiconductor microsensor technology? Describe with suitable diagrams.

   What are the different film deposition processes in this technology and how are they implemented?

7. Discuss with suitable diagrams, the techniques of diffusion and ion implantation as are required in microsensor systems. In which stage of the sensor development are these methods of processing used?

    If the number of bombarding ions per unit area in ion implantation method is 10^4, wafer depth is 200 μm, and the layer to be formed by doping is 20 μm below the exposed surface, calculate the distribution of ions in the wafer.

8. Draw the flowchart and discuss the steps in reactive plasma etching. How does it differ from the conventional dry chemical etching? What is ion-milling etcher?

9. What is so special in microelectromechanical system that future sensor production technology is expected to revolve around this only? Explain microengineering and micromachining.

    Describe briefly, the four micromachining techniques and compare their advantages and uses.

10. Sketch the constructional features of a FET using a single-walled carbon nanotube and describe its operation. Which special technology is used to put the tube on its source and drain ends?

    What is GMR effect? Where is it being used presently? What special advantage does it offer compared to the earlier components used in that equipment?

Chapter 9

Sensors—Their Applications

9.1 INTRODUCTION

An apt name of this chapter could be 'Sensors for miscellaneous applications'. Sensors elucidated here are only highlighted in terms of their applications as many of these are discussed in details in the earlier chapters. Their adaptation to specific applications has been given due consideration. In doing so, a little analysis of the sensor system has been made in some cases. Obviously not all fronts of application can be covered in the present text though the ones presented offer a wider exposure.

9.2 ON-BOARD AUTOMOBILE SENSORS (AUTOMOTIVE SENSORS)

Sensors for automobiles, that is, automotive on-board sensors come with some special constraints and features that include environment, reliability, cost, and resources and innovations.

Engine is the heart of the automobile which is exposed to vibration, dust, electrical noise, extreme temperature variations, and hence, the sensors used too get exposed to all these conditions even though the performance level has to be kept unabated. One such sensor is the flow-rate sensor as discussed in the following subsection. The engine compartment of an automobile has a temperature varying from -40 to $150°C$ and vibrational acceleration ranging from $3g$–$30g$. Exposure to water, oil, mud, electromagnetic interference, and the like are also to be taken into consideration for better performance.

Reliability in itself is important criterion for production to be trouble-free and for maintaining accuracy at least over ten years. Besides, safety can be ensured in this way—at least partially.

Reduced cost reduces price but it should not be done at the cost of safety and accuracy. Innovative designs are therefore often required for application in automotive conditions.

In present day automobile systems, sensing is required to be done majorly for (i) engine control, (ii) manouvering control, (iii) room and operational comfort control, (iv) safety and reliability, and (v) fuel consumption control.

9.2.1 Flow-rate Sensors

For the conventional engines with carburettor, such sensors are not necessary as the air-to-fuel ratio is self-adjustable here. For the upcoming electronic fuel injection engines, these sensors are made use of as the air volume input to the engine is estimated by flow-rate or pressure sensing. The estimation is done on the basis of engine revolution and the negative pressure measured at its intake. As an advancement, for engines using carburettor also the use of sensors is being considered now and the proposed sensors for this application are (a) ultrasonic flowmeters where the flow-rate is measured by measuring the difference between the speeds of sound both upstream and downstream. But the success of such sensors is yet limited because of large flow-rate changes and temperature variation. As in other areas, microsensors are making inroads into the automobile systems as well. Solid state sensors developed through semiconducting technology are used for sensing air and fuel flows. The underlying principle is to use a heating element in the form of a transistor or a 'semiconductor' resistance bridge. With the advancement of micromachining technology (described in the previous chapter), these can now be realized on silicon substrates. The schemes are shown in Figs. 9.1(a) and (b).

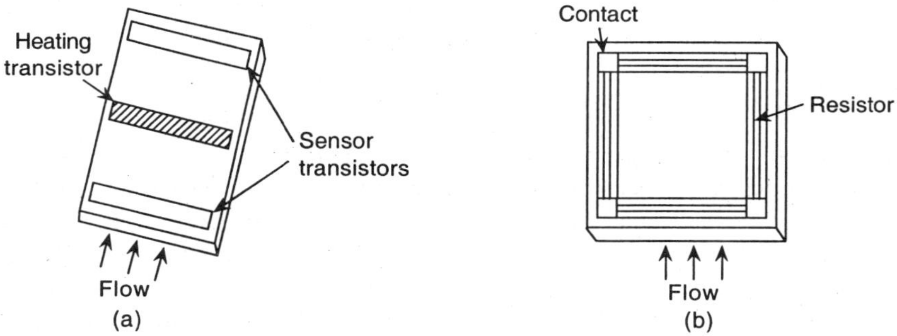

Fig. 9.1 Examples of semiconductor flow sensors.

9.2.2 Pressure Sensors

Pressure is a very important parameter in on-board automobiles and pressures of intake manifold, engine oil, brake oil, tyres, room atmosphere, and so on need be measured. The conventional diaphragms and bellows elements in association with strain gauge, LVDT, and capacitive elements are already in use for pressure measurement. The semiconductor capacitive devices and SAW devices are being increasingly used from which high frequency output can be derived for easier signal processing. With applied pressure, a crystal oscillator changes its frequency and is being tried now as also the PZT types devices. Such a pressure sensor processed by the semiconductor technology and MEMS is shown schematically in Fig. 9.2. For negative pressure sensing for the intake manifold, a semi-smart sensor is used.

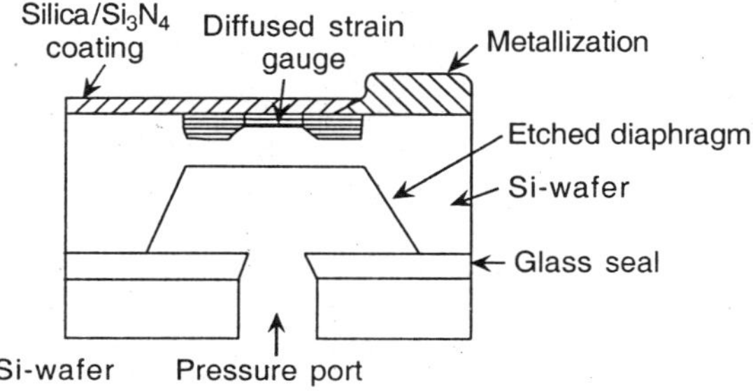

Fig. 9.2 Schematic of a pressure sensor.

Amplification, calibration, and temperature compensation are internally made for in the sensor using IC technology. A basic sensor unit uses a silicon diaphragm and a vacuum chamber is created. The sensor unit is depicted in Fig. 9.3. The sensing silicon chip is resistive in nature and is obtained by adding appropriate impurity to the diffused design.

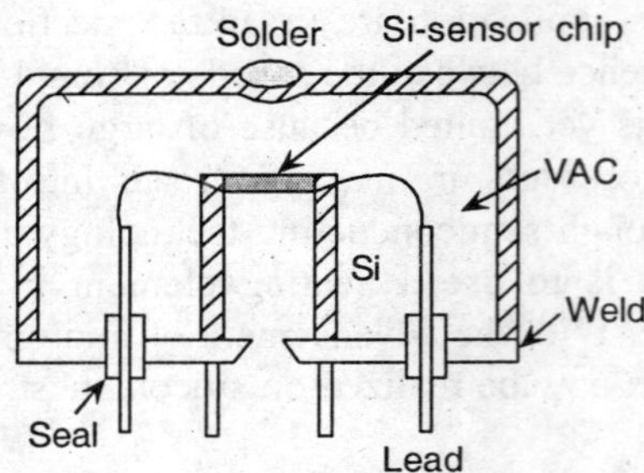

Fig. 9.3 Negative pressure sensing silicon chip.

9.2.3 Temperature Sensors

Temperature sensors for motor vehicles are of various kinds depending on the place where they are used. Temperature switches are pretty common that must be fast. High temperature sensitive sensors are also very much in demand and there are traditional types which are used for oil and water. Temperature ranges from −40–180°C in engine coolant, oil basin, and gear box while for the exhaust gas at the exhaust pipe it may rise upto about 1200°C; the temperature inside the car is less than 80°C, and so on. Therefore, sensors need to be carefully chosen to offer maximum output and reliability under these conditions.

In engine coolant and oil, bimetal elements are commonly used as thermorelays having hysteresis < 5°C for a temperature range of −40°–140°C, and switching current 6 A at a rated supply of 12 V.

For conversion of coolant, air, oil, and gas temperatures in automobiles into analog signal output for indication or for use by the computer/processor for regulation purposes, negative temperature coefficient thermistors are mostly used. The thermistor range lies between 50–150°C with 1 kΩ resistance at 25°C. However, for cylinder head, RTD (platinum resistance) appears to be a better choice. It is the design in either case that has got to be special to fit into the feature of the vehicle. Figure 9.4 shows a typical RTD used in the vehicle, it is an RTD 600 variety with a range of −40–300°C that possesses an accuracy of ±2°C.

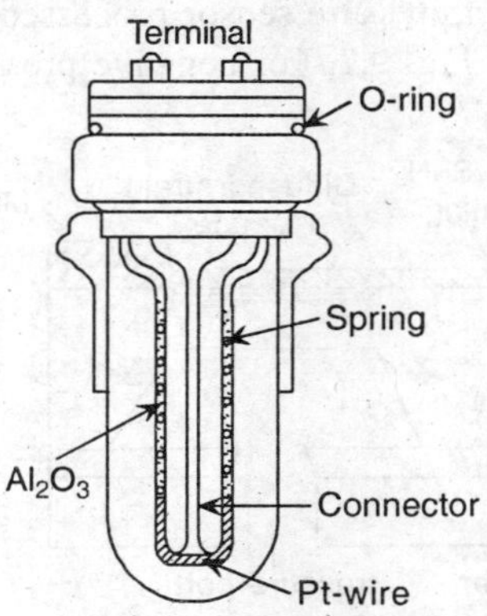

Fig. 9.4 RTD as used in automobiles.

For coolant, engine oil, and air experiencing temperature range of −40–140°C, quartz temperature sensors are now quite frequently used. These have an accuracy of ±1°C. The thermometer houses the electronic processing unit which is a surface-monitored device mounted on a ceramic hybrid system. It can also be adjusted by laser trimming system. The frequency of quartz changes with temperature and this property is exploited for detection. Figure 9.5 shows the basic schematic of such a system.

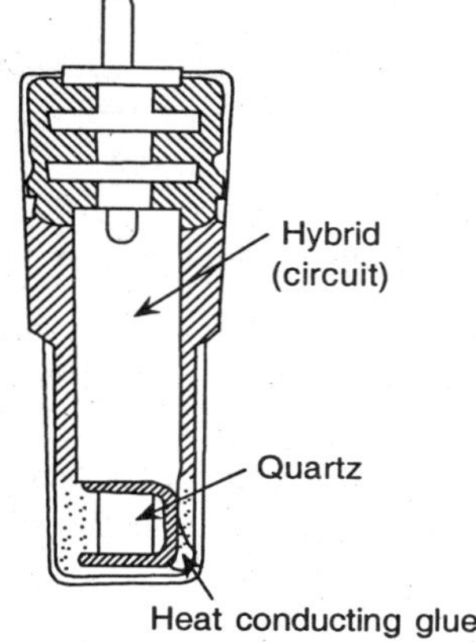

Fig. 9.5 Hybrid IC temperature sensor.

For exhaust gas temperature sensing, Ni–Cr/Ni thermocouples are used upto 1200°C which has a sensitivity of about 40 μV/°C. Thermocouples are gradually getting strong foothold in auto-monitoring because of their active nature.

9.2.4 Oxygen Sensors

Oxygen is a very important parameter for autoexhausts and gas concentration inside the automobile. Appropriate oxygen sensors are installed in the emmision controlled systems to reduce the toxic exhaust and improve fuel consumption. Zirconia and titania sensors are being used for detecting air-to-fuel ratio for quite some time now. Niobium oxide sensors have also recently been introduced and are observed to show better performance. Sensors for checking smoke, humidity, and odour are fixed inside the automobiles and limiting current oxygen sensors are still extensively used in automobile exhaust. The saturation current, called the limiting current in these sensors is proportional to the ambient oxygen. Figures 9.6(a), (b), and (c) present three typical designs of such a sensor.

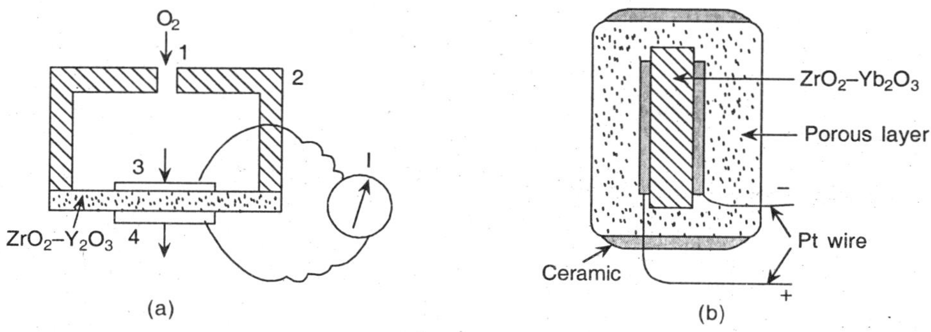

Fig. 9.6 Cont.

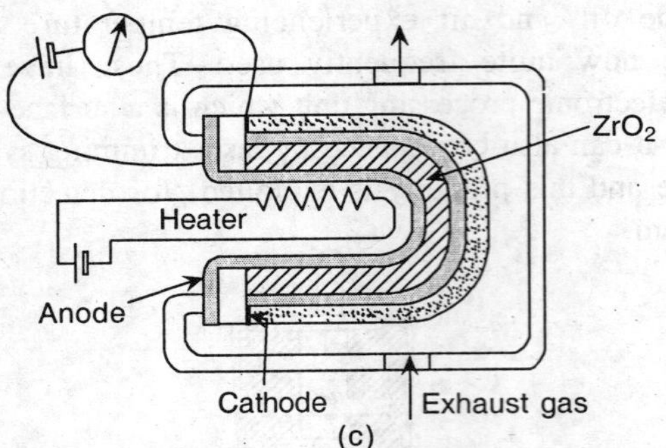

Fig. 9.6 Three typical designs of oxygen sensor.

In Fig. 9.6(a), the cathode is exposed through a pin hole made inside a cover that limits the diffusion of gas. The size of the pin hole is so chosen as to limit the rate in the transfer of oxygen at the cathode. The voltage–current relationship with oxygen concentration as a parameter for this device is shown in Fig. 9.7(a). For oxygen sensors to operate, temperature should be around 700°C which is easily available in the exhaust. The relation between the limiting current I_l and the molar concentration of oxygen, C, is given by

$$I_l = \left(\frac{4FD\,\alpha C_g}{l} \right) \ln \left(\frac{1}{1 - C/C_g} \right) \tag{9.1}$$

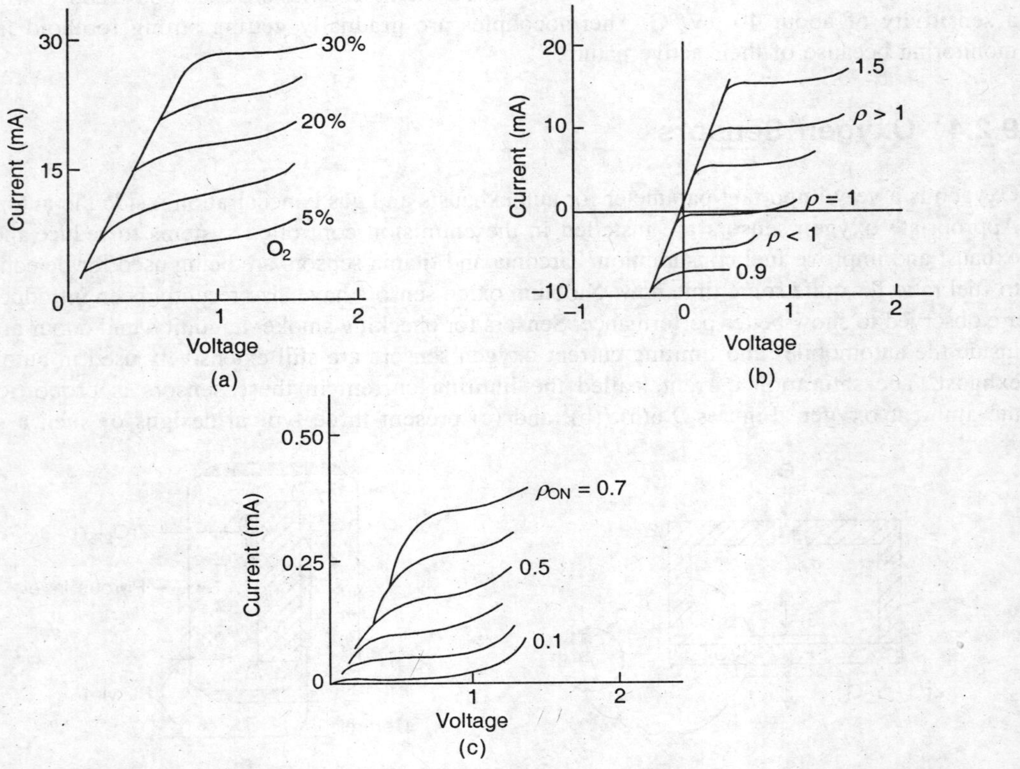

Fig. 9.7 (a) Current–voltage characteristics with oxygen as a parameter, (b) with ρ as a parameter, (c) with ρ_{ON} as a parameter.

where

$$D = \text{diffusion constant,}$$
$$\alpha = \text{the cathode area,}$$
$$C_g = \text{total molar concentration of gases,}$$
$$l = \text{length of the pin hole, and}$$
$$F = \text{Faraday constant.}$$

Obviously $C/C_g \ll 1$, so that

$$\ln\left(\frac{1}{1 - C/C_g}\right) = \ln 1 - \ln\left(1 - \frac{C}{C_g}\right) = \frac{C}{C_g} + \left(\frac{C}{C_g}\right)^2 + \left(\frac{C}{C_g}\right)^3 \cdots$$

$$= \frac{C}{C_g}$$

giving

$$I_l = \frac{4FD\,\alpha\,C}{l} = KC \tag{9.2}$$

The design to limit the diffusion rate by a pin hole has been redesigned with a porous layer used for diffusion, as shown in Fig. 9.6(b) and (c). In Fig. 9.6(b), it is shown that a zirconia disc about 0.3 cm diameter is used as an electrolyte which is made as a film; platinum electrodes are bonded to this electrolyte. Both the cathode and the anode are coated with a porous layer of minerals, for example spinel, for limiting diffusion of oxygen. The anode is protected against toxicity and thermal shock. The design shows that the sensor has small size, simple structure, and low time constant. Further improvement of this design is shown in Fig. 9.6(c) where a cylindrical heater has been incorporated for raising the temperature range to 610–700°C; the outside of the sensor is exposed to the exhaust gas. The inside, on the other hand, is exposed to air and the design resembles the conventional zirconia cell. Figure 9.7(b) shows the characteristics of such a sensor with air-to-fuel ratio (ρ) as a parameter.

There is another variety of temperature sensors where a thin film platinum cathode, a zirconia electrolyte slice, and a platinum anode are all deposited on a porous Al_2O_3 substrate by sputtering and on the other side, a platinum is deposited as a heater. Controlling the substrate porosity during the design stage itself limits oxygen transport by the diffusion process. The V–I characteristics are similar to those of Fig. 9.7(b) but calibration is in oxygen to nitrogen ratio, $\rho_{O/N}$ in volume percentage. These are shown in Fig. 9.7(c).

9.2.5 Torque and Position Sensors

The power trains by which an automobile is run consists of the engine itself, the transmission link, differential gear, axle, and wheels. The torque generated in the engine is distributed to the wheels through power trains. A torque sensor for each component at appropriate position of the power train provides quick and precise response to power controls. Non-contact sensor is found to be suitable for practical adaptability, particularly in its miniature form. Such a sensor is now available and works on the magnetoresistive effect and which can be installed in main bearing and hence, mean output torque can be detected for multicylinder engine with a single sensor. The design is packaged approximately within $12 \times 8 \times 16$ mm small package in which both exciting and pick-up cores of laminated silicon steel and permalloy respectively are housed wound with 200–400 turns of wire. The sensor area exposed to the shaft at a gap of 0.2 mm is about 12×3 mm.

Position sensing is another important aspect in automobiles for detecting shaft position, engine speed, throttle position, potentiometer position, and so on. Here also non-contact sensors receive preference. The semiconducting sensors such as Hall and magnetoresistive ones and the other varieties such as ferromagnetic, electromagnetic pick up, optical modular device, Wiegand wire, and capacitive modular device are also considered suitable. Of the electromagnetic pick-up variety, proximity sensors are most common because of high resolution and low cost offered by them although its output varies with changes in rotational speed. Integrated magnetic sensor using ferromagnetic resistive element is also being increasingly used. It has high sensitivity at low magnetic field and is comparatively less sensitive to temperature variations. Figure 9.8 diagrammatically represents such a sensor. The sensing unit is the NiCo thin film deposited by electron-beam evaporation method. The IC contains a signal processing unit with a differential amplifier and other processing units leading to digital output. Figure 9.9 shows its mounting arrangement.

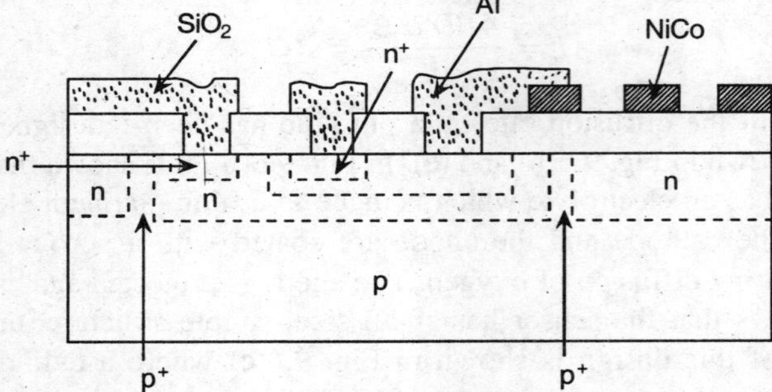

Fig. 9.8   Integrated magnetic sensor.

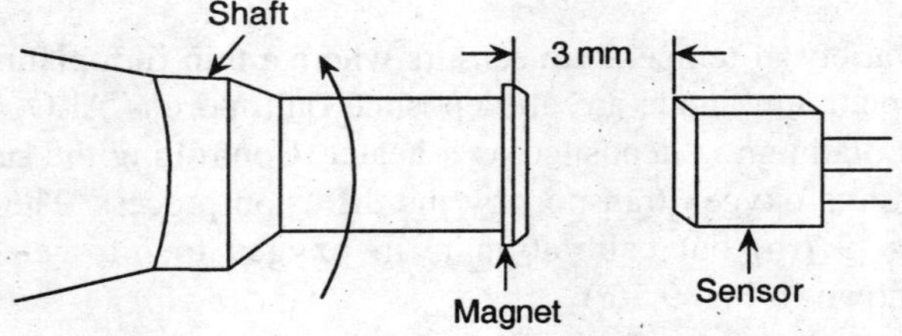

Fig. 9.9   Mounting of proximity sensor.

9.3   HOME APPLIANCE SENSORS

Home comfort depends to a great extent on the availability of automatic home appliances such as cleaner, refrigerator, washing machine, microwave ovens, and so forth. Semiconductor technology has grown fast over a last few decades leading to development of microminiaturized processors, circuits, and of course sensors therefore enhancing the capabilities of the home appliances in terms of automation, safety, and efficiency. Smart operation of the home appliances depends largely on appropriate sensors which have made the equipment more convenient, energy-economic, and safe.

Basically, the sensors are used in electronic control of the appliances and when coupled with microcomputers, all these requirements are almost fully met. Therefore, the basic requirements for the sensors for home appliances can now be revisited as they must have low cost, small size, light weight, better reliability, and easy handling.

The sensors used in home appliances are nothing new though the tendency is to miniaturize them retaining the reliability and efficiency. Sensors so used belong to all categories, that is, mechanical, chemical, magnetic, temperature, and radiation types—the last two types having major applications.

In the mechanical category, Silicon pressure sensors, metal diaphragms, and potentiometers are used in carpet cleaners, while bellows element types are being used in refrigerators. Potentiometers are also used in washing machines.

In the chemical type, humidity sensors have considerable applications in microwave ovens, clothes dryers, air conditioners, dehumidifiers, and also in VCR cameras, while gas sensors are used in ovens, and exhaust fans.

Magnetic sensors are widely used in electronic gadgets in entertainment. There are Hall sensors in VCR cameras, stereo sets and tape recorders. Magnetoresistive sensors are used in VCR cameras, and tape recorders.

In the temperature sensing category, thermistors are extensively used in ovens, cooking ranges, refrigerators, dishwashers, dryers, dehumidifiers, air-conditioners, exhaust fans, CD players, and stereoplayers. They are also used in electric carpet and blanket cleaners for temperature control. In contrast, PTC devices are used in cookers and electric cooking ranges; thermocouples are used in gas ovens; bimetallic elements find use in gas ovens and rice cookers while infrared sensors are employed in microwave ovens.

Radiation sensors, that is, photodiodes, and phototransistors are used as the major elements in refrigerators, washing machines, air-conditioners, TV sets, CD players, stereo players, and video disc players. Photoresistors such as CdS are used in TV sets while VCR camera uses charge control device (CCD) image sensors and MOS image sensors.

The pyroelectric IR sensor used in microwave ovens comes, in general, in TO package. It consists of a $LiTaO_3$ pyroelectric element on a silicon base plate and is irradiated through a silicon window. The sensor appears as shown in Fig. 9.10. Its responsivity is 200–300 V/W, NEP is less than 2 nW/Hz, response time is around 0.2 s, temperature range is –20°C to 100°C, and with silicon window the spectral response is 2–15 μm.

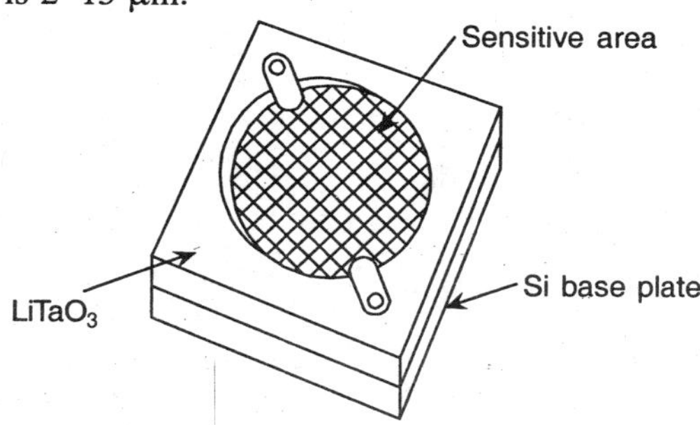

Fig. 9.10   Basic pyroelectric IR sensor.

In a microprocessor-controlled washing machine, water level is sensed using optics principles that comprise units like a light emitting diode (LED), a photodiode/phototransistor, and a light slit. The light slit is moved by the water level. This type of sensor is also used in rinsing chambers for the detection of degree of rinsing which provides information about the concentration of residual detergents. Here, two such systems are coupled—one for reference and the other for measuring. The outputs from the phototransistors of the two sets are compared for achieving a standard condition. The sensor is schematically shown in Fig. 9.11.

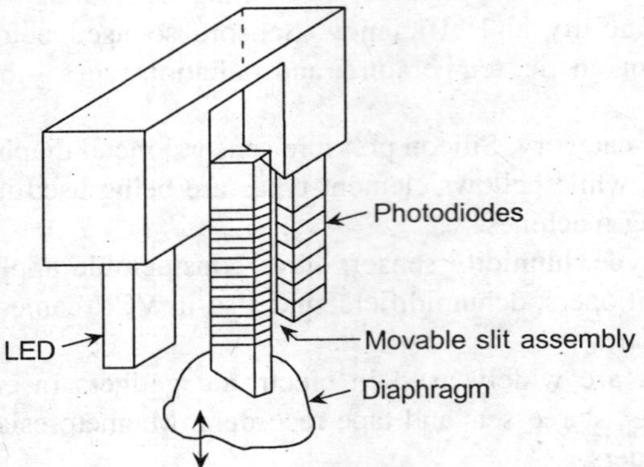

Fig. 9.11   Optical sensor for water level detection.

The sensor used for spin-dry system in washing machines is a PZT ceramic sensor. It is based on the principle that when water drips on to the surface of the sensor, voltage developed in the sensor becomes less with more impinging force of water on it. As the clothes are dried, voltage also increases. PZT, as discussed in an earlier chapter, is a solid solution of lead zirconate ($PbZrO_3$) and lead titanate ($PbTiO_3$)—it also belongs to pervoskite structures. Its piezoelectric property depends on Ti/Zr (T/Z) ratio. Most ceramic piezoelectric transducers belong to this group. Variations in properties have been obtained by partial replacement of Pb^{2+} by other divalent cations (such as Ba, Ca, and Sr) and Ti^{4+} and Zr^{4+} by tetravalent cations. Figure 9.12 gives the structure of a cubical pervoskite. The PZT ceramics are manufactured (see Chapter 8) in the same way as thermistors, from mixed oxide powders while sintering is done at 1200–1300°C at normal atmospheric pressure. Excess lead oxide (PbO) should be used, as a precaution, in a refractory enclosure during sintering as PbO has a high partial pressure. The grain-size and porosity of the ceramic are controlled by hot pressing which in turn controls the electrical property.

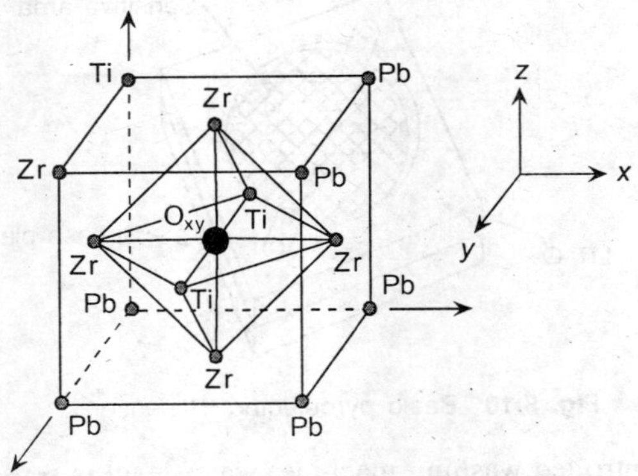

Fig. 9.12   The pervoskite cube.

Photodiode–LED assembly has also been used for frost detection in refrigerators. With frost, the light intensity received by the photodetector is reduced as in the case of a rinsing system. An alternative system for this use is piezocrystal oscillator and a PTC thermistor system. The crystal in an oscillator circuit vibrates at its natural frequency and with frost formation, its resonance frequency changes. PTC thermistor heats the crystal for making it frost free.

The video cassette recorders, for precise control of the servomotor that drives the playback and recording heads, use Hall and magnetoresistive sensors. For control of the cylinder head and its start and stop operations, photosensors are made use of. The video tape has to be protected from dew drops formed on the cylinder head of a VCR for which special humidity sensors have been developed as dew detectors. The detecting film is a hydrophilic acrylic polymer on which carbon particles are sprinkled.

Homes are moving towards being automated and for that alongwith the variety of sensors available, new sensors need be developed. Fortunately, the development is already underway. Three important categories in home automation are (i) house control, (ii) energy control/optimization, and (iii) home security. A block schematic of a home automation system with various sensors is shown in Fig. 9.13.

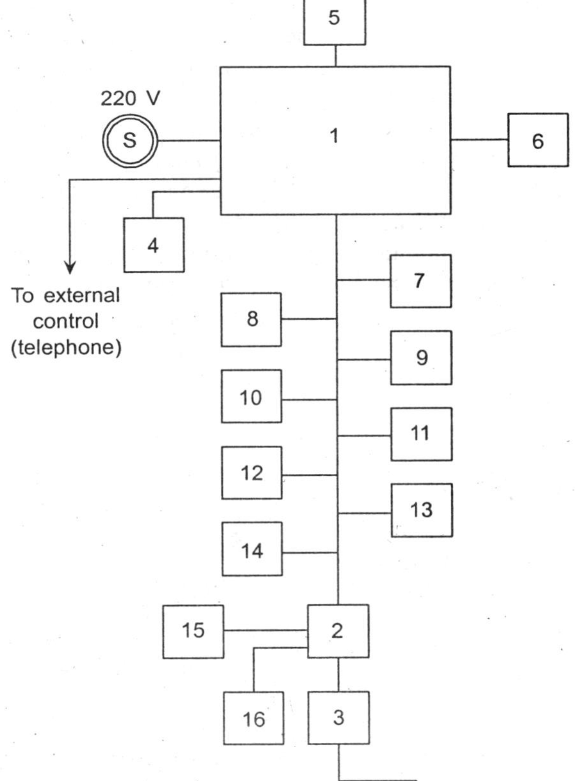

Fig. 9.13 Block schematic of a house automation system: (1) main controller, (2) and (3) secondary controllers, (4) air conditioners, (5) current sensor, (6) light control, (7) smoke control, (8) ventilation, (9) security, (10) gas, (11) thermal/electrical keys, (12) bath etc., (13) earthquake protection, (14) electrical keys (15) air adaptor, (16) light (secondary).

9.4　AEROSPACE SENSORS

Sensors in aerospace applications have to withstand wide variation in physical conditions. Density of air changes from a free molecule to normal state, fluid flow temperatures vary from 20–350 K. At supersonic speed, gases lose their near ideal behaviour. In air, sensors are subjected to varying acceleration, rain, ice, and other environmental changes.

Parameters which are to be monitored for optimization are not new and include pressure, flow and flow direction, temperature, aircraft speed, fluid velocity, strain, thrust, force, acceleration, and so forth. Sensors that are used are not special though their installation, data acquisition, and processing are to be carefully done.

9.4.1　Static Pressure Sensors

For altitude and time rate of change of altitude, that is, vertical speed static pressure is to be monitored. For this, probes are used which are Pitot tubes of appropriate design and require to be aligned accurately. One such probe for static and total pressure measurement is shown in Fig. 9.14. Here p_s denotes the static pressure and p_t denotes the total pressure. The holes for placing probes for p_s and p_t are to be properly chosen with reference to their sizes and location. Besides length, nose shape and cross-section of the probe are also equally important. There are differences in the actual pressure p_s and the indicated static pressure p_{si}, specially for supersonic flow. Correction and evaluation of p_s is possible by Mach number versus p_{si}/p_{ti} and p_{si}/p_s curves. Figure 9.15 shows the technique for the type of probe of Fig. 9.14. Using the curve of the ratio of the indicated pressure versus mach number (M), M is first determined, then using this M, from curve p_{si}/p_s, p_s is evaluated. The curves are dependent also on γ, the ratio of specific heats or the isentropic exponent. Reduction in γ (by, say, 7%) lowers the upper curve (by $\approx$ 5%) and raises the p_{si}/p_{ti} curve by approximately 1%.

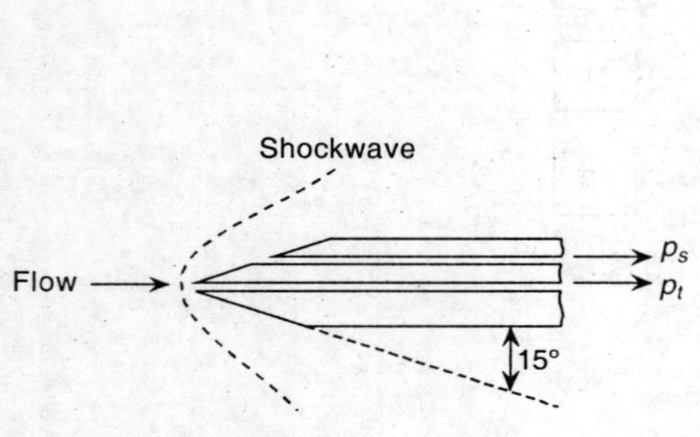

Fig. 9.14　Probe for static and total pressure sensing.

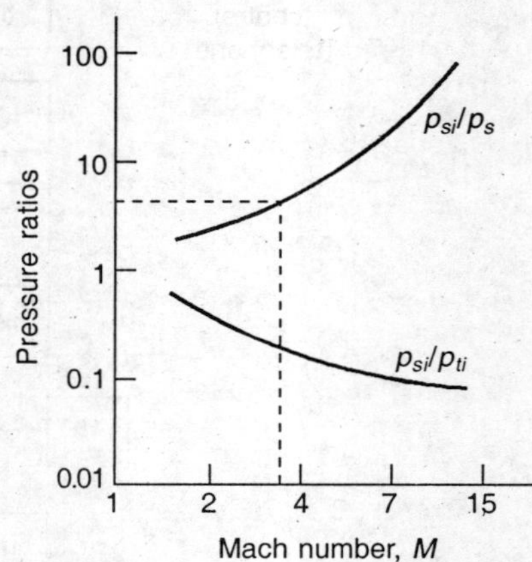

Fig. 9.15　Technique for evaluation of the desired pressure.

Misalignment of the probes incurs considerable error in static pressure. Angle with the flow channel of a 10° cone Pitot is shown in Fig. 9.16. Prandtl tube and wedge are two other pressure probes which also suffer from this error. The errors are also shown for 8° wedge and 30° Prandtl tube in Fig. 9.16.

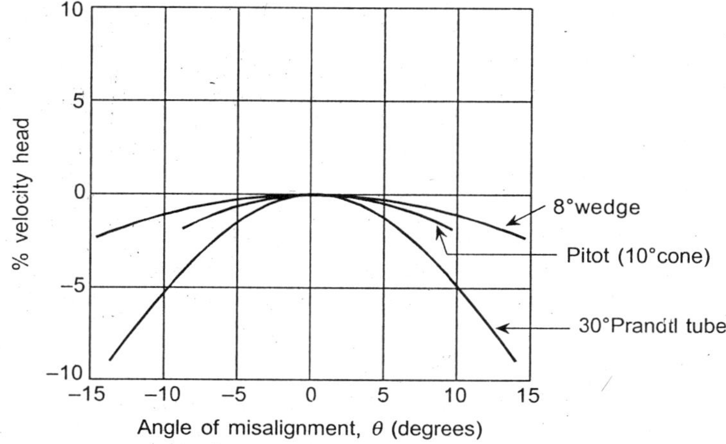

Fig. 9.16 Error curves for three different probes.

For Pitot and other tubes, the term impact pressure is often used in practice. It is the difference of p_t and p_s and is thus, identical with the dynamic head or dynamic pressure, $\rho v^2/2$ (see equation for Pitot tube), where v is the velocity and ρ is the fluid density. Target meters using strain-gauge sensors are also used for dynamic pressure measurement specially in the studies of turbulence in local flow velocities of jet engine drafts where frequency is of the order of 5–10 kHz.

9.4.2 Temperature Sensing

Static temperature is the simple streamline temperature needed to establish the acoustic speed and hence, the gas velocity from the knowledge of Mach number M. This is denoted as T_s. On the other hand, total temperature T_t is the one that the gas acquires if it is isentropically stagnated. For measurement of temperature, the probes usually chosen are RTD and/or thermocouple. In the aerospace terminology, these are called 'temperature sensitive elements' (TSE). Here again, if T_t is measured, T_s is deducible from T_t, M and γ.

Obviously for measurement of T_t, stagnation should be complete otherwise what is measured is called the indicated T_t, that is, T_{ti} and the recovery ratio $(\rho_r T_{ti})/T_t$ is a measure of success of stagnating. Another parameter called recovery factor, $\rho_f = (T_{ti} - T_s)/(T_t - T_s)$ is, however, the theoretical measure of this success.

For high temperature measurement, thermocouple is the first choice. Type K upto 850°C and types B, R, or S thermocouples upto about 1250°C are used. For even higher temperature (upto about 2300°C), $W_{95}Re_5$ and $W_{74}Re_{26}$ are recommended specially in nonoxidizing atmosphere.

As gas density changes with temperature, a pneumatic probe in the form of an aspirator is used to measure temperature by measuring gas density specially if the velocity at the aspirator entrance is small. If the flow is sonic, it measures the product of density and pressure p_t. By stopping aspiration, p_t is measured and density can then be found out and from that temperature.

Static temperature of hot gases such as the exhaust of a jet engine or hydrogen–oxygen rocket, is obtained by the optical absorption–emission method. Here, actually the radiation from H_2O is utilized and accuracy is improved by simultaneous measurements at several wavelengths by spectral scanning. There are other optical methods such as (a) line and band reversal method,

(b) spectral intensity distributions within a molecular band, (c) Raman spectroscopy, and many more wherein the temperature is correlated with the spectral distribution pattern of scattered radiation. In all these techniques, only the measurements of relative intensities are involved.

For measurement of cryogenic temperatures, around that of liquid hydrogen (20 K), metallic resistance thermometers are used. Examples are gold–irridium alloy, platinum, and so on. These are well-known and have been discussed earlier.

For measuring temperatures of solid components of aircrafts and rockets, common methods such as thermocouple are still resorted to. For measuring turbine blade temperature, a narrow-band or monochromatic radiation pyrometer is used. This avoids the uncertainty in blade surface emittance. Two wavelength pyrometry can be used for improving performance reducing uncertainty.

9.4.3 Fluid Velocity Sensors

Fluid velocity can be divided into two parts (i) local linear velocity and (ii) bulk velocity or bulk flow-rate. The former is measured by Pitot tube, hot wire anemometer and/or laser Doppler velocitymeter (LDV), where the latter, LDV, is of more interest in aerospace applications. Bulk mass flow-rate can be measured by gyroscopic, Coriolis, moment of momentum or calorimetric sensors. Besides, for such flow-rates other sensors such as vortex type, orifice type, flow-nozzles and turbine types are also used. A common variety is the orifice type velocity meter, shown in Fig. 9.17 where it is shown that the flow is restricted through an orifice creating a differential pressure at two convenient points, both upstream and downstream which is related to bulk velocity of flow by simple equation. Often empirical, this equation is obtainable through rigorous analysis. The relation is given as

$$q = k\sqrt{\frac{P}{RT}}\sqrt{\Delta p} \tag{9.3}$$

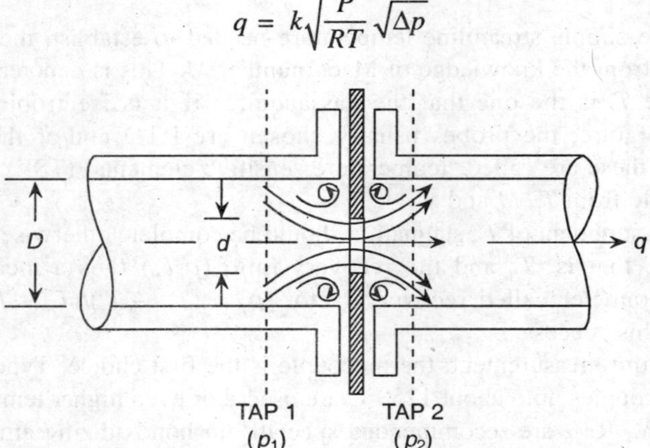

Fig. 9.17 The orifice flow sensor.

where,

q = volume flow-rate,
Δp = differential pressure = $p_1 - p_2$,
P = absolute pressure,
R = gas constant, and
T = absolute temperature.

The relation follows the energy conservation principle through Bernoulli's equation. However, k, a constant, is seen here which takes care of the losses and is dependent on dimensions of the pipe and orifice. The shape and size of the orifice and the flow-rate play an important role in its choice. The value of k is available in charts or tables with d/D as a variable parameter.

Solid state flow-rate sensors have already been considered. These are manufactured using semiconductor technology.

9.4.4 Sensing Direction of Air-flow

It is important to determine the direction of air-flow in aerospace services. In cone type probes or probes resembling cone or wedge, two holes on the inclined surfaces that lie diametrically opposite to each other are used. The difference of the pressures of these two taps is measured which is more or less proportional to the angle between the axis of the probe and the direction of air-flow. There are variations in the design both of the probe (specially its nose) and hole positions. Figure 9.18 shows a typical design of a device used for this purpose.

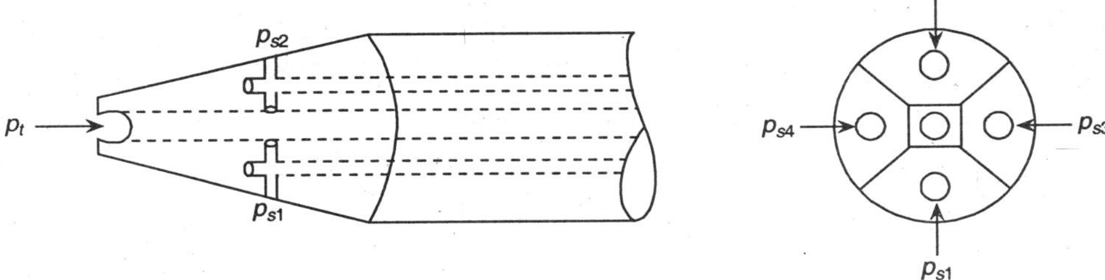

Fig. 9.18 Air velocity and direction of flow sensor.

It can be shown that the pitch and yaw angles θ_p and θ_y are given respectively by

$$\theta_p \propto \frac{p_{s1} - p_{s2}}{p_{s1} + p_{s2}} \tag{9.4}$$

$$\theta_y \propto \frac{p_{s4} - p_{s3}}{p_{s4} + p_{s3}} \tag{9.5}$$

the average pressure being given by

$$p_{sa} = p_{s1} + p_{s2} + p_{s3} + p_{s4} \tag{9.6}$$

9.4.5 Measuring Air-speed on Aircrafts

Air-speed on aircraft can be computed from the measurement of total pressure (p_t), temperature (T_t), and static pressure (p_s) using the equations of ideal gas. If m is molecular weight or relative molar mass, R_0 is the gas constant, and γ is the isentropic exponent (being close to 1.4 at high altitudes), then for subsonic flow velocity

$$v_{ss} = \sqrt{\left\{1 - \left(\frac{p_{ti}}{p_{si}}\right)^{-\frac{\gamma-1}{\gamma}}\right\}\left\{\frac{2\gamma T_t R_0}{m(\gamma-1)}\right\}} \tag{9.7}$$

and for the normal shockwave upstream of the craft, it is

$$v_{sw} = \sqrt{\left\{\frac{2\gamma(\gamma-1)}{(\gamma+1)^2}\right\}\left\{\frac{(R_0 T_t)/m}{1-\left(\dfrac{p_{ti}}{p_{si}}\right)^{-\frac{\gamma-1}{\gamma}}}\right\}} \qquad (9.8)$$

Total and static pressures are measured by a probe by extending it from the aircraft nose. Sometimes the static pressure is measured by two orifices located on the side of the fuselage, the two are placed diametrically opposite to each other to give average pressure of yaw and pitch. Temperature T_t is measured by an RTD.

At a high Mach number, deviations from the computed values are observed using Eqs. (9.7) and (9.8) mainly because

(a) the equations use ideal gas laws,
(b) change in specific heats with temperature occurs at high temperatures due to increased rotational and vibrational energies,
(c) a time lag appears for equilibrium between the above energies near the shock front,
(d) at very high temperatures, ionization and dissociation of molecules occur, and so on.

Obviously, adequate correction has to be applied for these changes.

9.4.6 Monitoring Strain, Force, Thrust, and Acceleration

Aerospace research and studies involve measurement of strain, force, thrust, acceleration, and so on for operation, innovation, and safety considerations. Both static and total pressures, and other parameters are measured using the usual sensors such as strain gauges specially for strain and force. Appropriate positioning of the gauges and compensation for temperature variations are to be taken care of. For dynamic strain alone with fluctuation frequency of several Hertz, the gauge material is chosen to be fatigue-free or to withstand high fatigue. Load cells are also made of strain gauges bonded to a spring whose deflection is sensed. Load cells are used for weighing aircrafts and for measurement of thrust forces.

Engine thrust of a rocket is sensed or determined in the test centre from the integral of dynamic pressure which is measured by an array of total head (pressure) tubes at the exhaust. Alternatively, the force exerted in a tank receiving the engine exhaust is measured by load cell; here also dynamic pressure is area-integrated.

Acceleration measurement in an aircraft is very important during acrobatic movements or gusts when stresses in the structures increase and require to be obtained by correlating stress with acceleration. Standard seismic element devices are used for the purpose but the location of the device in the aircraft is carefully chosen. One such point is near the CG(centre of gravity) of the aircraft, another is the main spar or rigid airframe.

9.5 SENSORS FOR MANUFACTURING

Manufacturing, basically is a controlled process or system—the key to the control being the sensors used in automated manufacturing. Diagrammatic representation between sensors used in such manufacturing and various automation levels in a plant is shown in Fig. 9.19. Increased interconnection has been employed because of the invasion of the computer in manufacturing and special software based on process data available from signals delivered by the sensors at various stages and levels. The concept is known as computer-integrated manufacturing (CIM). Obviously, the prime task of the sensor in this is to run the automated production processes smoothly by compensating for disturbances, tolerances of workpieces, and environmental conditions as and when required. Quality control is also an important area where sensorial assistance is now largely sought. Intelligent sensors are gradually becoming essential as they can interface with each other through organized software processing of electronic signals.

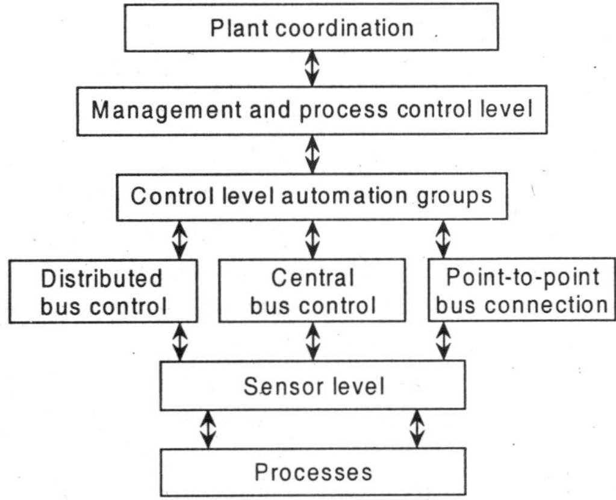

Fig. 9.19 Interaction diagram between sensors and levels of operations.

9.5.1 Sensors

Sensors used in production processes have to perform functions which are not conventional process control functions. The chart in Fig. 9.20 depicts the sensor functions briefly. They are not always as distinct as indicated in the diagram but may be performing in combination on demand. Most of the sensors used have been considered earlier but for robotic actions, specific sensors are applied in production engineering. Sensors used in such actions are discussed in this subsection with the actions for which they are meant for.

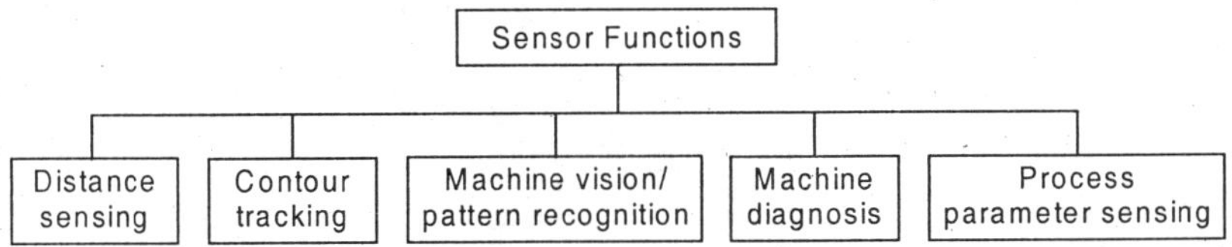

Fig. 9.20 Functional block diagram of the sensor.

Distance sensing: This can be done by (i) tactile sensors, (ii) electrical sensors such as inductive and capacitive, (iii) optical sensors using IR, UV, visible, and laser radiations, and (iv) acoustic sensors using ultrasonic principle.

Contour-tracking: This is a kind of scanning process and is performed by using (i) electrical sensors such as inductive and capacitive ones and (ii) optical sensors—mostly laser-based scanners.

Machine vision/Pattern recognition: Here also, tactile arrays and ultrasonic scanning serve some useful purpose. Besides optical systems with binary vision, grey level vision and stereovision are widely used.

Machine diagnosis: Well-known sensors are used to measure pressure, force, torque, speed (both linear and rotational), temperature, frequency, and a lot of other electrical parameters for obtaining indirect diagnostic data.

Process parameters: Parameters as mentioned in the preceding paragraphs are measured under different environmental conditions.

Distance sensing

During processing, the workpiece and the tool face the possibility of collision. Therefore, the distances between the two for various operations need be monitored. In some processing operations, the distance between the two should be maintained constant as in laser cutting.

Sensors for distance measurement are of two types, namely (i) contact type and (ii) non-contact type. The latter type is gaining ground because the sensors in this type are free from wear and tear.

Contact type distance sensors are common metrological instrument components such as pins, gauge blocks, dial gauges, and many others. Switches and buttons with potentiometric or inductive pick-up are also used. In the non-contact type distance sensors, inductive, capacitive, acoustic, and optical techniques are adopted. The inductive pick-ups are designed and named proximity sensors. Single coil and multi-coil designs are also common. Multi-coil designs allow to measure the distance in two coordinates. Example of an inductive proximity sensor used in distance measurement is schematically shown in Fig. 9.21. The middle coil, coil 2, is fed with ac of appropriate frequency allowing it to produce an ac magnetic field in its own proximity. Coil 1 and coil 3 symmetrically positioned with respect to coil 2 are also electrically energized with phase opposition with respect to supply of coil 2 in absence of any metallic body approaching the set up (coils 1 and 3). With any metallic body approaching, as shown, the magnetic field distributions to coils 1 and 3 change and a signal is generated which can be seen to be proportional to the distance and angle between the body and the coil (s).

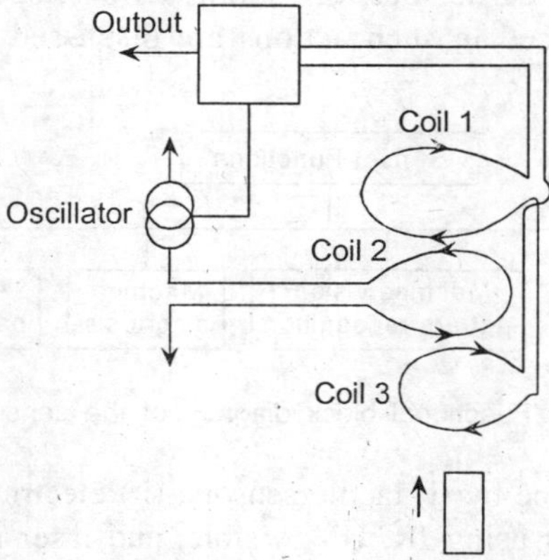

Fig. 9.21 Example of an inductive proximity sensor.

Ultrasonic sensors housed in robot gripper utilize the period between the reflected pulse (echo) and the original pulse sent by transmitter for distance measurement. Figure 9.22 shows this scheme. In this, h_r refers to the distance of reference plane and h_j is the height of the job attached to it. Two pulses, one from the plane and the other from the top of the job, are shown in Fig. 9.23. Focussed ultrasonic beam pulse after reflection undergoes change in the pulse height depending on the distance from where the reflection occurs. Figure 9.23 shows this change. For measurement of time, digital counting technique can be used.

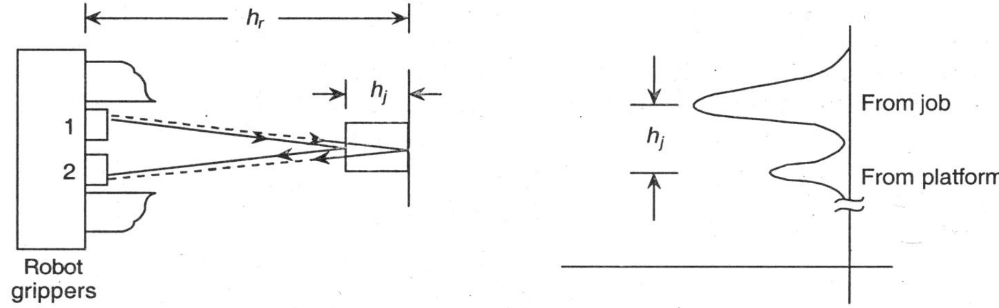

Fig. 9.22   Reflected pulse type distance sensor: (1) transmitter, (2) receiver.

Fig. 9.23   Evaluation of the distance.

In optoelectronic technique of distance sensing, a laser beam is focussed on the approaching body and its reflection is then detected by a properly aligned photodiode after being converged by a lens. This can detect either (i) the intensity of light, or (ii) the angle of approach. The latter technique is known as *optical triangulation*. The detector system consists of about 1000 diodes arranged in an array which can help to enhance resolution. The scheme is shown in Fig. 9.24.

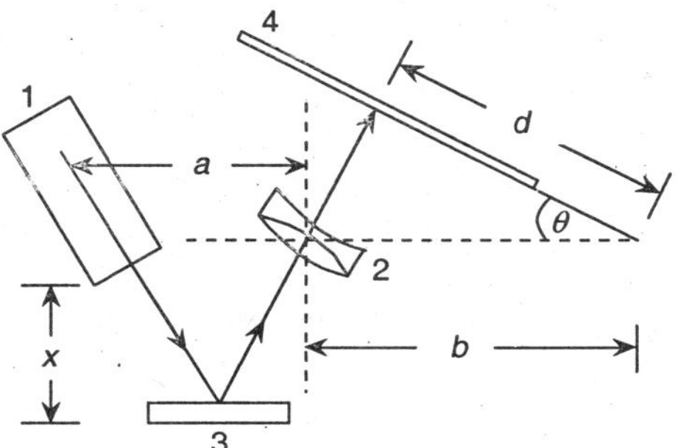

Fig. 9.24   Diode array sensor: 1. laser source, 2. focussing lens, 3. job, 4. diode array.

Obviously, distance d of the detecting diode is important here. From the figure,

$$x = \frac{ad \sin \theta}{b - d \cos \theta} \qquad (9.9)$$

The angle θ, b, and a are fixed while x is a function of d. For the method to be successful, the approaching surface should have good polish for reflection. The technique can have a resolution of the order of tens of μm.

Triangulation principle is also used in contour tracking using preview laser scanner. The job profile is obtained by scanning it with a narrow laser beam and sensing the reflection from the job-piece with an array of diode detectors.

Machine vision is an intelligent sensing system. It involves scanning the object with a video camera whose output is converted to digital by an ADC for image processing, and feature computing and identification. Then comparison with model, called pattern recognition, is performed for the desired output. The system obviously requires a very sensitive viewing of the object with adequate resolution and discrimination. Images are obtained by (i) ultrasonic transducer scanner, (ii) X-ray scanner. However, for robotics, the tactile arrays which generate electrical signals from pressure sensing have been of special significance. The transducers so used may be conducting rubber type, the capacitance type, or piezoresistive type. The rubbing pressures produce change in resistance in the conducting rubber type transducers while capacitance changes in the second even by touching. In the piezoresistive transducer, the normal piezoresistive action takes place. The scanned output obtained from a multiplexer may be stored. Response time of each of such sensors is less than 1 ms per 100 units in an array. Figure 9.25 shows such a sensor system.

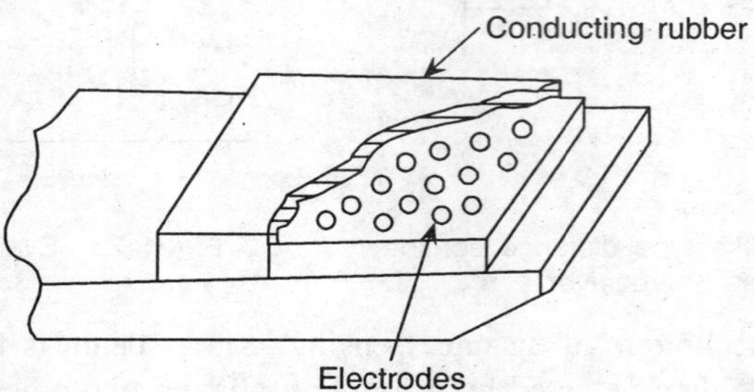

Fig. 9.25 Piezoresistive array sensor.

For machine diagnosis, the techniques applied are (i) process parameter monitoring, (ii) power consumption by the machine and edges of work-pieces (their condition), (iii) force and torque sensing, and (iv) change in the noise of the machine in operation. The first technique is not very straightforward. In force and torque sensing, strain gauges are extensively used. Noise sensing, however, has become an important technique with the advancement of device technology. Noise sensors are, in general, capacitive type. Often, ceramic pieces of PZT material consisting of lead zirconate and lead titanate are used for the purpose. One such scheme is shown in Fig. 9.26. Acoustic pressure is transmitted to ceramic beam with the conical diaphragm. Holes in the diaphragm are provided for equalizing average pressure. Back up plate prevents sag. Very small capacitive microphones have also been developed for sound intensity measurement. Their appropriate placement is a very important aspect. Figure 9.27 gives a schematic arrangement of the system where dynamic sound pressure p_d can be measured with static pressure p_s acting as 'buffer'.

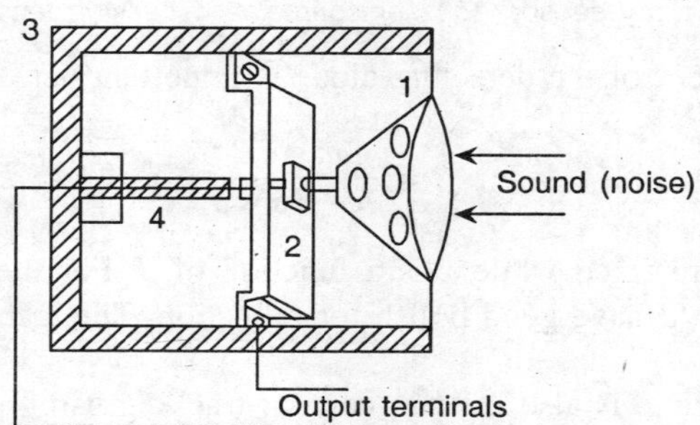

Fig. 9.26 PZT sensor acoustic pick-up system: 1. conical diaphragm with back-up plate, 2. PZT ceramic beam, 3. metallic support, 4. insulator.

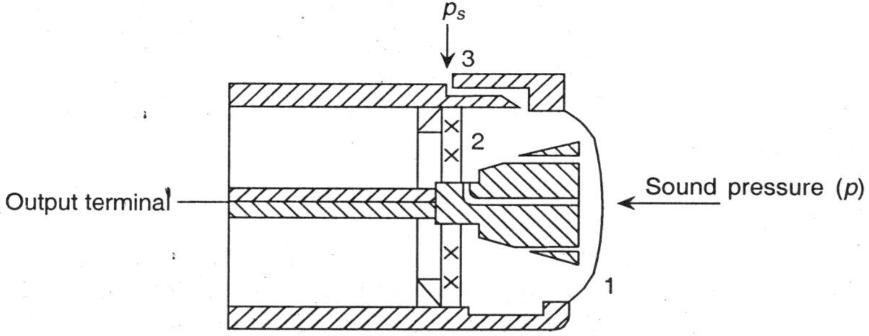

Fig. 9.27   Acoustic pressure sensor for static and dynamic pressure sensing:
1. diaphragm, 2. insulator, 3. air-leak.

9.6   MEDICAL DIAGNOSTIC SENSORS

With the advent of sensor technology involving different materials, processing, and size and shape, clinical studies are gradually being replaced by instrumental studies in medical science with the sensors serving as the interface systems between the instrument hardware and the biological systems. This requires that the sensors used have very stringent specifications.

The most important consideration is whether the sensor is invasive or noninvasive. Gradations have been made here, for example, in invasive type minimally invasive category has been defined where surgical implantation is not needed. Special sensors may be placed remote from the body.

When sensors are to be implanted and/or made indwelling type, they should have minimal effect on the biological system. This requires choice of materials for the sensors which is to be nontoxic and system compatible. It is to be remembered that often the sensor is in a package, so compatibility means that the packaging should also match the tissues in contact such that irritation and inflammation do not develop and measurement is not affected by any means. Noninvasive biosensing electrodes are often used on the skin surface leading to manifestation of allergy from the adhesive tapes for some patients. This is found to change with stretch conditions and is considered to be a mechanical mismatch.

A very important consideration for indwelling and implanted sensors packaged generally from ploymers is that the internal aqueous body environment containing chlorides and enzymes must not affect the sensor packages and the sterility of the sensors must persist.

9.6.1   Sensors

Conventional sensors and microsensors are being adopted more and more now in biomedical activities. Many of these are based on (i) radiation—electromagnetic and acoustic, (ii) force and pressure, (iii) temperature, (iv) electromagnetic variables, (v) chemical and electrochemical principles, (vi) variables related to blood flow, and (vii) kinematic and geometric etc.

Radiation

In the electromagnetic range of radiation, infrared radiation detection of human body by scanning is now quite common diagnostic technique for low–deep circulation abnormalities. These abnormalities produce a thermal image different from the normal case. This technique, known as *thermography,* now uses sensors such as photodetectors, LDR's and bolometers.

Photodiode detectors are also used in phototherapy which uses visible light to convert bilirubin in a new-born baby into naturally excreting materials.

In the shorter wave techniques, X-ray methods are very common and useful. In recent years, conventional ionization detectors or Geiger–Muller counters are replaced by radiation-to-electrical signal sensors, specifically Si-based semiconductor sensors, as these can easily be adopted in tomography, digital subtraction radiography, low intensity fluoroscopy, and so on. Scintillation detectors of special designs are also used for this purpose.

Ultrasonic sensors, that is, piezoelectric sensors are now quite effectively used in measurement of blood pressure, heartbeat of foetus, and grown up people. The conventional carbon microphone in stethoscope is now being replaced with new sensors that convert sound into electrical output. These are used for sensing breathing sound, gastrointestinal sound, and so forth.

Biomechanics

It is an area where sensors for measurement of force and pressure are required. Force sensors are basically load cells which are properly adopted to derive data from models that may provide informations regarding material properties of bone, skin, muscle, tendon, and the like. In direct measurement of human specimens, these sensors provide data for various activities like movements in sports, walking, and others.

Tactile sensors consisting of arrays of force sensors are mostly used in robotics. Thin film multi-element force sensors are also being used for measuring force in patients having difficulty in gripping.

Temperature

Temperature in medical diagnosis is now measured by electronic thermometers using infrared sensors or non-invasive skin-bound sensors which utilize resistive elements such as bolometers or pyroelectric devices.

Electromagnetic variables

In the recent past, electromagnetic mapping of the human body has been done quite extensively from which a lot of information can be extracted. In such mapping, resistance metal strips are used as electrodes. Basically, chemical gradients and membrane potentials are converted into electrical voltage signals by these electrodes. The strip is coupled to the body by an electrolyte layer. Silver–silver chloride form a good combination although toxic effect is not precluded under certain body conditions.

Studies of heart, muscle, brain, stomach, and position of eye are made possible by measuring bioelectric potentials using electrode systems which can be implanted in the tissue or tied to it surgically. These results are obtained as electrocardiograms, electromyograms, electroencephalograms, nerve action potential diagrams, electrogashograms, electrooptograms and so on. Measurement of resistance of tissues by measuring the current with a constant voltage applied between a pair of electrodes has also been used as a diagnostic tool particularly for diagnosis of blood volume in tissue (plethysmographic) or other fluid volume for monitoring breathing and more.

Chemical and electrochemical sensors

In medical/biomedical applications, these sensors are easily adopted. For example, enzyme electrodes are used for biochemical substrates, immunosensors for immunological reactions, polarography for body electrolytes and blood gases, and so on. Electrochemical sensors are more common. One such sensor is the Clark cell which consists of two electrodes for amperometric measurement where an oxygen permeable membrane is used. It is used for the measurement of partial pressure of oxygen in body fluids as well as tissues. Oxygen diffusing through the membrane is reduced following the reaction

$$O_2 + 2H_2O + 4e^- \Leftrightarrow 4OH^- \tag{9.10}$$

Enzyme electrodes have received more attention in recent years. The principle is to immobilize the enzyme in the sensor. Enzyme is a complex of proteins, it must be immobilized retaining its biochemical activity and kept in that condition for quite some time. Basically, a particular substrate (glucose, for example) enters a biochemical reaction either to consume or generate some substance which is detected by a chemical sensor (for example, membrane sensor).

Variables related to blood flow

Blood pressure and blood flow are two very important biological parameters which need be accurately determined for different parts of the body (for flow specifically). Blood pressure is determined by the conventional technique but in recent years, non-invasive electronic techniques are being used. Microsensor technology has been used to develop miniature pressure sensors which in probe form are packaged and introduced into the circulatory system.

Comparatively, blood flow measurement is more of an invasive method and microsensors have been developed which can be used for the purpose. Electromagnetic flow meters and ultrasonic Doppler methods are also being used for the purpose. In the ultrasonic Doppler method, the ultrasound is reflected off the cellular component of the blood; by measuring the Doppler shift in frequency of this reflected sound, both velocity and direction can be ascertained.

Kinematic and geometric

Sensors of linear and angular displacement, area, and volume related to biological systems have also been developed. Sensors are based on special type of strain gauges and are used for various detecting parameters such as position of joints, extra growth, bladder area, tumour size, and so forth.

Kinematic sensors are used to study movements of limbs and gait. Basically, accelerometers of very small sizes are attached to the limbs and the signals obtained therefrom are integrated for velocity and displacement and the computer is used for recreating gait for the patient.

9.7   SENSORS FOR ENVIRONMENTAL MONITORING

Entire living world is now at risk on counts of health and normal survivality due to hazards arising out of biological, chemical, and radiation effects on the environment which not only work locally but are also likely to affect around the globe through transportation. These hazards are thus, critical/serious environmental problems and to assess the extent to which they can affect human and other living entities, measurement of certain selective parameters are needed. Environmental monitoring is not possible to be done in a simple way by measuring temperature of a hot body—in fact, a few steps are involved in the process of monitoring. As environment is affected by pollution, the pollutants are to be identified. The quantity/concentration of pollutants in specific collected 'sample' need be determined.

As said, hazard occurring at a place is not endemic to that place alone and it is spread. The three main ways that cause this spread are (i) atmosphere, (ii) surface water, and (iii) ground water. The manner in which these hazards affect human/living being can be given by simple chart as shown in Fig. 9.28.

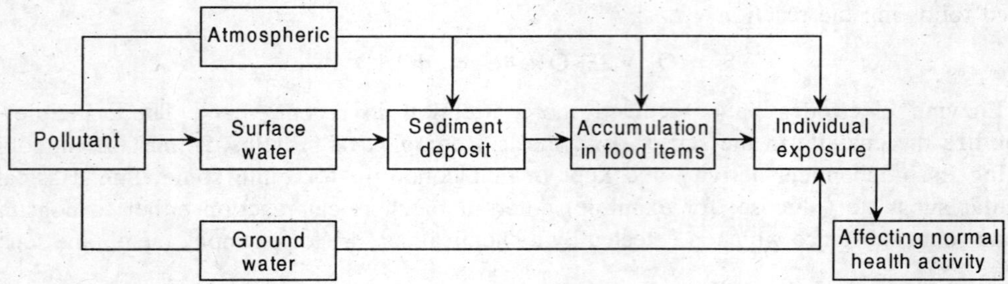

Fig. 9.28   The eco-hazard.

Monitoring the environment pollution again involves three steps, namely the (i) collection of sample representative enough of the environmental pollution content, (ii) pre-treatment of the sample using extraction, separation and so on, and (iii) analysis for identification and quantification of analytic pollutant in sample and expressing it in proper level of concentration.

Sampling (collection and preparation of a sample) is the major player in the three-stage process. Depending on the time and situation, the sampling techniques vary. Similarly, analysis techniques differ depending on the type of sample. The sensors/instrumentation in the analyzers are nothing new in general but their matching with the pollutant and source characteristics are important.

9.7.1   Pollution Hazards

Biological effects of the hazards of pollution on humans are manifested in the excreted wastes in general that may be considered for analysis. Of these, common ones are urine, stool, exhaled breath, sputum and so on. However, blood is an important medium and often nails, hair and fat too get affected. If people are exposed to chemical hazards, by analyzing the above samples, toxic levels can be found and compared to standard maximum that can be allowed. For example, maximum level for lead in blood is 0.5 mg/l and maximum concentration of creatinine in urine is 0.15 mg/g. The analysis is still done by standard chemical techniques or in some cases, where samples are properly available, by spectroscopic techniques. Selective sensors are not yet available widely but microsensors and chemical sensors using silicon and polymers are recently being experimented with.

As has been mentioned already, hazards evolve from various sources like (i) radiation—both ionizing and nonionizing, (ii) biological, (iii) chemical. Their monitoring is the first requirement and then control becomes mandatory. Ionizing radiation includes α-particles, β-particles, X-rays, neutrons, gamma-rays which are capable of biological mutation. Nonionizing radiation includes IR, UV, radiowave and microwave, and extremely low frequency (ELF) (within 300 Hz) radiation. Exposure of these radiations over long periods is also hazardous while IR radiation is known to cause injury to ophthalmic organs such as cornea, retina ($\lambda < 320$ nm), and to a certain extent skin (760 nm $\leq \lambda \leq 1400$ nm). UV radiation is known to damage skin (cancerous), sunburn (erythema) and occular organs. Radio and microwave exposure may lead to cardiovascular nervous, and haematopoitic functions. ELF causes disharmony in reproductive system including cancer.

Biological hazards cause physiological and psychological diseases which are well-known by now. Chemical agents affect water, air, and soil leading to long-term biological afflictions.

9.7.2 Sensing Environmental Pollution

For ionizing radiation, different sensors are used depending on the characteristics of ionization. For low penetration, α-particles (He^{++}) ionization and proportional counters are recommended, scintillation detectors using ZnS as scintillators are also used. Semiconductor detectors have also begun to be used lately.

For β-particles (e^+, e^-), which are slightly more penetrating than α-particles, Geiger counters such as proportional counters. Scintillation counters with solid and liquid scintillators are used besides semiconductor detectors.

γ-rays and X-rays are even more penetrating radiations. These are sensed by NaI and/or $Bi_4Ge_3O_{12}$ scintillators. On the higher wavelength sides, sometimes Geiger–Muller and proportional counters are used. Thermoluminescent and semiconductor detectors are also being used in increased numbers. Li-loaded thermoluminescent detectors and p-n junction diodes are used for neutron radiation detectors. Gas-filled detectors using ^3He and BF_3 are also used for detecting X- and γ-rays.

For sensing non-ionizing radiations thermopiles, bolometers, and diodes are used. Thermopiles are used for microwave, the bolometers are used for radio-frequency, and the diodes are used for ELF, whereas for detecting IR radiation bolometers, photoconductors, Schottky barrier diodes and pyroelectric ($BaTiO_3$) detectors are extensively used. For visible and ultraviolet radiations, photovoltaic cells, photodiodes, thermocouples/thermopiles and Schottky diodes are used.

In semiconductor detectors, it is the charge transport property of the materials such as Ge, Ge(Li), Si, and Si(Li) that is being utilized. These sensors can be packaged and made handy for site measurement as portable units. Over the recent years, new scintillation materials have also been used for the purpose. As such, fluorescent emissions from solid organic and inorganic phosphors are also used. Photodiode scintillators and planar silicon detectors have also appeared in the arena. Cryogenic bolometers working in pulse mode are known to provide high resolution.

While biological polluting agents are very difficult to be sensed and quantify because of the 'omnipresence' of microorganisms in the environment, special 'culture' and microscopic studies have now become common. Counting 'colonies' after culture in a prescribed medium is a traditional practice though time consuming. In recent times, immunoassays have been used with success for microbiological and clinical analysis. Immunoassays can be performed by microsensor implementation.

9.7.3 Ecological Studies of Air

Air pollutants form a major group of identifiable and quantifiable 'parameters'. These include CO, CO_2, Ci, SO_2, NO, NO_2, NO_x (in general), H_2S, hydrides, NH_3, mercury vapour, O_3 and many more. Besides, suspended particles of duct origin or air-effluents such as carbon particles also harm ecology.

Sensing techniques vary with pollutants, their amount and state of occurrence. In many cases, reaction with reagents suggests analysis processes which are purely chemical techniques. Some of these techniques have been adopted in instrumental form. For example, many electroactive pollutants such as NO_x, H_2S, HCOH, SO_2, CO and so on are detected by amperometric or potentiometric methods when appropriate sensing cells are developed with electrode catalysts and membrane permeability. This technique can measure the presence of reactants upto concentration of a hundred ppb.

Flame ionization and photoionization are commonly used for detection of organic samples. Burning the sample in flame or photoionization with UV rays produces ions which are stripped in the form of current. In gas, chromatographic system flame ionization is quite extensively used for better selectivity. Photoionization produces high sensitivity.

Metal oxide semiconductors like SnO_x or metals in filament form such as platinum produce current when CO, H_2S, or hydrocarbons are oxidized on them. Selectivity, however, is not good with this method, although sensitivity is quite high (upto 100 ppm).

Spectrophotometry is extensively used in measurement and detection of pollutants in air and water. Absorption spectrometry is more of use than emission type. Absorption is proportional to gas concentration and the wavelength ranges of the spectral medium change with materials. Thus, for NO_2, visible range is quite common. For SO_2, O_3, and Hg vapour, UV radiation is recommended while IR radiation is used for organic and dipolar inorganic samples.

Aerosol photometry is another sensing mechanism used for particulate pollutants. Forward scattering that causes attenuation of incident radiation is used for the purpose. Particle size is important for detection, as also is the presence of particles in $\mu g/m^3$ which should not exceed $100{-}150$ $\mu g/m^3$.

Elemental analysis including emission spectroscopy for metals appears to be a better choice. Of the various types of such techniques, inductively coupled plasma atomic emission type spectroscopy is more in use though graphite furnace atomic absorption spectroscopy is also sometimes used for monitoring ecological pollutants.

REVIEW QUESTIONS

1. Sketch the bridge type and the transistor type semiconductor flow sensors, and a semiconductor pressure sensor as used in automobiles. How and where are they installed and how do they function?

2. Describe three types of oxygen sensors used in automobiles (on-board) comparing their advantages and operations with the help of V–I characteristics.

    Show that in the range of operation, the limiting current in the device is proportional to concentration of oxygen in the gas.

3. Draw the sketch of a pyroelectric IR sensor as used in microwave oven. What is the material used for developing this sensor?

    How is water level sensed in washing machines? Sketch a sensor and explain its operation. How does ceramic PZT sensor work in the spin-dry system of a washing machine?

4. How is static pressure measured in aerospace studies? How is it dependent on total pressure, isentropic ratio, and Mach number? Explain with graph. How does misalignment of the probes affect measurement? Show with curves, how the per cent errors change with angle of misalignment for three different types of probes.

5. Describe the technique of computation of air speed on aircraft by measuring the static pressure, total pressure, and temperature. How far is this computation valid? Discuss its deviation with reference to Mach number, ratio of specific heats, temperature variation, and consideration of gas laws.

6. Draw a block diagram to show how sensors interact with the automated manufacturing process. Describe distance sensing in this context.

   Draw the sketch of a laser beam operated system of distance sensing and explain its operation. What types of detectors are used here?

7. How are acoustic systems used in machine diagnostic systems? What variations are used in the sensors for the purpose? What are their functioning modes? Sketch an acoustic sensor that can identify both static and dynamic pressures?

8. Describe, on what principles do the microsensors work in biomedical systems? Which electromagnetic ranges of frequencies are important for biomedical studies and how are they detected?

9. How are environmental hazards spread? Draw a block diagram to explain the same. How has instrumentation improved the studies of ecology? Give a few examples.

Bibliography

1. Baker, H.D., E. Ryder, and N.H. Baker, *Temperature Management in Engineering,* John Wiley, New York, Vol. I, 1953, Vol. III, 1961.

2. Baker, R.C., *Development in Flow Measurement,* Applied Science Publishers, London, 1982.

3. Baltes, H.P., and R.S. Popovic, Integrated Semiconductor Magnetic Field Sensors, *Proc. IEEE,* Vol. 74, pp. 1107–1132, 1986.

4. Bard, A.J., and L.R. Faulkner, *Electrochemical Methods,* John Wiley, New York, 1980.

5. Barth, P.W., Silicon Sensors meet Integrated Circuits, *IEEE Spectrum,* Vol. 18, p. 33, Sept. 1981.

6. Bates, R.G., *Determination of pH: Theory and Practice* (2nd ed.), John Wiley, New York, 1973.

7. Bean, H.W. (Ed.), *Fluid Meters* (6th ed.), ASME, 1971.

8. Bentley, J.P., *Principles of Measurement Systems,* Longman, London, 1983.

9. Betts, D.B., *The Spectral Response of Radiation Thermopiles,* Journal of Science Institutes, Vol. 42, pp. 243, 1965.

10. Bockrath, M., et al., Single Electron Transport in Ropes of Carbon Nanotubes, *Science,* 275, p. 1922, 1997.

11. Boll, R. (Ed.), *Soft Magnetic Materials,* Heyden, London, 1979.

12. Bozorth, R.M., *Ferromagnetism,* Van Nostrand, Princeton, 1951.

13. Braggins, D., and J. Hollimgum, *The Machine Vision Source Book,* IFS Publications, London, 1986.

14. Braun, R.D., *Introduction to Instrumental Analysis,* McGraw Hill, New York, 1987.

15. Bryer, D.W., and R.C. Pankhurst, *Pressure Probe Methods for Determining Wind Speed and Flow Direction,* HMSO, London, 1971.

16. Chen, J., M.A. Reed, A.M. Rawlett, and J.M. Tour, Large on-off Ratios and Negative Differential Resistance in a Molecular Electronic Device, *Science,* 286, pp. 1550–1552, 1999.

17. Chien, C.L., and C.R. Westgate (Eds.), *The Hall Effect and Its Applications,* Plenum Press, New York, 1980.

18. Clevite Corporation, *Bulletin: Piezoelectric Division,* 1965.

19. Cobbold, R.S.C., *Transducers for Biomedical Measurements,* John Wiley, New York, 1974.

20. Collins, P.G., A. Zettle, H. Bando, A. Thess, and R.E. Smalley, Nanotube Nanodevices, *Science,* 278, p. 100, 1998.

21. Considine, D.M., and G.D. Considine (Eds.), *Process Instruments and Controls Handbook* (3rd ed.), McGraw Hill, Singapore, 1984.

22. Cook, B.E., *Electronic Noise and Instrumentation Control,* Vol. 12, pp. 326–335, 1979.

23. Doebelin, E.O., *Measurement Systems: Application and Design,* McGraw Hill, New York, 1990.

24. Eisenman, G., *Glass Electrodes for Hydrogen and Other Cations,* Marcel Dekker, New York, 1967.

25. Eller, E., This is Piezoelectricity, *IEEE Student Journal,* Vol. 4, pp. 2, 1966.

26. Ferris, S.A., J.M. Ivison, and D. Walker, *The Magnetoresistor as a Displacement Transducer Element,* Journal for Physical and Electrical Science Instrumentation, Vol. 3, p. 639, 1970.

27. Fleming, D.G., W.H. Ko, and M.R. Newman (Eds.), *Indwelling and Implantable Pressure Transducers,* CRC Press, Cleveland, 1977.

28. Freiser, H. (Ed.), *Ion Selective Electrodes in Analytical Chemistry,* Vols. I and II, Plenum Press, New York, 1978, 1980.

29. Fri, F.J. (Ed.), *Ultrasound: Its Applications in Medicine and Biology,* Elsevier, Amsterdam, 1978.

30. Garshelis, I.J., Displacement Transducer using Impedance Variations due to Core Torsions, *Trans. IEEE,* Vol. MAG 16, pp. 704–706, 1980.

31. Geddes, L.A., *Electrodes and the Measurement of Bioelectric Events,* John Wiley, New York, 1972.

32. Ghosh, D., and D. Patranabis, Linearization of Transducers through a Generalized Software Technique, *Meas. Sci. Tech.,* Vol. 2, pp. 102–105, 1991.

33. Ghosh, D., and D. Patranabis, Software-based Linearization of Thermistor type Nonlinearity, *IEEE Proceedings, G.,* Vol. 139, pp. 339–342, 1992.

34. Gopel, W., J. Hesse, and J.N. Zemel (Eds.), *Sensors: A Comprehensive Survey;* T. Grandke and W.H.Ko (Eds.), Vol. I: *Fundamental and General Aspects,* 1989; R. Boll and K.J. Overshott (Eds.), Vol. 4: *Thermal Sensors,* 1989, VCH Velagsgeselschaft mbH, D. 6940, Weinheim, FDR.

35. Gracey, W., *Measurement of Aircraft Speed and Altitude,* John Wiley, New York, 1980.

36. Guo, T., et al., Catalytic Growth of Single-walled Nanotubes by Laser Vaporization, *Chemical and Physical Letters,* 243, pp. 49, 1995.

37. Heiks, R.R., and R.W. Ure, *Thermoelectricity,* Interscience Publishers, New York, 1961.

38. Hinton, W.R., Piezoelectric Crystal Constants, *Electronic Engineering,* Vol. 24, pp. 76, 1952.

39. Iijima, S., Helical Microtubes of Graphitic Carbon, *Nature,* 354, pp. 56, 1991.

40. Jung, T.A., et al., Controlled Room Temperature Positioning of Individual Molecules: Molecular Flexure and Motion, *Science,* 271, p. 181, 1996.

41. Khandpur, R.S., *Handbook of Analytical Instruments,* Tata McGraw-Hill, New Delhi, 1989.

42. Lawrence Berkeley Laboratory, *Instrumentation for Environmental Monitoring,* Vol. I, Chapter 3, John Wiley, New York, 1983.

43. Lion, K.S., *Instrumentation in Scientific Research,* McGraw Hill, New York 1959.

44. Lion, K.S., *Elements of Electrical and Electronic Instrumentation,* McGraw Hill, Kogakusha, Tokyo, 1975.

45. Lin, C.C., et al., *Application of Nanocrystalline Porous Tin Oxide Thin Film for CO-sensing, Sensors and Actuators,* B 52, pp. 188–194, 1998.

46. Loxton, R., and P. Pope, *Instrumentation: A Reader,* Open University Press, Milton Keynes, Philadelphia, 1986.

47. Ma, T.S., and S.S.H. Hasan, *Organic Analysis Using Ion-selective Electrodes,* Academic Press, New York, (Vols. 2), 1982.

48. McDowell, M.E., *The Identification of Man made Environmental Hazards to Health,* Macmillan, London, 1987.

49. Meijer, G.C.M., and A.W. Van Herwaarden, *Thermal Sensors,* Institute of Physics Publishing, Bristol, 1994.

50. Mohri, K., Review of Recent Advances in the Field of Amorphous Metal Sensors and Transducers, *Trans. IEEE,* Vol. MAG 20, pp. 942–947, 1984.

51. Moseley, P.T., and B.C. Tofield (Eds.), *Solid State Gas Sensors,* Adam Hilger, Bristol, UK, 1987.

52. Nakra, B.C., and K.K. Chaudhary, *Instrumentation, Measurement and Analysis,* Tata McGraw-Hill, New Delhi, 1985.

53. Noltink, B.E. (Ed.), *Jones Instrument Technology* (4th ed.), Vols. 1–4, ELBS Butterworth, Kent, 1987.

54. Nomura, T., et al., Battery-operated Semiconductor CO-sensors using Pulse Heating Method, *Sensors and Actuators,* B52, pp. 90–95, 1998.

55. Norton, H.N., *Sensor and Analyzer Handbook,* Prentice Hall, Inc., Englewood Cliffs, N.J., 1982.

56. Neubert, H.K.P., *Instrument Transducers,* Oxford University Press, London 1968.

57. Patranabis, D., *Principles of Industrial Instrumentation* (2nd ed.), Tata McGraw-Hill, New Delhi, 1996.

58. Patranabis, D., and P.C. Sen, A Linear Temperature Frequency Converter, *Journal for Instrumentation Engineers,* Vol. 53ET2, pp. 74, 1972.

59. Patranabis, D., S. Ghosh, S. Bandyopadhyay, and S. Sarkar, New Improved Bipolar and Bilateral Voltage to Current Converters, *Trans. IEEE,* Vol. IM-37, pp. 58–61, 1988.

60. Patranabis, D., and D. Ghosh, A Novel Software-based Transducer Linearizer, *Trans. IEEE,* Vol. IM-38, pp. 1122–1126, 1939.

61. Price, W.J., *Nuclear Radiation Detection,* McGraw Hill, New York, 1949.

62. Putley, E.H., *The Hall Effect and Semiconductor Physics,* Dover Press, New York, 1960.

63. Quinn, T.J., *Temperature,* Academic Press, London, 1983.

64. Sasada, I., et al., Non-contact Torque Sensors using Magnetic Heads and Magnetostrictive Layer on the Shaft, *Trans. IEEE,* Vol. MAG-22, pp. 406–408, 1936.

65. Segall, S., *Environmental Health* (Ed. P.W. Purdom): Academic Press, New York, 1971.

66. Seippel, R.G., *Transducers, Sensors, and Detectors,* Reston Publishing, Reston, 1983.

67. Sinha, S.K., *Reliability and Life Testing,* John Wiley, New York, 1985.

68. Skoog, D.A., *Principles of Instrumental Analysis* (3rd. ed.), Saunders College Publishing, Holt, Reinhart and Winstton, Philadelphia, 1985.

69. Society of Automobile Engineers of Japan, *A Handbook of Automotive Engineering,* Vol. 3, 1983.

70. Stein, P.K., *Measurement Engineering,* Stein Engineering Services Inc., Arizona, 1970.

71. Summer, W., *Photosensistors,* Chapman and Hall, London, 1957.

72. Sydenham, P.H., *Transducers in Measurement and Control,* Adam Hilger, Bristol, 1980.

73. Thomson, W., *The Glass Electrode,* Kent Technical Review, pp. 16, May 1972.

74. Tipping, F., *Measurement of Oxygen Content in Gases, Measurement and Control,* Vol. 3, p. 145, 1970.

75. Usher, M.J., *Sensors and Transducers,* Macmillan, London, 1985.

76. Wallmark, T.J., A New Semiconductor Photocell using Metal Photoeffect, *Proc. IRE,* Vol. 45, pp. 474–483, 1957.

77. Watrasiewicz, B.M., and M.J. Rudd, *Laser Doppler Measurements,* Butterworth, London, 1976.

78. Watson, J.A., and J.R. Miller (Eds.), *Transistor Circuit Design,* McGraw-Hill, New York, 1963.

79. Wecker, G.P., Operation of p-n Junction Photodetectors in a Photon Flux Integrating Mode, *IEEE Journal on Solid State Circuits,* Vol. SC-2, pp. 65–73, 1967.

80. Welsby, V.G., *The Theory and Design of Inductance Coils,* McDonald, London 1960.

81. Wolff, E.G., *Oxide Thermistors for Use to 2500°K,* Review of Science and Instrumentation, Vol. 40, pp. 544, 1969.

82. Wood, D., *Optoelectronic Semiconductor Devices,* Prentice Hall, New York 1994.

83. World Health Organization, *Global Pollution and Health,* Yale Press, London, 1987.

84. Yamazoe, N., et al., Grain-size Effects on Gas Sensitivity of Porous SnO_2-based Elements, *Sensors and Actuators,* B3, pp. 147–155, 1991.

85. Zieren, V., and B.P.M. Duynden, Magnetic Field Sensitive Multicollector n-p-n Transistors, *Trans. IEEE,* Vol. ED-19, pp. 83–90, 1982.